FLORA OF THE DZHIZAK PROVINCE, UZBEKISTAN

Tojibaev Komiljon, Beshko Natalya, Batoshov Avazbek,
Azimova Dilnoza, Yusupov Ziyoviddin, Tao Deng, Hang Sun

中国林业出版社
China Forestry Publishing House

The Dzhizak Province is a large administrative region of the Republic of Uzbekistan located in the central part of the country, between two large rivers, Syrdarya and Zeravschan. There are represented all altitudinal zones and the majority of ecosystems and vegetation types of Central Asia. The territory of the Dzhizak Province is divided into two physiographical parts. The northern plain part includes the Kyzylkum Desert, the Aydar-Arnasay lake system, and the Hungry Steppe depression in the middle reaches of the Syrdarya River. The southern mountainous part includes Turkestan Range and its western spurs, Malguzar and Nuratau mountains. The highest and lowest points are 4029 m and 240 m above sea level, respectively. This area is characterized by rich plant diversity with a large number of endemic, endangered species, and species of global economic importance. The *Flora of the Dzhizak Province, Uzbekistan* has been compiled for the first time. It includes 1965 species of 613 genera and 105 families of vascular plants, among which 51 species are red-listed at the national level. The book contains following information of species: life form, habitat, altitudinal zone, distribution within province, conservation status, and economic use. Photographs of selected plant species and landscapes are provided.

Authors: Tojibaev Komiljon, Beshko Natalya, Batoshov Avazbek, Azimova Dilnoza and Yusupov Ziyoviddin from the Institue of Botany, Academy Sciences of the Republic of Uzbekistan; Tao Deng and Hang Sun from the Kunming Institute of Botany, Chinese Academy of Sciences.

图书在版编目(CIP)数据

乌兹别克斯坦吉扎克省植物 = Flora of the Dzhizak Province,Uzbekistam：英文/(乌兹)托基比夫·科米尔杰恩等著. -- 北京：中国林业出版社，2020.6

（"一带一路"沿线植物多样性保护研究）
ISBN 978-7-5219-0591-5

Ⅰ.①乌… Ⅱ.①托… Ⅲ.①植物—生物多样性—生物资源保护—乌兹别克—英文 Ⅳ.①Q948.536.2

中国版本图书馆CIP数据核字(2020)第089026号
北京版权局著作权合同登记号：01-2020-2460

中国林业出版社·自然保护分社（国家公园分社）

策划编辑：肖静
责任编辑：肖静

出版发行	中国林业出版社（100009　北京市西城区德内大街刘海胡同7号）https://www.forestry.gov.cn/lycb.html　电话：(010) 83143577
发　行	中国林业出版社
印　刷	河北京平诚乾印刷有限公司
版　次	2020年9月第1版
印　次	2020年9月第1版
开　本	889mm×1194mm　1/16
印　张	32.75
字　数	400千字
定　价	500.00元

未经许可，不得以任何方式复制或抄袭本书的部分或全部内容。

©版权所有，侵权必究。

FOREWORD

Dzhizak Province is the large region in the central part of the Republic of Uzbekistan, which has a unique history, great political, economic and spiritual status, and this region has a great importance for the development of science, culture and tourism.

Dzhizak Province is characterized by outstanding diversity of natural landscapes and wildlife, including a large number of endemic, rare species, and species of global economic importance. There are represented different ecosystems and cultural landscapes: deserts and lakes, foothills and mountains, croplands and pastures, gardens and vineyards. The territory has colourful natural landscapes. In the mountains of Zaamin, Bakhmal and Forish districts with an exotic relief and temperate climate, there are juniper, almond and walnut forests, and valuable medicinal and ornamental plants.

The plant diversity of the Dzhizak Province is extremely rich, and its detailed study is very important. This book provides detailed information about almost 2000 wild-growing plant species of the Dzhizak Province, as well as their images. This is the fruitful result of extensive joint research performed by Uzbek and Chinese botanists.

Uzbekistan and China are neighbors and partners along the belt and road. In recent years, the Academy of Sciences of the Republic of Uzbekistan has been working more and more closely with the Chinese Academy of Sciences. In particular, the Institute of Botany, Academy of Sciences of the Republic of Uzbekistan and the Kunming Institute of Botany, Chinese Academy of Sciences, within the framework of projects devoted to the study and protection of plant diversity, sustainable use of plant resources in the two countries, have made good achievements in joint scientific researches, personnel training and other aspects. The Global Allium Garden with Kunming center and Tashkent center has been created in collaboration of Chinese and Uzbek sides. These unique living collections conserve and display the outstanding global diversity of *Allium* species, provide resource support for systematic and thorough research on this group of plants, knowledge dissemination, education, environmental protection, *etc*.

Academician Yuldashev Bekhzod Sadikovich
The President of Academy of Sciences of the Republic of Uzbekistan

PREFACE

The Republic of Uzbekistan is located in the heart of the Eurasian continent and occupies a position of great bio-geographical importance. The territory of Uzbekistan characterized by rich biological and landscape diversity belongs to the Iran-Turanian region in the Tethyan (Ancient Mediterranean) floristic subkingdom of Holarctic kingdom, which is one of major centers of plant diversity in the world (Takhtajan, 1986). Ecosystems of Uzbekistan are included in the list of key ecological regions of our planet (Olson & Dinerstein, 2002). Flora of vascular plants of Uzbekistan is rich with a large number of crop wild relatives, relicts, endemic and endangered species. Considerable diversity of plants can be explained by the geographic position and huge area of the country, as well as wide spectrum of physiographic factors and complicated geological history. The unique centers of origin and diversity of many genera are revealed on the territory of the country. However, it is necessary to note that the ecosystems of the country have been influenced by human activities since ancient times; anthropogenic impact has increased considerably in recent decades.

Floristic surveys in Uzbekistan were started more than 150 years ago, and the first edition of *Flora of Uzbekistan* was published in 6 volumes (1941–1962). This treatment includes 4,148 species of vascular plants (3,663 of them are native for Uzbekistan, and 485 are naturalized, alien and cultivated taxa). However, at the present stage of our knowledge, this fundamental synopsis and the data on distribution of many plant species were significantly outdated. Currently, non-published database of the flora of Uzbekistan counts more than 4,375 species; thus, at least 712 wild-growing species were added to the national checklist after issue of the first edition of the *Flora of Uzbekistan* (Sennikov et al., 2016).

Detailed study of plant diversity became extremely important in the last decades in connection with current trends and approaches in plant sciences and biodiversity conservation. Protection and sustainable use of biodiversity are priority directions of state ecological policy in Uzbekistan. The *National Strategy and Action Plan on CBD* (1998) includes identification, sampling and monitoring of flora and vegetation (*The Fifth National Report*, 2015). Compilation of national and regional floristic checklists has a special significance, as well as the assessment of populations of economic valuable and endangered plants does so. The Flora of Uzbekistan Project aimed at taxonomical revision of the national flora was started in 2012, and the first two volumes of the second edition of

Flora of Uzbekistan were published (Sennikov, 2016, 2017). Synopses of flora were compiled for some large bio-geographical and administrative regions of the country.

In this book, the results of inventory of flora of the Dzhizak Province, a large administrative region of the Republic of Uzbekistan, are represented for the first time.

The Dzhizak Province has been chosen as the study area because this territory is located in the central part of the country and characterized by rich plant diversity with a large number of endemic and rare species. There are represented all altitudinal zones and the majority of ecosystems of Central Asia, including deserts, foothills, mountains, highlands and wetlands.

The synopsis of the flora of the Dzhizak Province has been compiled on the basis of the data of our field surveys, analysis of TASH herbarium collections, publications, and scientific reports. We hope that this book will become a reliable source for further studies and nature conservation planning, and will be useful for botanists, ecologists, staff of nature reserves and national parks, students, educators, and amateur-naturalists.

The study has been performed within the framework of the Uzbek state research projects I5-FA-0-17440 and BA-FA-F5-010 and supported by the Second Tibetan Plateau Scientific ExPedition and Research (STEP) Program (2019QZKK0502), the Strategic Priority Research Program of Chinese Academy of Sciences (XDA20050203), the International Partnership Program of Chinese Academy of Sciences (151853KYSB20160021) as well as the International Partnership Program of Chinese Academy of Sciences (151853KYSB20180009). We would like to extend special thanks to the Institute of Botany of Academy Sciences of the Republic of Uzbekistan, the Bureau of International Cooperation of Chinese Academy of Sciences (CAS), the Kunming Institute of Botany of CAS and the CAS Key Laboratory for Plant Diversity and Biogeography of East Asia for supporting the study and to the China Forestry Publishing House for publishing this book.

CONTENTS

FOREWORD	/ 3
PREFACE	/ 4
STUDY AREA	/ 7
INVENTORY OF THE FLORA	/ 24
POLYPODIOPHYTA [FERNS]	/ 33
GYMNOSPERMAE [GYMNOSPERMS]	/ 39
ANGIOSPERMAE [ANGIOSPERMS]	/ 47
LILIOPSIDA [MONOCOTS]	/ 47
MAGNOLIOPSIDA [EUDICOTS]	/ 135
REFERENCES	/ 499
INDEX OF LATIN NAMES	/ 507

STUDY AREA

The Dzhizak Province (Jizzax Viloyati in Uzbek) has been established in 1973. It is a large administrative region of the Republic of Uzbekistan located in the central part of the country. The total area is 21.210 km^2 (approximately 5% of the country), 23% are occupied with arable lands, 2.2% with gardens and vineyards, 36% with pastures, and 7.7% are forested. The average population density is 54 people/km^2. The province is divided into 12 administrative districts: Arnasay, Bakhmal, Dustlik, Farish, Gallaaral, Mirzachul, Pakhtakor, Sharaf Rashidov, Yangiabad, Zaamin, Zafarabad and Zarbdar. The Dzhizak Province borders with the Syrdarya Province to the east and north-east, with the Navoi Province to the west, with the Samarkand Province and Tajikistan to the south, and with Kazakhstan to the north. The economy of the province is primarily based on agriculture; cotton and wheat are the main crops, and extensive irrigation is used. Approximately 80% of population lives in rural areas. The industry, mining operations and transportation infrastructure also are well-developed (www.jizzax.uz; www.stat.uz).

The study area is located between two large rivers, Syrdarya and Zeravschan. The territory of the Dzhizak Province is divided into two physiographical parts (Figure A). The northern plain part includes the desert of south-eastern Kyzylkum, the Aydar-Arnasay lake system located in the large saline depression, and the Hungry Steppe depression in the middle reaches of the Syrdarya River. The southern mountainous part includes Turkestan Ridge and its western spurs, Malguzar and Nuratau mountains. The natural boundary between plain and mountainous part of the Dzhizak Province runs approximately at 380–400 m.s.l.[①] in the west (piedmonts of the Nuratau Ridge), and 480–500 m.s.l. in the east (piedmonts of the Turkestan Ridge). The highest and lowest points are 4029 m (Mt. Shaukartau, Turkestan Ridge) and 240 m.s.l. (the shore of Aydar-Arnasay lake system), respectively (Aramov, 2012).

The plain part of region has been transformed significantly due to human activity. Up to the middle of the 20th Century, the Aydar salt marsh and a number of small salty lakes were located in the large drainless depression in the south-eastern edge of Kyzylkum. With the beginning of intensive agricultural development of the Hungry Steppe, this depression used to evacuate the drainage water from irrigated lands. In 1969, the water level was extremely high; more than 21210 km^2 of

---

① m.s.l.=meter(s) above sea level.

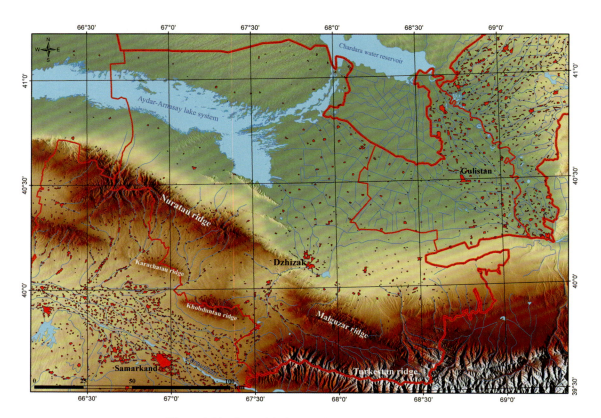

Figure A Physiographical map of the Dzhizak Province

water was discharged from the Chardara Reservoir, the Aydar salt marsh was flooded completely, and a large drainless salty Aydar-Arnasay lake system was formed. At the present time, this anthropogenic water body holds the 4th position in Central Asia in order of size, after the Aral Sea, Balkhash and Issyk-Kul. In the past, an alluvial-proluvial plain on the left bank of the Syrdarya River called the Hungry Steppe or Mirzachul was an almost waterless desert with ephemeroid vegetation and fragments of salt marshes. At present, almost the entire territory of Hungry Steppe is occupied by irrigated crop lands and settlements.

The climate is continental (Csa and BWk according to the Köppen climate classification), with high amplitudes of daily, monthly and annual air temperature, low precipitation and high intensity of solar radiation. In general, the winter is moderately cold, and summer is hot and dry.

There are significant differences among the climates of different natural regions of the province related to their geomorphology. The continentality of climate and contrast seasonal changes is much more expressed in the desert areas than in the mountains. The vast arid plain of Kyzylkum, located in the north and north-west, and huge mountain ranges in the south and southeast have a great influence on the climate of the Dzhizak Province.

Within the Dzhizak Province, the annual and monthly air temperature decreases from the south to the north and from the plain to the mountains. In the plain and piedmonts, the mean annual temperature ranges from 12°C to 15°C, the mean temperature of July is +27–+29°C, and the mean temperature of January ranges from –4.4°C in the nortern part of the province to –0.1–0°C in piedmonts. The average duration of the no frost period in plain and foothills ranges from 170 to 230 days. The warmest area of the province is the piedmont plain of the Turkestan, Malguzar and Nuratau ridges (the mean annual temperature reaches 15.1°C, and the mean temperature of July reaches 29.4°C). The coldest area is the alpine zone of the Turkestan range (the mean annual temperature does not exceed 5°C, and the mean temperature of July is 11–16°C). The lowest temperature recorded in the province is –37°C, and the highest temperature is 48°C (Alibekov & Nishanov, 1978; Williams & Konovalov, 2008).

The amount of precipitation and humidity increases from the northwest to the southeast and from the plain to the mountains. The mean annual precipitation is 150–200 mm in the plain, and 200–300 mm in foothills. In the montane and alpine zones, this indicator ranges from 300–400 mm (Nuratau Mts.) to 400–500 mm (Turkestan Ridge). Precipitation in the desert and foothill zones occurs mainly in the winter and spring period; the maximum is in March–April, and the dry period is July–September. In the mountains, the rainfall takes place throughout the year, but the maximum is in April–May (Alibekov & Nishanov, 1978; Williams & Konovalov, 2008).

The Dzhizak Province is distinguished by outstanding diversity of flora and vegetation determined by its geographical location, natural conditions and wide variety of landscapes and habitats. The territory of the province is represented by all altitudinal zones, from the plain to highlands, the majority of Central Asian vegetation types and most types of ecosystems including sandy and clay desert, saline lands, wetlands, insular relic mountains, foothills, montane and alpine ecosystems. In accordance with phytogeographical division of Uzbekistan, the territory of the Dzhizak Province belongs to Nuratau and Kuhistan dictricts of Mountain Central Asian Province, and Middle Syrdarya and Kyzylkum districts of the Turan Province (Tojibaev et al., 2016; Tojibaev et al., 2017). The study area is divided into following phytogeographical regions: Nuratau, Nuratau Relic Mountains and Aktau regions of the Nuratau district, North Turkestan and Malguzar regions of the Kuhistan dictrict, Mirzachul region of the Middle Syrdarya district, and Kyzylkum region of the Kyzylkum district (Figure B).

Northern and north-western part of the Dzhizak Province belongs to the Kyzylkum region of the Kyzylkum district. This area includes the aeolian plain of the Southeastern Kyzylkum with fixed and semi-fixed sands, and the Aydar-Arnasay lake system with saline wetlands and small plots of tugay vegetation. The sand desert is covered with psammophilous woodlands of saxaul, sand acacia and *Calligonum* species, psammophilous shrubs, semishrubsand herbs (Zakirov, 1971, 1973, 1976, 1984).

Before 1993-1995, the shore of the Aydar-Arnasay lake system was fringed with a wide belt of tugay vegetation (reeds and riparian woodlands of bloomy poplar, willow, oleaster and tamarisk). Since 1993, the Chardara Reservoir was overflowed with waters of the Syrdarya River several times. A huge volume of water was drained into the Aydar-Arnasay lake system, the water surface level was drastically increased, and large areas were flooded. Currently, tugay vegetation is represented mainly with local plots of reeds, camel-thorn and tamarisks.

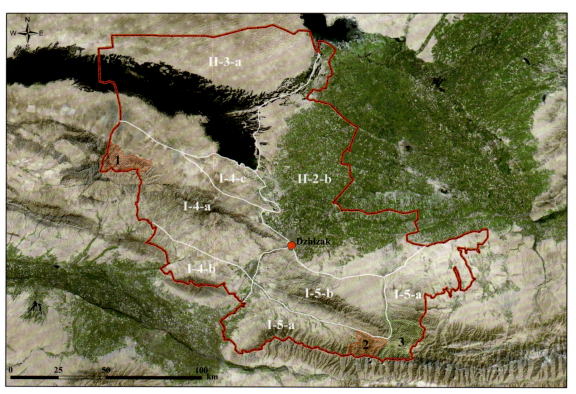

Figure B Phytogeographical regions of the Dzhizak Province (according to Tojibaev et al., 2017)

Notes: I. Central Asian Mountain Province: I-4 Nuratau district (I-4-a Nuratau, I-4-b Aktau, I-4-c Nuratau Relic Mountains). II. Turan Province: II-2 Middle-Syrdarya district (II-2-b Mirzachul), II-3 Kyzylkum (II-3-a Kyzylkum). Protected areas: 1—Nuratau Nature Reserve, 2—Zaamin Nature Reserve, 3—Zaamin National Park.

The central and north-eastern part of the Dzhizak Province belongs to the Mirzachul region of the Middle Syrdarya district. The Mirzachul or Hungry Steppe is an alluvial-proluvial plain on the left bank of the Syrdarya River in its middle reaches between the throat of the Fergana valley in the east, the Chardara water reservoir, Aydar-Arnasay lake system and salt river Kly in the west, and foothills of Turkestan and Malguzar ridges on the south. At present, this region is densely populated and occupied by anthropogenic landscapes. Small areas of natural ecosystems with ephemeral-ephemeroid, saline and tugay vegetation can be found on the eastern shore of the Aydar-Arnasay lake system, in the piedmonts of the Turkestan and Malguzar ridges, and along the irrigation network (Zakirov, 1971, 1973, 1976, 1984).

The region of Nuratau Relic Mountains include the insular ridges, Pistalitau (557 m.s.l.), Khanbandytag (476 m.s.l.), Egarbelistag (618 m.s.l.) and Balyklitau (581 m.s.l.), located at the southern edge of the Kyzylkum, to the north of the Nuratau ridge. They stretch parallel to the Nuratau Mountains from south-east to north-west. The largest of them is the Pistalitau ridge approximately 40 km long (Aramov, 2012). These small ridges are covered with sagebrush communities and xerophytic shrublands. The general appearance of vegetation and dominant species is similar to that in the lower belts of the Nuratau and other mountain ridges of the North-Western Pamir-Alay (Batoshov & Beshko, 2013, 2015; Batoshov, 2016; Tojibaev et al., 2017).

The territory of the Nuratau region includes the northern branch of the Nuratau Mountains (or North Nuratau Ridge) and is divided into two parts separated by the Saurbel pass: the Nuratau and Koytash ridges. The length of this middle-altitude mountain chain is approximately 200 km, and the highest peak named Khayat-bashi is 2169 m.s.l. (Aramov, 2012).

The Nuratau ridge was formed in the late Proterozoic and early Paleozoic as a result of the Baikal and the Caledonian orogeny. It is one of the oldest mountains of Central Asia, almost unaffected by the Alpine orogeny. During transgressions of the Tethys Sea (Cretaceous–Paleogene), it was an isolated island. These circumstances contributed to the evolution of the autochthonous flora. In general, the vegetation has xerophytic character with an original type of vertical zonality determined by geographical position and natural conditions of the territory, but the majority of mountainous Central Asian vegetation types (except for alpine communities) are represented here. The piedmont plain named the Farish Steppe, and foothills of the Nuratau ridge (400–800 m.s.l.) are covered with ephemeroid–sagebrush and phlomis–ephemeroid communities. The stony slopes in the upper part of the foothills are occupied by petrophytic ephemeral–sagebrush vegetation and xerophytic shrublands. The montane zone (from 800 m.s.l. up to the crest of the ridge) has mosaic vegetation represented by tall grass and tall herb communities (so-called savannoids), forb-sagebrush, semishrub and dwarf shrub communities (so-called phryganoids), bunch grass steppes, spiny herbs and pulvinates (so-called mountain xerophytes), shrubs and open woodlands. This area is the most western point of the distribution of forests of *Juniperus polycarpos* var. *seravschanica*. In the past, juniper woodlands grew in the Nuratau Mountains, but they were cut down almost entirely by the beginning of the 20th century and were replaced with scrublands and tall grass communities. Small populations and solitary trees of *Zeravshan juniper* were still preserved in some remote places (Zakirov P., 1969, 1971; Zakirov K., 1971, 1973, 1976, 1984; Tojibaev et al., 2017).

The Nuratau ridge is one of the three isolated sites of the relict walnut-fruit forests in Uzbekistan. The walnut-fruit stands occupy the valleys of the largest streams on the northern slope of the Nuratau ridge. Currently, they represent an original cultural landscape (so-called forest-garden) that has been transformed extensively by local people over the centuries-old horticultural activities. These gallery woodlands with 0.7–0.8 degree of density are composed by *Juglans regia* L., *Malus sieversii* (Ledeb.) M. Roem, *Prunus armeniaca* L., *Morus alba* L., *Acer semenovii* Regel & Herder, *Crataegus turkestanica* Pojark., *etc*. Walnut-fruit forests of the Nuratau ridge accounts approximately 16.7 percent of the national area.

In central part of Nuratau ridge, there is situated the Nuratau Nature Reserve (17.752 hectares, among which 2.303 are forested), a strictly protected area of Category I IUCN established in 1975. Flora of the Nuratau Nature Reserve numbers 835 species, 33 of them are red-listed at the national level (Beshko et al., 2013).

Aktau phytogeographical region includes the southern branch of the Nuratau Mountains, which consists of several small ridges. Within the study area, it is represented with the north slopes of Khobduntau and Karachatau ridges (1672 and 1101 m.s.l., respectively) situated in the south-western part of the Dzhizak Province (Aramov, 2012). Vegetation of of this territory is similar with foothills and lower part of montane zone of the Nuratau ridge.

The southern and south-eastern part of the Dzhizak Province belongs to the North Turkestan phytogeographical region. This area covers the northern slope of the Turkestan ridge with the complete profile of altitudinal zones. The zone of foothills lies here between of 480–500 and 1000–1300 m.s.l, montane zone is situated between 1000–1300 and 2800–3000 m.s.l., and upper part of the ridge (up of 2800–3000 m.s.l.) belongs to the alpine zone. The highest peak of the Dzhizak Province, Mt. Shaukartau (4029 m.s.l.) is located here (Aramov, 2012). This is the most humid area of the Dzhizak Province. The flora and vegetation of this area is distinguished by the significant diversity due to the strongly rugged terrain, large amplitude of elevations, and wide range of soil and climatic conditions. There are represented the semi-arid version of all piedmont and mountainous types of landscapes of Central Asia and almost all mountain types of vegetation. The piedmont plain and foothills are characterized by ephemeral-sagebrush and low herb ephemeral–ephemeroid communities, often with saltworts. Tall grass savannoids and xerophytic open woodlands replace them at 1000–1300 m.s.l. In river valleys, there are small plots of walnut–fruit forests. A thick belt of juniper forests begins at 1500–1700 m.s.l. In some areas, the juniper trees form stands with a stock density of 0.6–1.0. *Juniperus polycarpos* var. *seravschanica* (Kom.) Kitam., *J. semiglobosa* Regel, and *J. pseudosabina* Fisch. & C. A. Mey. dominates in the lower, middle, and upper part of the juniper belt, respectively. Despite the rich diversity of deciduous trees and shrubs represented with maple, hawthorn, Sievers apple, cherry-plum, honeysuckle, and other species, broad-leaved woodlands not play a notable role in the landscape.The vegetation of highlands is comprised of juniper elfin wood, bunchgrass steppes, mountain xerophytes and alpine meadows (Demurina, 1975; Konnov, 1973, 1990; Zakirov K., 1971, 1973, 1976, 1984; Esankulov, 2012; Khassanov et al., 2013; Tojibaev, 2017).

The Zaamin Nature Reserve is situated here, it is one of the oldest nature reserves in Central Asia established in 1926 under the name Guralash Reserve. At the present time, the area of the nature reserve is 26.840 hectares (22.137 hectares are forested). The check-list of flora of the Zaamin Nature Reserve includes 1216 plant species (Khassanov et al., 2013). In the east, the reserve borders with the territory of the Zaamin National Park (23.894 hectares) established in 1976. A synopsis of the flora of the national park is still incomplete (Beshko et al., 2013a).

The Malguzar phytogeographical region includes the Malguzar ridge (2620 m.s.l.), a north-western spur of the Turkestan ridge stretching about 80 km and separated from the Turkestan ridge by the Zaaminsu River valley in the east and the Sanzar river valley in the south (Aramov, 2012). The gorge called Tamerlane's Gate is the border between the Malguzar and Nuratau ridges. The northern slope of the ridge faces the Hungry Steppe, which is almost completely occupied by arable land. The vegetation of Malguzar and the northern slope of the Turkestan ridge is similar (Demurina, 1975; Zakirov K., 1971, 1973, 1976, 1984). There are arid and semi-arid piedmont and mountain landscapes, and most mountain vegetation types (except for highlands). As on the Turkestan ridge, the foothills are covered with ephemeroid and ephemeroid–sagebrush communities. The vegetation of the montane zone is represented with savannoids, fescue steppes, xerophytic shrubs, juniper woodlands with small plots of broad-leaved stands. The natural boundary between foothills and montane zone lies at the same elevations as in the northern slope of the Turkestan ridge. The differences between the flora and vegetation of the Malguzar and North Turkestan phytogeographical regions (e.g., the absence of alpine plants and communities) are explained by the relatively low altitude and xeric conditions of the Malguzar ridge (Demurina, 1975; Azimova, 2017; Tojibaev et al., 2017).

Uzbekistan is the most densely populated country in Central Asia, and the ecosystems of the Dzhizak Province have been influenced by human activities since ancient times. Anthropogenic impact has increased considerably since the second half of the 20th century. Indigenous vegetation communities have been transformed extensively, except for remote desert and mountain areas. In the Hungry Steppe, natural ecosystems have been replaced almost completely by arable lands and settlements. In the piedmonts, large areas are covered by rainfed crops and fallow lands with ruderal vegetation. In the mountainous part of the Province, the main threats influencing the plant diversity are excessive grazing and clear-cutting of trees and shrubs. Forests and open woodlands are under serious threat from over-exploitation, over-grazing and desertification. In this connection, the study and analysis of historical and current distribution of plant species and communities within this large region of the country is extremely important.

Landscapes of South-eastern Kyzylkum and sandy desert with a giant Umbelliferae *Ferula foetida* (photography by N. Yu. Beshko)

Landscapes of South-eastern Kyzylkum and psammophilous woodlands of white saxaul *Haloxylon persicum* (photography by N. Yu. Beshko)

Aydar-Arnasay lake system (photography by N. Yu. Beshko)

Aydar-Arnasay lake system, a view from the ridge Pistalitau (photography by N. Yu. Beshko)

The insular ridge Balyklitau (photography by N. Yu. Beshko)

Piedmonts of the ridge Nuratau (photography by N. Yu. Beshko)

Landscapes of the Nuratau ridge (photography by N. Yu. Beshko)

Landscapes of the Nuratau ridge and the rock massif Parrandas (photography by N. Yu. Beshko)

Landscapes of the Nuratau ridge and spiny almond communities in foothills (photography by N. Yu. Beshko)

Landscapes of the Nuratau ridge and the walnut-fruit gallery forest (photography by N. Yu. Beshko)

Landscapes of the Nuratau ridge and tall grass communities and xerophytic shrublands (photography by N. Yu. Beshko)

Landscapes of the Nuratau ridge and mountain xerophytes (photographby N. Yu. Beshko)

Landscapes of the Turkestan ridge and juniper forests (photography by N. Yu. Beshko)

Landscapes of the Malguzar ridge (photography by N. Yu. Beshko)

Landscapes of Mirzachul (Hungry Steppe) and *Papaver pavoninum* on the fallow lands (photography by N. Yu. Beshko)

Landscapes of Mirzachul (Hungry Steppe) and saline lands (photography by N. Yu. Beshko)

INVENTORY OF THE FLORA

First botanical expeditions in the territory that currently belongs to the Dzhizak Province were conducted in the Sanzar river basin and surroundings of the town Dzhizak by A. Fedtschenko (1866), P. Capus and G. Bonvalot (1881), A. Regel (1882), and V. I. Lipsky (1890). Many naturalists and officials, who worked in Central Asia in the late 19th century, also collected herbarium specimens and contributed to the accumulation of floristic data.

The systematic and purposeful study of the flora and vegetation of Central Asia began in the 1910s. Several significant botanical expeditions were carried out in the first half of the 20th century and in 1950–1970s on the Turkestan, Malguzar and Nuratau Ranges and in the Hungry Steppe. The studies of this period have been focused on applied purposes, mainly on an investigation of the vegetation cover and resource plants. The most important collections in this region were made by B. A. Fedtschenko, E. P. Korovin, V. P. Drobow, M. G. Popov, I. I. Sprygin, M. V. Kultiassov, B. S. Zakrzczewsky, S. N. Kudrjashev, P. A. Gomolitsky, M. M. Sovetkina, E. M. Demurina, E. E. Korotkova, O. E. Knorring, A. I. Vvedensky, L. L. Bulgakova, L. I. Nazarenko, P. K. Zakirov, and R. V. Kamelin. *Essay on the Vegetation of the Pistalitau Mountains* by M. V. Kultiassov (1923), *Plant Formations of the Nurata Valley* (1923) and *Vegetation of Middle Asia and South Kazakhstan* (1934; 1961; 1962) by E. P. Korovin, *Vegetation of the Khobduntau and Karachatau Mountains* by S. N. Kudrjashev (1930), *Vegetation of the Guralash Nature Reserve and Zaamin* Forestry by M. G. Popov and N. V. Androssov (1937), *Flora and Vegetation of the Zeravschan River Basin* by K. Z. Zakirov (1955, 1962), *Vegetation Cover of the Nuratau Mountains* (1969) and *Botanical Geography of Low Mountains of Kyzylkum and Nuratau Ridge* (1971) by P. K. Zakirov, *Vegetation of the Western Part of Turkestan Ridge and Its Spurs* by E. M Demurina (1975), *To the Knowledge of the Flora of Nuratau Mountains* (1973) and *The Kuhistan District of Mountainous Middle Asia* (1979) by R.V. Kamelin should be cited as the most important publications of the second half of the 20th century dedicated to the plant diversity of this region. Several new studies have been performed since 2000 (Beshko, 2000a, 2000b, 2011; Tirkasheva, 2011; Botirova, 2012; Esankulov, 2012; Khassanov et al., 2013; Batoshov, 2016; Azimova, 2017), a number of new species have been described from this territory and dozens of new records were found. Analysis of obtained data allows us to give a summary of floristic studies in the Dzhizak Province.

The synopsis of flora of the Dzhizak Province of the Republic of Uzbekistan has been compiled for the first time as a result of analysis of a huge volume of data of our long-term field surveys, revision of herbarium material stored in TASH, LE and MW, publications, and reports (Azimova, 2017; Batoshov & Beshko, 2013, 2015; Batoshov, 2016; Belolipov, 1973; Beshko, 2000a, 2000b, 2011; Beshko & Azimova, 2013, 2014; Beshko et al., 2013, 2014; Botirova, 2012; Botschantzev et al., 1961; *Conspectus Florae Asiae Mediae* (1963–1993); Demurina, 1975; Esankulov, 2012; *Flora of Tajikistan* (1957–1991); *Flora of Uzbekistan* (1941–1962); Kamelin, 1973a, 1973b, 1979, 1990; Khassanov, 2015; Khassanov et al, 2013; Konnov, 1973, 1990; Korovin, 1923; Kudrjashev, 1930; Kultiassov, 1923; Popov & Androsov, 1937; Sennikov, 2016, 2017; Tirkasheva, 2011; Tojibaev & Beshko, 2007, 2015; Tojibaev et al., 2015; Zakirov, 1955, 1962; Zakirov & Burygin, 1956; Zakirov, 1969, 1971).

The *Flora of the Dzhizak Province, Uzbekistan* contains information about plant species arranged in the following order: currently accepted scientific name, commonly-used synonymy (optional), life form, habitat, altitudinal zone, distribution in the Province, character of use, conservation status (for nationally or globally red-listed species).

Families and genera of ferns in the checklist are arranged in accordance with the modern classification that was recently proposed by the Pteridophytes Phylogeny Group (Christenhusz et al., 2011a), Gymnosperms are arranged according to Christenhusz et al. (2011b), flowering plants are arranged in accordance with the APG system (APG IV, 2016). Species are arranged in alphabetic order. Accepted names of species are provided in accordance with The Plant List database (www.theplantlist.org), databases and recently published treatments of different taxonomical groups (Akhani et al., 2007; Al-Shehbaz et al., 2014; Clayton et al., 2006; Degtjareva et al., 2013; Downie et al., 2010; Egorova, 1999; http://www.e-monocot.org/; Hill, 2017; Horandl & Emadzade, 2012; www.ildis.com; Khassanov & Rakhimova N. 2012; Lammers, 2007; Lazkov, 2016; Lopez-Vinyallonga et al., 2009, 2011; Podlech & Zarre, 2013; Salmaki et al., 2012; Sennikov, 2016, 2017; The Global Compositae Checklist, 2009; The Gymnosperm Database, 2017; The World Umbellifer Database, 1999; Tsvelev, 1983; Wiegleb et al., 2017). Citation of authorship of taxa corresponds to the *Authors of Plants Names* (Brummitt & Powell, 1992) and the International Plant Names Index (www.ipni.org).

Altitudinal distribution is provided with the following elevation zones: plains, foothills, montane zone, and alpine zone (their natural boundaries were described above). The distribution of species within the Dzhizak Province was indicated according to the scheme of phytogeographical regions of Uzbekistan (Tojibaev et al., 2017).

Character of economical use of plants is done in accordance with published data (*Plant Resources of the U. S. S. R.,* 1984–1996; Larin, 1956; Gammerman & Grom, 1976; Nikitin, 1983; Kurmikov & Belolipov, 2012, *etc.*).

IUCN RL is the abbreviation for the IUCN Red List (www.iucnredlist.org). UzbRDB is the abbreviation for the *Red Data Book of Uzbekistan* (2009), and categories of threatened species used in that book are employed as follows: 0 (extinct species) — corresponds to EX or EW categories of the IUCN Red List, 1 (endangered, disappearing species) — meets CR or EN categories of IUCN, 2 (rare species) — meets VU category of IUCN and 3 (declining species) — corresponds to NT category of IUCN.

The checklist presented below includes 1965 species of 613 genera and 105 families of vascular plants occurring in the wild in the study area including naturalized aliens. Alien taxa are marked with an asterisk. The assessment of alien status is based on analysis of publications and online databases (Nikitin, 1983; IUCN/ISSG, 2014; CABI, 2017).

There are 44 species of trees and 116 shrubs, among which 11 species can have a habit of tree or shrub; 108 species of semi-shrubs; 1062 perennial herbs, among which 61 biennial, 516 annual; 46 herbaceous plants can have different life forms (annual, biennial or perennial). The number of plant species recorded for different phytogeographical regions of the Dzhizak Province is shown in Table 1. The leading families and genera are represented in Table 2, and ratio of life forms is shown in Table 3.

Only 5 species of the flora of the Dzhizak Province are assessed as globally threatened (CR, EN, VU) on the IUCN Red List (www.iucnredlist.org), and 3 species are assessed as Near Threatened (Table 4). Only one species among them (*Lonicera paradoxa* Pojark.) is red-listed at the national level. The status of *Platycladus orientalis* (L.) Franco is uncertain because the species was naturalized in different Asian countries in ancient times, but many botanists regarded *Platycladus* in Uzbekistan as native and Tertiary relic. The list of plants included in the *Red Data Book of Uzbekistan* is represented with 51 species, most of them (42 species) occur within protected areas (Table 5).

Following 20 species of the flora of the Dzhizak Province are endemic to the Nuratau Mountains: *Acantholimon nuratavicum* Zakirov ex Lincz., *Acantholimon subavenaceum* Lincz., *Acantolimon zakirovii* Beshko, *Arctium pallidivirens* (Kult.) S. López, Romaschenko, Susanna & N. Garcia [*Anura pallidivirens* (Kult.) Tscherneva], *Cousinia botschantzevii* Juz. ex Tscherneva, *Dianthus helenae* Vved., *Dracocephalum nuratavicum* Adylov, *Eremurus nuratavicus* A. P. Khohkr., *Helichrysum nuratavicum* Krasch., *Jurinea zakirovii* Iljin, *Lagochilus olgae* Kamelin, *Lagochilus proskorjakovii* Ikramov, *Lappula nuratavica* Nabiev & Zakirov, *Lepidium olgae* (R. M. Vinogr.) Al-Shehbaz & Mummenhoff (*Stubendorffia olgae* R.Vinogradova), *Oxytropis pseudorosea* Filim., *Parrya nuratensis* Botsch. & Vved., *Phlomis nubilans* Zakirov, *Phlomoides anisochila* (Pazij & Vved.) Salmaki, *Salvia submutica* Botsch. & Vved., *Thymus subnervosus* Vved., Nabiev & Tulyag. One species, *Ferula helenae* Rakhm. & Melibaev, is endemic to the insular low mountains located to the north of the Nuratau ridge.

One species, *Eremurus chloranthus* Popov, is endemic to the Turkestan ridge. Two species are endemic to the Malguzar ridge: *Allium levichevii* F. O. Khass. & N. Sulejm. and *Oxytropis kamelinii* Vassilcz. Following 13 species occurring in Turkestan and Malguzar ridges are endemic to the Kuhistan phytogeographical district: *Allium alexeianum* Regel, *Astragalus belolipovii* Kamelin ex F. O. Khass. & N. Sulajm., *Astragalus russanovii* F. O. Khass., Sarybaeva & Esankulov, *Dianthus subscabridus* Lincz., *Euphorbia rosularis* Fed., *Ferula ovczinnikovii* Pimenov, *Jurinea helichrysifolia* Popov ex Iljin, *Oxytropis capusii* Franch., *Oxytropis seravschanica* Gontsch. ex Vassilcz. & B. Fedtsch., *Primula iljinskii* Fed., *Scutellaria schachristanica* Juz., *Serratula lancifolia* Zakirov, *Tanacetopsis urgutensis* (Popov ex Tzvelev) Kovalevsk. One species, *Bryonia melanocarpa* Nabiev, is endemic to the South-eastern Kyzylkum.

The identification of economically valuable species is an important result of the inventory of the flora of vascular plants of the Dzhizak Province. The most numerous group among them are fodder plants (558 species); 446 species are medicinal plants (officinal and used in folk medicine); 407 species are meliferous, 388 ornamental, 154 essential oil, 123 dye, 136 food plants (including wild-growing vegetables, fruits, berries, nut-bearing, spicy-aromatic species, *etc.*); 89 species can be used for afforestation and soil stabilization. At the same time, various ruderal, segetal and pastoral weeds (355 species), as well as poisonous plants (99 species) and alien species (114) are widespread in the Dzhizak Province. This fact indicates that ecosystems of study area have been noticeably transformed by anthropogenic impact.

Table 1. The number of species recorded for different phytogeographical regions of the Dzhizak Province

Phytogeographical regions	Number of species
Mountain Central Asian Province	
North Turkestan	1463
Malguzar	1255
Nuratau	1139
Nuratau Relic Mountains	614
Aktau	541
Turan Province	
Mirzachul	590
Kyzylkum	456

Table 2. Leading families and genera of the flora of the Dzhizak Province

Families	Number of genera	Number of species	Percentage (%)	Genera	Number of species	Percentage (%)
Asteraceae	82	281	14.32	*Astragalus*	107	5.45
Fabaceae	26	210	10.70	*Allium*	37	1.89
Poaceae	58	173	8.82	*Cousinia*	34	1.73
Brassicaceae	57	113	5.76	*Artemisia*	27	1.38
Lamiaceae	29	95	4.84	*Silene*	25	1.27
Apiaceae	41	91	4.64	*Ranunculus*	22	1.12
Amaranthaceae	29	86	4.38	*Oxytropis*	21	1.07
Caryophyllaceae	25	83	4.23	*Taraxacum*	20	1.02
Rosaceae	18	66	3.36	*Ferula*	19	0.97
Boraginaceae	21	58	2.96	*Carex*	18	0.92
Ranunculaceae	18	57	2.91	*Galium, Polygonum, Salsola, Stipa*	16	0.82

Table 3. The ratio of life forms of the flora of the Dzhizak Province

Life form	Number of species	Percentage (%)
Trees	44	2.24
Shrubs	116	5.91
Trees or shrubs	11	0.56
Semi-shrubs	108	5.50
Perennial herbs	1064	54.17
Biennial herbs	61	3.11
Annual herbs	516	26.22
Annual, biennial or perennial herbs	46	2.34

Table 4. List of plants of the flora of Dzhizak Province included in the *IUCN Red List*

No	Species	Category	Distribution in the Dzhizak Province (phytogeographical regions and protected areas)
1	*Prunus bucharica* (Korsh.) B. Fedtsch. ex Rehder [*Amygdalus bucharica* Korsh.] (Rosaceae)	VU B2ab(iii,v)	Nuratau, Aktau, North Turkestan, Malguzar. Nuratau Reserve, Zaamin Reserve, Zaamin National Park. Dominant species of open woodlands widely distributed in mountainous part of the province
2	*Prunus armeniaca* L. [*Armeniaca vulgaris* Lam.] (Rosaceae)	EN B2ab(iii)	Nuratau, North Turkestan, Malguzar. Nuratau Reserve, Zaamin Reserve, Zaamin National Park
3	*Fraxinus sogdiana* Bunge (Oleaceae)	NT	Nuratau, North Turkestan, Malguzar. Nuratau Reserve, Zaamin Reserve, Zaamin National Park
4	*Juglans regia* L. (Juglandaceae)	NT	Nuratau, North Turkestan, Malguzar. Nuratau Reserve, Zaamin Reserve, Zaamin National Park
5	*Lonicera paradoxa* Pojark. (Caprifoliaceae)	EN B2ab(iii,v) UzbRDB 1	North Turkestan
6	*Malus sieversii* (Ledeb.) M. Roem (Rosaceae)	VU A2cde	Nuratau, Aktau, North Turkestan, Malguzar. Nuratau Reserve, Zaamin Reserve, Zaamin National Park
7	*Platycladus orientalis* (L.) Franco (Cupressaceae)	NT	Nuratau. Nuratau Reserve. One giant "holy" tree in the valley Madjrum ca. 2000 years old
8	*Pyrus korshinskyi* Litv. (Rosaceae)	CR B2ab(iii,v)	Malguzar. Only one location is known

Table 5. List of plants of the flora of the Dzhizak Province included in the Red Data Book of Uzbekistan

No	Species	Category of UzbRDB	Distribution in the Dzhizak Province (phytogeographical regions and protected areas)
1	*Acantholimon nuratavicum* Zakirov ex Lincz. (Plumbaginaceae)	2	Nuratau. Nuratau Reserve
2	*Aconitum talassicum* Popov (Ranunculaceae)	3	North Turkestan, Malguzar. Zaamin Reserve, Zaamin National Park
3	*Allium isakulii* R. M. Fritsch & F. O. Khass. (Amaryllidaceae)	2	Nuratau, Malguzar. Nuratau Reserve
4	*Allium praemixtum* Vved. (Amaryllidaceae)	1	Nuratau, Malguzar. Nuratau Reserve
5	*Allochrusa gypsophiloides* Regel (Caryophyllaceae)	3	Nuratau, Malguzar, North Turkestan
6	*Arctium haesitabundum* (Juz.) S. López, Romasch., Susanna & N. Garcia [*Cousinia haesitabunda* Juz.] (Asteraceae)	1	Nuratau, Malguzar
7	*Arctium pallidivirens* (Kult.) S. López, Romaschenko, Susanna & N. Garcia [*Anura pallidivirens* (Kult.) Tscherneva] (Asteraceae)	1	Nuratau, Malguzar. Nuratau Reserve
8	*Astragalus belolipovii* Kamelin ex F. O. Khass. & N. Sulajm. (Fabaceae)	1	North Turkestan, Malguzar. Zaamin Reserve
9	*Astragalus kelleri* Popov (Fabaceae)	2	Nuratau, Aktau, Nuratau Relic Mountains, Malguzar. Nuratau Reserve
10	*Astragalus knorringianus* Boriss. (Fabaceae)	2	Nuratau, Aktau, Nuratau Relic Mountains, Malguzar, North Turkestan. Nuratau Reserve, Zaamin Reserve, Zaamin National Park
11	*Astragalus leptophysus* Vved. (Fabaceae)	2	Nuratau, North Turkestan. Nuratau Reserve
12	*Bryonia melanocarpa* Nabiev (Cucurbitaceae)	2	Kyzylkum
13	*Cicer grande* (Popov) Korotkova (Fabaceae)	2	Nuratau. Nuratau Reserve
14	*Colchicum kesselringii* Regel (Colchicaceae)	3	Nuratau, Malguzar, North Turkestan. Nuratau Reserve, Zaamin Reserve, Zaamin National Park
15	*Cousinia dshizakensis* Kult. (Asteraceae)	2	Nuratau, Aktau, Malguzar, North Turkestan. Nuratau Reserve, Zaamin Reserve, Zaamin National Park
16	*Eremurus chloranthus* Popov (Asphodelaceae)	0	North Turkestan. Zaamin Nature Reserve. Only type specimen collected in 1926 is known

(continued)

No	Species	Category of UzbRDB	Distribution in the Dzhizak Province (phytogeographical regions and protected areas)
17	*Eremurus lactiflorus* O. Fedtsch. (Asphodelaceae)	2	Nuratau. Nuratau Nature Reserve
18	*Eremurus nuratavicus* A. P. Khokhr. (Asphodelaceae)	2	Nuratau. Nuratau Nature Reserve
19	*Eremurus robustus* (Regel) Regel (Asphodelaceae)	3	Nuratau, Malguzar, North Turkestan. Nuratau Reserve, Zaamin Reserve, Zaamin National Park
20	*Ferula fedtschenkoana* Koso-Pol. (Apiaceae)	1	Malguzar, North Turkestan. Zaamin Reserve, Zaamin National Park
21	*Ferula moschata* (H. Reinsch) Koso-Pol. [*Ferula sumbul* (Kauffm.) Hook. f.] (Apiaceae)	2	Nuratau, Malguzar, North Turkestan. Nuratau Reserve, Zaamin Reserve, Zaamin National Park
22	*Gladiolus italicus* Mill. (Iridaceae)	1	North Turkestan. Only one herbarium specimen collected in 1934 is known
23	*Helichrysum nuratavicum* Krasch. (Asteraceae)	2	Nuratau. Nuratau Reserve
24	*Jurinea zakirovii* Iljin (Asteraceae)	1	Nuratau. Nuratau Reserve
25	*Lagochilus inebrians* Bunge (Lamiaceae)	2	Nuratau, Aktau, Malguzar. Nuratau Reserve
26	*Lagochilus olgae* Kamelin (Lamiaceae)	2	Nuratau. Nuratau Reserve
27	*Lagochilus proskorjakovii* Ikramov (Lamiaceae)	1	Nuratau. Nuratau Reserve
28	*Lappula nuratavica* Nabiev & Zakirov (Boraginaceae)	2	Nuratau. Nuratau Reserve
29	*Lepidium olgae* (R. Vinogradova) Al-Shehbaz & Mummenhoff [*Stubendorffia olgae* R. Vinogradova] (Brassicaceae)	2	Nuratau. Nuratau Reserve
30	*Lepidolopha nuratavica* Krasch. (Asteraceae)	1	Nuratau. Nuratau Reserve
31	*Lipskya insignis* Nevski (Apiaceae)	3	Malguzar
32	*Lonicera paradoxa* Pojark. (Caprifoliaceae)	1	North Turkestan
33	*Oxytropis pseudorosea* Filim. (Fabaceae)	2	Nuratau. Nuratau Reserve
34	*Paeonia tenuifolia* L. [*P. hybrida* Pall.] (Paeoniaceae)	3	North Turkestan. Zaamin Nature Reserve. Only one location is known
35	*Phlomoides anisochila* (Pazij & Vved.) Salmaki (Lamiaceae)	1	Nuratau. Nuratau Reserve
36	*Phlomis nubilans* Zakirov (Lamiaceae)	2	Nuratau. Nuratau Reserve

(continued)

No	Species	Category of UzbRDB	Distribution in the Dzhizak Province (phytogeographical regions and protected areas)
37	*Parrya olgae* (Regel & Schmalh) D. A. German & Al. Shehbaz [*Pseudoclausia olgae* (Regel & Schmalh.] (Brassicaceae)	2	Nuratau, Malguzar, North Turkestan. Nuratau Reserve, Zaamin Reserve, Zaamin National Park
38	*Parrya sarawschanica* (Regel & Schmalh.) D. A. German & Al-Shehbaz [*Pseudoclausia sarawschanica* (Regel & Schmalh.) Botsch.] (Brassicaceae)	2	Nuratau. Nuratau Nature Reserve
39	*Pseudosedum campanuliflorum* Boriss. (Crassulaceae)	1	North Turkestan, Malguzar. Zaamin National Park
40	*Salsola titovii* Botsch. (Chenopodiaceae)	1	Aktau
41	*Salvia submutica* Botsch. & Vved. (Lamiaceae)	2	Nuratau. Nuratau Reserve
42	*Saxifraga hirculus* L. (Saxifragaceae)	2	North Turkestan. Zaamin Reserve, Zaamin National Park
43	*Serratula lancifolia* Zakirov (Asteraceae)	1	Malguzar
44	*Seseli turbinatum* Korovin (Apiaceae)	2	Nuratau. Nuratau Reserve
45	*Silene paranadena* Bondarenko & Vved. (Caryophyllaceae)	2	Nuratau, Malguzar. Nuratau Reserve
46	*Tulipa affinis* Botschantz (Liliaceae)	2	Nuratau, Aktau, Nuratau Relic Mountains, Malguzar, North Turkestan. Nuratau Reserve, Zaamin Reserve, Zaamin National Park
47	*Tulipa dasystemon* Regel (Liliaceae)	2	North Turkestan, Malguzar. Zaamin Reserve, Zaamin National Park
48	*Tulipa korolkowii* Regel (Liliaceae)	2	Nuratau, Aktau, Nuratau Relic Mountains, Malguzar, North Turkestan. Nuratau Reserve, Zaamin Reserve, Zaamin National Park
49	*Tulipa lehmanniana* Mercklin (Liliaceae)	3	Kyzylkum
50	*Tulipa micheliana* Hoog (Liliaceae)	2	Nuratau, Nuratau Relic Mountains, Malguzar. Nuratau Reserve
51	*Vitis vinifera* L. (Vitaceae)	2	Nuratau. Nuratau Reserve

POLYPODIOPHYTA [FERNS]

Family 1. Equisetaceae

Genus 1. Equisetum L.

1. Equisetum arvense L.
Perennial. Banks of rivers, wet places. Plain, foothills, montane zone. Nuratau, Malguzar, North Turkestan. Medicinal.

2. Equisetum ramosissimum Desf.   Figure 1
Perennial. Banks of rivers, wet places, meadows. Plain, foothills, montane zone. Nuratau, Aktau, Nuratau Relic Mountains, Malguzar, North Turkestan, Mirzachul. Medicinal, fodder.

Figure 1 Equisetum ramosissimum (photography by Natalya Beshko)

Family 2. Ophioglossaceae

Genus 2. Botrychium Sw.

3. Botrychium lunaria (L.) Sw.
Perennial. Wet meadows, moraines. Montane and alpine zones. North Turkestan.

Family 3. Marsileaceae

Genus 3. Marsilea L.

4. Marsilea quadrifolia L.
Perennial. Standing and slowly flowing water, rice fields. Plain. Kyzylkum, Mirzachul.

Family 4. Salviniaceae

Genus 4. Salvinia Micheli

5. Salvinia natans (L.) All.
Perennial. Standing and slowly flowing water. Plain. Kyzylkum, Mirzachul.

Family 5. Pteridaceae

Genus 5. Cheilanthes Sw.

6. Cheilanthes persica (Bory) Mett. ex Kuhn.   Figure 2
Perennial. Rocks. Montane zone. Nuratau, Malguzar, North Turkestan.

Figure 2 Cheilanthes persica (photography by Natalya Beshko)

Genus 6. Adiantum L.

7. Adiantum capillus-veneris L.   Figure 3
Perennial. Wet rocks and boulders near waterfalls, stream banks. Montane zone. Nuratau, Malguzar, North Turkestan. Ornamental, medicinal.

Figure 3 Adiantum capillus-veneris (photography by Natalya Beshko)

Family 6. Aspleniaceae

Genus 7. Cystopteris Berh.

8. Cystopteris fragilis (L.) Bernh.   Figure 4
Perennial. Shady wet places. Montane zone. Nuratau, Aktau, Malguzar, North Turkestan.

Figure 4 Cystopteris fragilis (photography by Natalya Beshko)

Genus 8. Asplenium L.

9. Asplenium ruta-muraria L.   Figure 5
Perennial. Rocks. Montane zone. Nuratau, Malguzar, North Turkestan. Ornamental, medicinal.

Figure 5 Asplenium ruta-muraria (photography by Natalya Beshko)

10. Asplenium trichomanes L.
Perennial. Rocks. Montane zone. Nuratau, Malguzar, North Turkestan. Ornamental, medicinal.

Genus 9. Ceterach Gorsault.

11. Ceterach officinarum Willd.   Figure 6
Perennial. Rocks. Montane zone. Malguzar, North Turkestan. Ornamental, medicinal.

Figure 6 Ceterach officinarum (photography by Natalya Beshko)

Family 7. Polypodiaceae

Genus 10. Dryopteris Adans.

12. Dryopteris filix-mas (L.) Schott.
Perennial. Shady wet places. Montane zone. Malguzar, North Turkestan. Ornamental, medicinal.

GYMNOSPERMAE [GYMNOSPERMS]

Family 8. Ephedraceae

Genus 11. Ephedra L.

13. Ephedra botschantzevii Pachom.
Shrub. Stony slopes. Foothills, montane zone. Nuratau.

14. Ephedra equisetina Bunge   Figure 7
Shrub. Stony and fine-earth slopes, rocks, screes. Plain, foothills, montane and alpine zones. Nuratau, Aktau, Nuratau Relic Mountains, Malguzar, North Turkestan. Medicinal.

Figure 7 Ephedra equisetina (photography by Natalya Beshko)

15. Ephedra fedtschenkoae Paulsen
Shrub. Stony slopes. Montane and alpine zones. Nuratau, Malguzar, North Turkestan.

16. Ephedra foliata Boiss. ex C. A. Mey. [*Ephedra kokanica* Regel]   Figure 8
Shrub. Stony and fine-earth slopes. Foothills, montane zone. Nuratau, Nuratau Relic Mountains, Malguzar, North Turkestan.

Figure 8 Ephedra foliata (photography by Natalya Beshko)

17. Ephedra intermedia Schrenk & C. A. Mey.   Figure 9

Shrub. Stony and fine-earth slopes. Foothills, montane zone. Nuratau, Aktau, Nuratau Relic Mountains, Malguzar, North Turkestan. Medicinal.

Figure 9 Ephedra intermedia (photography by Natalya Beshko)

18. Ephedra regeliana Florin   Figure 10

Shrub. Gravelly and stony slopes. Montane and alpine zones. Nuratau, Malguzar, North Turkestan.

Figure 10 Ephedra regeliana (photography by Natalya Beshko)

19. Ephedra strobilacea Bunge   Figure 11

Shrub. Sandy and clay desert, relic mountains. Plain, foothills. Nuratau Relic Mountains, Kyzylkum. Fodder, afforestation.

Figure 11 Ephedra strobilacea (photography by Natalya Beshko)

Family 9. Cupressaceae

Genus 12. *Platycladus Spach[①]

20. *Platycladus orientalis (L.) Franco

Tree. A giant "holy" tree in the valley Madjrum, the Nuratau Nature Reserve, ca. 8 m in diam., ca. 2000 years old. Their status is uncertain because the species was naturalized in many Asian countries in ancient times, but botanists regarded *Platycladus* in Uzbekistan as native and Tertiary relic. Nuratau. Ornamental. IUCN NT.

Genus 13. Juniperus L.

21. Juniperus pseudosabina Fisch. & C. A. Mey. [*Juniperus turkestanica* Kom.]   Figure 12

Tree. Stony and fine-earth slopes. Montane and alpine zones. North Turkestan. Industrial, essential

---

① An asterisk means the alien taxon.

oil, medicinal, afforestation.

Figure 12 **Juniperus pseudosabina** (photography by Natalya Beshko)

22. Juniperus semiglobosa Regel   Figure 13

Tree. Stony and fine-earth slopes, screes. Montane and alpine zones. Malguzar, North Turkestan. Industrial, essential oil, medicinal, afforestation.

Figure 13 **Juniperus semiglobosa** (photography by Natalya Beshko)

23. Juniperus polycarpos var. **seravschanica** (Kom.) Kitam. [*Juniperus seravschanica* Kom.] Figure 14
Tree. Stony and fine-earth slopes, rocks, screes. Montane zone. Nuratau, Malguzar, North Turkestan. Industrial, essential oil, food (spice-aromatic), medicinal, afforestation.

Figure 14 Juniperus polycarpos var. **seravschanica** (photography by Natalya Beshko)

ANGIOSPERMAE [ANGIOSPERMS] LILIOPSIDA [MONOCOTS]

Family 10. Araceae

Genus 14. Lemna L.

24. Lemna gibba L.
Perennial. Standing and slowly flowing water. Plain. Mirzachul, Kyzylkum. Medicinal, fodder, food.

25. Lemna minor L.
Perennial. Standing and slowly flowing water. Plain. Mirzachul, Kyzylkum. Medicinal, fodder, food.

Genus 15. Arum L.

26. Arum korolkowii Regel   Figure 15
Perennial. Shady wet places, river valleys, stony slopes, rocks, screes. Montane zone. Nuratau, Aktau, Nuratau Relic Mountains, North Turkestan. Medicinal, poisonous, ornamental, dye.

Figure 15 Arum korolkowii (photography by Natalya Beshko)

Genus 16. Eminium Schott.

27. Eminium lehmanii (Bunge) O. Kuntze
Perennial. Sandy deserts. Plain. Nuratau Relic Mountains, Kyzylkum. Poisonous, ornamental.

Family 11. Alismataceae

Genus 17. Alisma L.

28. Alisma plantago-aguatica L.
Perennial. Banks of rivers, lakes and canals, shallows. Plain, foothills, montane zone. Nuratau, Nuratau Relic Mountains, Malguzar, North Turkestan, Mirzachul, Kyzylkum. Medicinal, poisonous.

Genus 18. Sagittaria L.

29. Sagittaria trifolia L.
Perennial. Banks of rivers, lakes and canals, shallows. Plain, foothills. Nuratau Relic Mountains, Malguzar, Mirzachul, Kyzylkum. Medicinal, fodder, ornamental.

Family 12. Butomaceae

Genus 19. Butomus L.

30. Butomus umbellatus L.
Perennial. Banks of rivers, lakes and canals, shallows. Plain, foothills, montane zone. Nuratau Relic Mountains, Mirzachul, Kyzylkum. Medicinal, fodder, meliferous, ornamental.

Family 13. Hydrocharitaceae

Genus 20. *Vallisneria L.

31. *Vallisneria spiralis L.
Perennial. Standing and slowly flowing water. Plain. Nuratau Relic Mountains, Malguzar, North

Turkestan, Mirzachul, Kyzylkum. Fodder (for geese and ducks), ornamental.

Genus 21. Najas L.

32. Najas marina L.
Annual. Standing and slowly flowing water. Plain. Nuratau Relic Mountains, Malguzar, Mirzachul, Kyzylkum.

33. Najas minor All.
Annual. Standing and slowly flowing water. Plain, foothills. Nuratau Relic Mountains, Malguzar, North Turkestan, Mirzachul, Kyzylkum.

Family 14. Juncaginaceae

Genus 22. Triglochin L.

34. Triglochin palustre L.
Perennial. Banks of rivers, lakes and canals, swamp meadows. Plain, foothills, montane zone. Nuratau Relic Mountains, Malguzar, North Turkestan, Mirzachul. Fodder, food.

Family 15. Potamogetonaceae

Genus 23. Potamogeton L.

35. Potamogeton crispus L.
Perennial. Standing and slowly flowing water. Plain, foothills. Nuratau Relic Mountains, Malguzar, North Turkestan, Mirzachul, Kyzylkum. Fodder (for geese, ducks and muskrats).

36. Potamogeton filiformis Pers.
Perennial. Standing and slowly flowing water. Plain, foothills, montane zone. Nuratau Relic Mountains, Malguzar, North Turkestan, Mirzachul, Kyzylkum. Fodder (for geese, ducks and muskrats).

37. Potamogeton lucens L.
Perennial. Standing and slowly flowing water. Plain, foothills, montane and alpine zones. North Turkestan, Mirzachul, Kyzylkum. Fodder (for geese, ducks and muskrats).

38. Potamogeton natans L.
Perennial. Standing and slowly flowing water. Plain, foothills. Nuratau Relic Mountains, Mirzachul,

Kyzylkum. Fodder (for geese, ducks and muskrats), medicinal, ornamental.

39. Potamogeton pectinatus L.
Perennial. Standing and slowly flowing water. Plain, foothills. Nuratau Relic Mountains, Mirzachul, Kyzylkum. Fodder (for geese, ducks and muskrats), medicinal, ornamental.

40. Potamogeton perfoliatus L.
Perennial. Standing and slowly flowing water. Plain, foothills. Nuratau Relic Mountains, Mirzachul, Kyzylkum. Fodder (for geese, ducks and muskrats), medicinal, ornamental.

41. Potamogeton pusillus L.
Perennial. Standing and slowly flowing water. Plain, foothills. Malguzar, Mirzachul, Kyzylkum. Fodder (for geese, ducks and muskrats), medicinal, ornamental.

Genus 24. Zannichellia L.

42. Zannichellia pedunculata Rchb.
Perennial. Standing and slowly flowing water, usually saline. Plain. Nuratau Relic Mountains, Mirzachul, Kyzylkum. Fodder (for geese, ducks and muskrats).

Family 16. Ruppiaceae

Genus 25. Ruppia L.

43. Ruppia maritima L.
Perennial. Standing and slowly flowing water. Plain. Mirzachul, Kyzylkum. Fodder (for geese, ducks and muskrats).

Family 17. Colchicaceae

Genus 26. Colchicum (Tourn.) L.

44. Colchicum kesselringii Regel   Figure 16
Perennial. Wet places, fine-earth, gravelly and stony slopes, places near melting snow. Plain, foothills, montane and alpine zones. Nuratau, Malguzar, North Turkestan. Medicinal, alkaloid-bearing, poisonous, ornamental, threatened species. UzbRDB 3.

Figure 16 Colchicum kesselringii (photography by Natalya Beshko)

45. Colchicum luteum Baker   Figure 17

Perennial. Fine-earth, gravelly and stony slopes, places near melting snow. Montane and alpine zones. Malguzar, North Turkestan. Medicinal, alkaloid-bearing, poisonous, ornamental, dye.

Figure 17 Colchicum luteum (photography by Natalya Beshko)

Genus 27. Merendera Ramond.

46. Merendera robusta Bunge   Figure 18

Perennial. Sandy and clay deserts, fine-earth slopes. Plain, foothills. Nuratau Relic Mountains, Kyzylkum. Medicinal, alkaloid-bearing, poisonous, ornamental.

Figure 18 Merendera robusta (photography by N. Yu. Beshko)

Family 18. Liliaceae

Genus 28. Gagea Salisb.

47. Gagea afghanica A. Terracc.
Perennial. Sandy soils. Plain. Nuratau Relic Mountains, Mirzachul, Kyzylkum. Ornamental.

48. Gagea bergii Litv.
Perennial. Sandy deserts. Plain. Kyzylkum. Ornamental.

49. Gagea calyptrifolia Levichev
Perennial. Fine-eart and stony slopes. Montane zone. North Turkestan. Ornamental.

50. Gagea capillifolia Vved.
Perennial. Shady places. Montane and alpine zones. North Turkestan. Ornamental.

51. Gagea capusii A. Terracc.   Figure 19
Perennial. Fine-eart and stony slopes. Montane zone. Nuratau, Malguzar, North Turkestan. Ornamental.

Figure 19 Gagea capusii (photography by Natalya Beshko)

52. Gagea chomutovae (Pascher) Pascher   Figure 20
Perennial. Fine-earth and stony slopes. Plain, foothills, montane zone. Nuratau, Aktau, Nuratau Relic Mountains, Malguzar, North Turkestan, Mirzachul. Ornamental.

Figure 20 Gagea chomutovae (photography by Natalya Beshko)

53. Gagea divaricata Regel
Perennial. Sandy soils. Plain. Nuratau Relic Mountains, Kyzylkum. Ornamental.

54. Gagea dschungarica Regel
Perennial. Fine-earth and stony slopes, places near melting snow. Montane and alpine zones. Malguzar, North Turkestan. Ornamental.

55. Gagea filiformis (Ledeb.) Kar. & Kir.
Perennial. Fine-earth and stony slopes. Montane and alpine zones. Malguzar, North Turkestan. Ornamental.

56. Gagea gageoides (Zucc.) Vved. Figure 21
Perennial. Shady wet places. Montane and alpine zones. Nuratau, Aktau, Nuratau Relic Mountains, Malguzar, North Turkestan. Ornamental.

Figure 21 Gagea gageoides (photography by N. Yu. Beshko)

57. Gagea graminifolia Vved.   Figure 22

Perennial. Clay deserts, piedmont plains, fine-eart slopes. Foothills. Nuratau, Aktau, Nuratau Relic Mountains, Malguzar, North Turkestan, Mirzachul. Ornamental.

Figure 22 Gagea graminifolia (photography by Natalya Beshko)

58. Gagea hissarica Lipsky   Figure 23

Perennial. Fine-earth and stony slopes, places near melting snow. Montane and alpine zones. Malguzar, North Turkestan. Ornamental.

Figure 23 Gagea hissarica (photography by N. Yu. Beshko)

59. Gagea kamelinii Levichev   Figure 24

Perennial. Fine-earth and stony slopes. Foothills, montane zone. Nuratau, Malguzar, North Turkestan. Ornamental.

Figure 24 Gagea kamelinii (photography by Natalya Beshko)

60. Gagea ova Stapf.

Perennial. Sandy and clay deserts, piedmont plains, fine-earth and stony slopes. Plain, foothills, montane zone. Nuratau, Aktau, Nuratau Relic Mountains, Malguzar, North Turkestan, Mirzachul, Kyzylkum. Ornamental.

61. Gagea liotardii (Sternb.) Schult. & Schult. f. [*Gagea emarginata* Kar. & Kir]   Figure 25
Perennial. Fine-earth and stony slopes. Montane and alpine zones. Malguzar, North Turkestan. Ornamental.

Figure 25 Gagea liotardii (photography by N. Yu. Beshko)

62. Gagea nabievii Levichev
Perennial. Fine-earth and stony slopes. Montane and alpine zones. Malguzar, North Turkestan. Ornamental.

63. Gagea olgae Regel   Figure 26
Perennial. Clay deserts, piedmont plains, fine-earth slopes. Plain, foothills, montane zone. Nuratau, Aktau, Nuratau Relic Mountains, Malguzar, North Turkestan, Mirzachul. Ornamental.

Figure 26 Gagea olgae (photography by Natalya Beshko)

64. Gagea reinhardii Levichev

Perennial. Fine-earth and stony slopes. Fothills, montane zone. Malguzar, North Turkestan. Ornamental.

65. Gagea reticulata (Pall.) Schult. & Schult. f. [*Gegea pseudoreticulata* Vved.] Figure 27

Perennial. Fine-earth and stony slopes. Foothills, montane zone. Nuratau, Malguzar, North Turkestan. Ornamental.

Figure 27 Gagea reticulata (photography by N. Yu. Beshko)

66. Gagea stipitata Merckl. ex Bunge Figure 28

Perennial. Sandy and clay deserts, piedmont plains, fine-earth and stony slopes. Plain, foothills, montane zone. Nuratau, Aktau, Nuratau Relic Mountains, Malguzar, North Turkestan, Mirzachul, Kyzylkum. Ornamental.

Figure 28 Gagea stipitata (photography by Natalya Beshko)

67. Gagea stolonifera Popov & Czugaeva

Perennial. Fine-earth and stony slopes. Foothills, montane zone. Nuratau, Malguzar, North Turkestan. Ornamental.

68. Gagea subtilis Vved.   Figure 29

Perennial. Stony slopes, rocks, shady places. Foothills, montane zone. Nuratau, Malguzar, North Turkestan, Mirzachul. Ornamental.

Figure 29 Gagea subtilis (photography by N. Yu. Beshko)

69. Gagea taschkentica Levichev

Perennial. Clay deserts, piedmont plains, fine-earth and stony slopes. Plain, foothills, montane zone. Nuratau, Nuratau Relic Mountains, Malguzar, North Turkestan, Mirzachul. Ornamental.

70. Gagea tenera Pascher

Perennial. Clay deserts, piedmont plains, fine-earth and stony slopes. Plain, foothills, montane zone. Nuratau, Aktau, Nuratau Relic Mountains, Malguzar, North Turkestan, Mirzachul. Ornamental.

71. Gagea turkestanica Pascher

Perennial. Stony slopes. Foothills. Malguzar, North Turkestan. Ornamental.

72. Gagea vegeta Vved.

Perennial. Fine-earth and stony slopes. Montane zone. Nuratau, Malguzar, North Turkestan, Mirzachul. Ornamental.

73. Gagea vvedenskyi Grossh.

Perennial. Fine-earth and stony slopes. Montane and alpine zones. Malguzar, North Turkestan. Ornamental.

Genus 29. Tulipa L.

74. Tulipa affinis Botschantz.   Figure 30

Perennial. Fine-earth, gravelly and stony slopes. Foothills, montane zone. Nuratau, Aktau, Nuratau Relic Mountains, Malguzar, North Turkestan. Ornamental. Threatened species, UzbRDB 2.

Figure 30 **Tulipa affinis** (photography by N. Yu. Beshko)

75. Tulipa borszczowii Regel

Perennial. Sandy deserts. Plain. Kyzylkum. Ornamental.

76. Tulipa buhseana Boiss.   Figure 31

Perennial. Sandy and clay deserts, piedmont plain, relic mountains. Plain. Nuratau, Nuratau Relic Mountains, Kyzylkum. Ornamental.

Figure 31 Tulipa buhseana (photography by Natalya Beshko)

77. Tulipa dasystemon Regel

Perennial. Fine-earth, gravelly and stony slopes. Montane and alpine zones. Malguzar, North Turkestan. Ornamental. Threatened species, UzbRDB 2.

78. Tulipa dasystemonoides Vved.

Perennial. Fine-earth, gravelly and stony slopes. Alpine zone. North Turkestan. Ornamental.

79. Tulipa korolkowii Regel

Perennial. Fine-earth, gravelly and stony slopes. Foothills, montane zone. Nuratau, Aktau, Nuratau Relic Mountains, Malguzar, North Turkestan. Ornamental. Threatened species, UzbRDB 2.

80. Tulipa lehmanniana Mercklin   Figure 32

Perennial. Sandy and clay deserts, relic mountains. Plain, foothills. Nuratau Relic Mountains, Kyzylkum. Ornamental. Threatened species, UzbRDB 3.

Figure 32 Tulipa lehmanniana (photography by N. Yu. Beshko)

81. Tulipa micheliana Hoog   Figure 33

Perennial. Fine-earth, gravelly and stony slopes. Plain, foothills, montane zone. Nuratau, Nuratau Relic Mountains, Malguzar, North Turkestan. Ornamental. Threatened species, UzbRDB 2.

Figure 33 Tulipa micheliana (photography by Natalya Beshko)

82. Tulipa turkestanica Regel
Perennial. Fine-earth, gravelly and stony slopes. Foothills, montane zone. Nuratau, Aktau, Nuratau Relic Mountains, Malguzar, North Turkestan. Ornamental.

Genus 30. Lloydia Salisb.

83. Lloydia serotina (L.) Reichenb
Perennial. Wet stony slopes, rocks, screes, moraines. Alpine zone. North Turkestan. Ornamental.

Genus 31. Fritillaria L.

84. Fritillaria olgae Vved.   Figure 34
Perennial. Fine-earth, gravelly and stony slopes, wet shady places. Montane and alpine zones. Nuratau, Malguzar, North Turkestan. Ornamental, medicinal.

Figure 34 Fritillaria olgae (photography by Natalya Beshko)

Family 19. Orchidaceae

Genus 32. Epipactis Adans.

85. Epipactis latifolia (L.) All.
Perennial. Swamp meadows, river valleys. Montane and alpine zones. Nuratau, North Turkestan. Ornamental.

Genus 33. Orchis L.

86. Orchis pseudolaxiflora Czerniak.
Perennial. Swamp meadows, river valleys. Montane zone. Nuratau, Malguzar, North Turkestan. Ornamental, medicinal.

Genus 34. Dactylorhiza Nevski

87. Dactylorhiza umbrosa (Kar. & Kir.) Nevski   Figure 35
Perennial. Swamp meadows, river valleys. Foothills, montane and alpine zones. Nuratau, Malguzar, North Turkestan. Ornamental, medicinal.

Figure 35 Dactylorhiza umbrosa (photography by Natalya Beshko)

Family 20. Ixioliriaceae

Genus 35. Ixiolirion Herb.

88. Ixiolirion tataricum (Pall.) Schult. & Schult. f.   Figure 36
Perennial. Fine-earth, gravelly and stony slopes, river vallys, piedmont plains, sandy and clay deserts. Plain, foothills, montane zone. Nuratau, Aktau, Nuratau Relic Mountains, Malguzar, North Turkestan, Mirzachul, Kyzylkum. Ornamental, medicinal.

Figure 36 Ixiolirion tataricum (photography by Natalya Beshko)

Family 21. Iridaceae

Genus 36. Crocus L.

89. Crocus korolkowii Maw & Regel   Figure 37

Perennial. Fine-earth and gravelly slopes, places near melting snow. Foothills, montane zone. Nuratau, Aktau, Malguzar, North Turkestan. Ornamental, medicinal, food (spice).

Figure 37 Crocus korolkowii (photography by Natalya Beshko)

Genus 37. Iris L.

90. Iris linifoliiformis (Khalk.) Tojibaev & Turginov

Perennial. Fine-earth, gravelly and stony slopes. Montane zone. Malguzar, North Turkestan. Ornamental.

91. Iris loczyi Kanitz

Perennial. Fine-earth, gravelly and stony slopes. Montane and alpine zones. North Turkestan. Ornamental.

92. Iris longiscapa Ledeb.   Figure 38

Perennial. Sandy and clay deserts. Plain. Nuratau Relic Mountains, Kyzylkum. Ornamental.

Figure 38 Iris longiscapa (photography by Natalya Beshko)

93. Iris maracandica (Vved.) Wendelbo   Figure 39
Perennial. Fine-earth, gravelly and stony slopes. Foothills, montane zone. Nuratau, Aktau, Nuratau Relic Mountains, Malguzar, North Turkestan. Ornamental.

Figure 39 Iris maracandica (photography by Natalya Beshko)

94. Iris narbutii O. Fedtsch.   Figure 40
Perennial. Fine-earth and gravelly slopes, clay deserts, piedmont plains. Plain, foothills. Nuratau, Aktau, Nuratau Relic Mountains, Malguzar, North Turkestan. Ornamental.

Figure 40 Iris narbutii (photography by Natalya Beshko)

95. Iris parvula (Vved.) T. Hall & Seisums

Perennial. Fine-earth, gravelly and stony slopes. Alpine zone. North Turkestan. Ornamental.

96. Iris songarica Schrenk   Figure 41

Perennial. Sandy and clay deserts, piedmont plains. Plain, foothills. Nuratau, Aktau, Nuratau Relic Mountains, Kyzylkum. Ornamental, poisonous.

Figure 41 **Iris songarica** (photography by Natalya Beshko)

97. Iris tadshikorum Vved.

Perennial. Fine-earth and stony slopes. Montane and alpine zones. Malguzar, North Turkestan. Ornamental.

Genus 38. Gladiolus L.

98. Gladiolus italicus Mill.

Perennial. Fine-earth slopes, meadows, fields. Foothills, montane zone. North Turkestan. Ornamental. Threatened species, UzbRDB 1.

Family 22. Asphodelaceae

Genus 39. Eremurus M. Bieb.

99. Eremurus anisopterus (Kar. & Kir.) Regel
Perennial. Sandy deserts. Plain. Kyzylkum. Ornamental.

100. Eremurus chloranthus Popov
Perennial. Alpine zone. North Turkestan. Ornamental. Threatened species, UzbRDB 0. Endemic to the Turkestan ridge.

101. Eremurus fuscus (O. Fedtsch.) Vved.
Perennial. Fine-earth, gravelly and stony slopes. Montane and alpine zones. Nuratau, Malguzar, North Turkestan. Ornamental, meliferous.

102. Eremurus inderiensis (Stev.) Regel
Perennial. Sandy deserts. Plain. Kyzylkum. Ornamental, meliferous.

103. Eremurus kaufmannii Regel   Figure 42
Perennial. Fine-earth, gravelly and stony slopes. Montane and alpine zones. Malguzar, North Turkestan. Ornamental.

Figure 42 Eremurus kaufmannii (photography by Natalya Beshko)

104. Eremurus lactiflorus O. Fedtsch.   Figure 43

Perennial. Gravelly and stony slopes, rocks, screes. Montane zone. Nuratau. Ornamental. Threatened species, UzbRDB 2.

Figure 43 **Eremurus lactiflorus** (photography by N. Yu. Beshko)

105. Eremurus nuratavicus A. P. Khohkr.   Figure 44

Perennial. Fine-earth, gravelly and stony slopes. Montane zone. Nuratau. Ornamental. Threatened species, UzbRDB 2. Endemic to the Nuratau ridge.

Figure 44 **Eremurus nuratavicus** (photography by N. Yu. Beshko)

106. Eremurus olgae Regel   Figure 45

Perennial. Fine-earth, gravelly and stony slopes. Foothills, montane zone. Nuratau, Aktau, Malguzar, North Turkestan. Ornamental, food, medicinal, dye, meliferous.

Figure 45 Eremurus olgae (photography by Natalya Beshko)

107. Eremurus regelii Vved.   Figure 46

Perennial. Fine-earth, gravelly and stony slopes. Montane zone. Nuratau, Malguzar, North Turkestan. Ornamental, food, medicinal, dye, melliferous.

Figure 46 Eremurus regelii (photography by Natalya Beshko)

108. Eremurus robustus (Regel) Regel   Figure 47

Perennial. Fine-earth, gravelly and stony slopes, rocks. Montane zone. Nuratau, Malguzar, North Turkestan. Ornamental, medicinal, melliferous. Threatened species, UzbRDB 3.

Figure 47 **Eremurus robustus** (photography by Natalya Beshko)

109. Eremurus sogdianus (Regel) Benth. & Hook. f.   Figure 48

Perennial. Gravelly and stony slopes. Foothills, montane zone. Nuratau, Aktau, Malguzar, North Turkestan. Ornamental.

Figure 48 **Eremurus sogdianus** (photography by Natalya Beshko)

110. Eremurus tianschanicus Pazij & Vved. Figure 49

Perennial. Fine-earth, gravelly and stony slopes. Montane zone. North Turkestan. Ornamental, meliferous.

Figure 49 Eremurus tianschanicus (photography by Natalya Beshko)

111. Eremurus turkestanicus Vved. Figure 50

Perennial. Fine-earth, gravelly and stony slopes. Montane zone. Nuratau, Malguzar, North Turkestan. Ornamental, meliferous.

Figure 50 Eremurus turkestanicus (photography by Natalya Beshko)

Family 23. Amaryllidaceae

Genus 40. Ungernia Bunge

112. Ungernia oligostroma Popov & Vved.   Figure 51
Perennial. Fine-earth, gravelly and stony slopes. Montane zone. Nuratau, Malguzar, North Turkestan. Ornamental, medicinal.

Figure 51 **Ungernia oligostroma** (photography by Natalya Beshko)

Genus 41. Allium L.

113. Allium alexeianum Regel   Figure 52
Perennial. Stony and gravelly slopes, screes. Montane and alpine zones. Malguzar, North Turkestan. Ornamental.

Figure 52 **Allium alexeianum** (photography by Natalya Beshko)

114. Allium altissimum Regel   Figure 53

Perennial. River valleys. Montane zone. Nuratau, Malguzar, North Turkestan. Ornamental, food.

Figure 53 Allium altissimum (photography by Natalya Beshko)

115. Allium barsczewskii Lipsky

Perennial. Fine-earth, gravelly and stony slopes. Montane and alpine zones. Nuratau, Malguzar, North Turkestan. Ornamental.

116. Allium caesium Schrenk   Figure 54

Perennial. Fine-earth, gravelly and stony slopes, river valleys. Foothills, montane zone. Nuratau, Nuratau Relic Mountains, Malguzar, North Turkestan. Ornamental, food.

Figure 54 Allium caesium (photography by Natalya Beshko)

117. Allium carolinianum DC.   Figure 55
Perennial. Fine-earth, gravelly and stony slopes. Alpine zone. North Turkestan. Ornamental.

Figure 55 **Allium carolinianum** (photography by Natalya Beshko)

118. Allium caspium (Pall.) M. Bieb.   Figure 56
Perennial. Sandy deserts. Plain. Nuratau Relic Mountains, Kyzylkum.

Figure 56 **Allium caspium** (photography by Natalya Beshko)

119. Allium clausum Vved.   Figure 57
Perennial. Fine-earth, gravelly and stony slopes. Montane zone. North Turkestan. Ornamental.

Figure 57 Allium clausum (photography by Natalya Beshko)

120. Allium cupuliferum Regel subsp. cupuliferum   Figure 58

Perennial. Fine-earth, gravelly and stony slopes. Foothills, montane zone. Nuratau, Aktau, Nuratau Relic Mountains, Malguzar. Ornamental.

Figure 58 Allium cupuliferum subsp. **cupuliferum** (photography by Natalya Beshko)

121. Allium cupuliferum subsp. nuratavicum R. M. Fritsch & Beshko   Figure 59

Perennial. Stony slopes. Montane zone. Nuratau. Ornamental.

Figure 59 Allium cupuliferum subsp. **nuratavicum** (photography by Natalya Beshko)

122. Allium drepanophyllum Vved.

Perennial. Stony and gravelly slopes. Foothills, montane zone. Nuratau, Aktau, Nuratau Relic Mountains. Ornamental.

123. Allium filidens Regel   Figure 60

Perennial. Piedmont plains, fine-earth, gravelly and stony slopes, rocks. Plain, foothills, montane zone. Nuratau, Aktau, Nuratau Relic Mountains, Malguzar, North Turkestan. Ornamental.

Figure 60 Allium filidens (photography by Natalya Beshko)

124. Allium griffithianum Boiss.   Figure 61

Perennial. Fine-earth slopes, piedmont plains. Plain, foothills, montane zone. Nuratau, Aktau, Nuratau Relic Mountains, Malguzar, North Turkestan. Ornamental.

Figure 61 Allium griffithianum (photography by Natalya Beshko)

125. Allium gusaricum Regel   Figure 62

Perennial. Stony slopes, screes, rocks. Montane zone. Nuratau, Malguzar, North Turkestan. Ornamental.

Figure 62 Allium gusaricum (photography by Natalya Beshko)

126. Allium inconspicuum Vved.

Perennial. Stony and gravelly slopes. Foothills. Nuratau, Nuratau Relic Mountains. Ornamental.

127. Allium isakulii R. M. Fritsch & F. O. Khass.   Figure 63

Perennial. Stony slopes. Foothills, montane zone. Nuratau, Malguzar. Ornamental. UzbRDB 2.

Figure 63 Allium isakulii (photography by Natalya Beshko)

128. Allium jodanthum Vved.   Figure 64

Perennial. Fine-earth and gravelly slopes. Montane and alpine zones. Nuratau, Malguzar, North Turkestan. Ornamental.

Figure 64 Allium jodanthum (photography by Natalya Beshko)

129. Allium karataviense Regel   Figure 65

Perennial. Stony and gravelly slopes, screes. Montane zone. Malguzar, North Turkestan. Ornamental.

Figure 65 Allium karataviense (photography by Natalya Beshko)

130. Allium kaufmannii Regel   Figure 66
Perennial. Swamp meadows, wet places. Montane and alpine zones. Malguzar, North Turkestan. Ornamental.

Figure 66 **Allium kaufmannii** (photography by Natalya Beshko)

131. Allium kokanicum Regel   Figure 67
Perennial. Stony slopes. Alpine zone. North Turkestan. Ornamental.

Figure 67 **Allium kokanicum** (photography by Natalya Beshko)

132. Allium komarowii Lipsky   Figure 68

Perennial. Screes. Montane zone. Malguzar, North Turkestan. Ornamental.

Figure 68 Allium komarowii (photography by Natalya Beshko)

133. Allium levichevii F. O. Khass., Esankulov & N. Sulejm

Perennial. Stony slopes. Montane zone. Malguzar. Ornamental.Endemic to the Malguzar ridge.

134. Allium longicuspis Regel

Perennial. Ravines, river valleys. Montane zone. Nuratau. Ornamental, food, medicinal.

135. Allium oreodictyum Vved.

Perennial. Stony slopes. Montane zone. Nuratau, North Turkestan. Ornamental.

136. Allium oreophiloides Regel

Perennial. Stony slopes, rocks. Alpine zone. North Turkestan. Ornamental.

137. Allium oreophilum C. A. Mey.   Figure 69

Perennial. Stony slopes. Alpine zone. North Turkestan. Ornamental.

Figure 69 Allium oreophilum (photography by Natalya Beshko)

138. Allium oschaninii O. Fedtsch.
Perennial. Rocks, stony slopes. Montane zone. Malguzar, North Turkestan. Ornamental, food.

139. Allium praemixtum Vved.   Figure 70
Perennial. Rocks, stony slopes, screes. Montane zone. Nuratau, Malguzar. Ornamental, food. Threatened species, UzbRDB 1.

Figure 70 Allium praemixtum (photography by Natalya Beshko)

140. Allium protensum Wendelbo

Perennial. Stony and gravelly slopes. Foothills. Nuratau Relic Mountains. Ornamental.

141. Allium sarawschanicum Regel   Figure 71

Perennial. River valleys, fine earth slopes. Montane zone. Malguzar, North Turkestan. Ornamental.

Figure 71 Allium sarawschanicum (photography by Natalya Beshko)

142. Allium stephanophorum Vved.

Perennial. Fine-earth and gravelly slopes. Montane zone. Nuratau, Malguzar, North Turkestan. Ornamental.

143. Allium stipitatum Regel   Figure 72

Perennial. Fine-earth slopes, wet places, river valleys. Montane zone. Nuratau, Aktau, Malguzar, North Turkestan. Ornamental, food.

Figure 72 Allium stipitatum (photography by Natalya Beshko)

144. Allium suworowii Regel   Figure 73

Perennial. Fine-earth slopes, river valleys, burials, gardens. Plain, foothills, montane zone. Nuratau, Aktau, Nuratau Relic Mountains, Malguzar, North Turkestan, Mirzachul. Ornamental, food.

Figure 73 **Allium suworowii** (photography by Natalya Beshko)

145. Allium taeniopetalum Popov & Vved.   Figure 74

Perennial. Fine-earth and stony slopes, shady places. Foothills, montane zone. Nuratau, Aktau, Nuratau Relic Mountains, Malguzar, North Turkestan. Ornamental.

Figure 74 **Allium taeniopetalum** (photography by Natalya Beshko)

146. Allium talassicum Regel

Perennial. Stony slopes. Montane zone. North Turkestan. Ornamental.

147. Allium turkestanicum Regel Figure 75
Perennial. Stony slopes, rocks. Foothills, montane zone. Nuratau, Malguzar. Ornamental.

Figure 75 Allium turkestanicum (photography by Natalya Beshko)

148. Allium verticillatum Regel Figure 76
Perennial. Piedmont plains, fine-earth and gravelly slopes. Foothills, montane zone. Nuratau, Aktau, Nuratau Relic Mountains, Malguzar, North Turkestan. Ornamental.

Figure 76 Allium verticillatum (photography by Natalya Beshko)

149. Allium xiphopetalum Aitch. & Baker
Perennial. Stony slopes. Foothills, montane zone. Nuratau, Aktau, Nuratau Relic Mountains, Malguzar, North Turkestan. Ornamental.

Family 24. Asparagaceae

Genus 42. Polygonatum Adans.

150. Polygonatum sewerzowii Regel   Figure 77
Perennial. Fine-earth and stony slopes, shady places. Montane zone. Malguzar, North Turkestan. Medicinal, food, poisonous, ornamental.

Figure 77 Polygonatum sewerzowii (photography by Natalya Beshko)

Genus 43. Asparagus L.

151. Asparagus brachyphyllus Turcz.   Figure 78

Perennial. Saline lands and depressions, sand deserts, alkaline meadows, lake shores. Plain. Kyzylkum. Medicinal, fodder.

Figure 78 **Asparagus brachyphyllus** (photography by Natalya Beshko)

152. Asparagus persicus Baker   Figure 79

Perennial. River valleys, rocks, fine-earth, gravelly and stony slopes. Foothills, montane zone. Malguzar, North Turkestan.

Figure 79 **Asparagus persicus** (photography by Natalya Beshko)

Genus 44. Fessia Speta

153. Fessia puschkinioides (Regel) Speta
Perennial. Fine-earth and stony slopes, places near melting snow. Montane and alpine zones. Malguzar, North Turkestan. Ornamental.

Family 25. Typhaceae

Genus 45. Sparganium L.

154. Sparganium stoloniferum (Buch.-Ham. ex Graebn.) Buch.-Ham. ex Juz.
Perennial. Banks of rivers, lakes and canals. Plain. Mirzachul, Kyzylkum.

Genus 46. Typha L.

155. Typha angustata Bory & Chaub
Perennial. Swamps, banks of rivers, lakes and canals. Plain. Mirzachul, Kyzylkum.

156. Typha angustifolia L.   Figure 80
Perennial. Swamps, banks of rivers, lakes and canals. Plain, foothills. Malguzar, North Turkestan, Mirzachul, Kyzylkum.

Figure 80 Typha angustifolia (photography by Natalya Beshko)

157. Typha latifolia L.

Perennial. Swamps, banks of rivers, lakes and canals. Plain, foothills, montane zone. Malguzar, North Turkestan, Mirzachul, Kyzylkum.

158. Typha laxmannii Lepech.

Perennial. Swamps, banks of rivers, lakes and canals. Plain. Mirzachul.

159. Typha minima Funck

Perennial. Swamps, banks of rivers, lakes and canals. Plain. Mirzachul.

Family 26.*Eriocaulaceae

Genus 47.*Eriocaulon L.

160.*Eriocaulon sieboldianum Sieb. ex Zucc.

Annual. Swamps, banks of rivers, lakes and canals, pebbles, rice fields. Plain, foothills, montane and alpine zones. Nuratau, North Turkestan, Mirzachul. Medicinal, weed.

Family 27. Juncaceae

Genus 48. Juncus L.

161. Juncus articulatus L.

Perennial. Swamps, banks of rivers, lakes and canals, pebbles. Plain, foothills, montane and alpine zones. Nuratau, Aktau, Malguzar, North Turkestan, Mirzachul. Fodder.

162. Juncus bufonius L.

Annual. Swamp meadows, banks of rivers, lakes and canals, pebbles, saline lands. Plain, foothills, montane and alpine zones. Nuratau, Malguzar, North Turkestan, Mirzachul. Fodder, medicinal, weed.

163. Juncus compressus Jacq.

Perennial. Banks of rivers and lakes, dry riverbeds, swamp meadows, fields, saline lands. Plain, foothills, montane and alpine zones. North Turkestan. Fodder, medicinal, industrial.

164. Juncus gerardii Loisel

Perennial. Swamp meadows, banks of rivers, lakes and canals, wet slopes, fallow lands. Plain, foothills, montane zone. Nuratau, Malguzar, North Turkestan, Mirzachul. Fodder, medicinal, industrial.

165. Juncus heptopotamicus V. I. Krecz. & Gontsch.

Perennial. Swamp meadows, banks of rivers, wet stony slopes, saline lands. Plain, foothills, montane and alpine zones. North Turkestan. Fodder.

166. Juncus hybridus Brot.
Annual. Banks of rivers and lakes, dry riverbeds, swamp meadows, fallow lands. Plain, foothills, montane and alpine zones. North Turkestan, Mirzachul, Kyzylkum. Fodder.

167. Juncus inflexus L.
Perennial. Swamp meadows, banks of rivers, lakes and canals, wet slopes. Plain, foothills, montane and alpine zones. Nuratau, Aktau, Malguzar, North Turkestan, Mirzachul. Fodder.

168. Juncus jaxarticus V. I. Krecz. & Gontsch.
Perennial. Swamp meadows, banks of rivers, lakes and canals, saline lands, depressions, fields. Plain, foothills, montane zone. Malguzar, Mirzachul, Kyzylkum. Fodder.

169. Juncus macrantherus V. I. Krecz. & Gontsch.
Perennial. Swamps, banks of rivers, wet slopes. Montane and alpine zones. North Turkestan. Fodder.

170. Juncus persicus subsp. **libanoticus** (J.Thiébaut) Novikov & Snogerup [*Juncus vvedenskyi* V. I. Krecz.]
Perennial. Swamp meadows, banks of rivers and lakes, wet slopes. Plain, foothills, montane zone. Nuratau. Fodder.

171. Juncus rechingeri Snogerup.
Annual. Swamp meadows, banks of rivers and lakes. Plain, foothills, alpine zone. Nuratau. Fodder.

172. Juncus sphaerocarpus Nees.
Annual. Swamp meadows, banks of rivers and lakes. Plain, foothills, montane and alpine zones. Nuratau. Fodder.

173. Juncus triglumis L.[*Juncus schischkinii* Krylov & Sumnev.]
Perennial. Swamp meadows, banks of rivers and lakes. Montane and alpine zones. Malguzar, North Turkestan. Fodder.

174. Juncus turkestanicus V. I. Krecz. & Gontsch.
Annual. Swamp meadows, banks of rivers and lakes. Plain, foothills, montane zone. Nuratau, North Turkestan. Fodder.

Genus 49. Luzula DC.

175. Luzula pallescens Sw.
Perennial. Riwer banks, swamps, alpine meadows. Montane and alpine zones. North Turkestan. Fodder.

176. Luzula spicata (L.) DC.
Perennial. Moraines, rocks, wet stony slopes, swamps, banks of rivers. Alpine zone. North Turkestan. Fodder.

Family 28. Cyperaceae

Genus 50. Schoenoplectus (Rchb.) Palla

177. Schoenoplectus litoralis (Schrad.) Palla [*Scirpus litoralis* Schrad.]
Perennial. Banks of rivers, lakes and canals. Plain. Nuratau Relic Mountains, Mirzachul, Kyzylkum.

178. Schoenoplectus triqueter (L.) Palla [*Scirpus triqueter* L.]
Perennial. Banks of rivers, lakes and canals. Plain. Mirzachul.

Genus 51. Scirpus L.

179. Scirpus triquetriformis (V. Krecz.) T. V. Egorova
Perennial. Banks of rivers, lakes and canals, near springs. Plain, foothills, montane zone. North Turkestan.

Genus 52. Bolboschoenus (Aschers.) Palla

180. Bolboschoenus maritimus (L.) Palla   Figure 81
Perennial. Banks of rivers, lakes and canals, near springs, swamp meadows, saline depressions. Plain, foothills. Nuratau, Nuratau Relic Mountains, Malguzar, North Turkestan, Mirzachul, Kyzylkum. Medicinal, fodder, industrial, weed.

Figure 81 Bolboschoenus maritimus (photography by Natalya Beshko)

181. Bolboschoenus popovii T. V. Egorova Figure 82

Perennial. Banks of rivers, lakes and canals, saline lands. Plain. Kyzylkum. Medicinal, fodder, industrial, weed.

Figure 82 Bolboschoenus popovii (photography by Natalya Beshko)

Genus 53. Holoschoenus Link.

182. Holoschoenus vulgaris Link.   Figure 83

Perennial. Banks of rivers, lakes and canals, pebbles, dry riverbeds, wet depressions, saline lands. Plain, foothills, montane zone. Nuratau, Malguzar, North Turkestan, Mirzachul, Kyzylkum.

Figure 83 Holoschoenus vulgaris (photography by Natalya Beshko)

Genus 54. Blysmus Panz. ex Schult.

183. Blysmus compressus (L.) Panz. ex Link.

Perennial. Swamp meadows, banks of rivers, lakes and canals. Foothills, montane and alpine zones. Nuratau, Nuratau Relic Mountains, Malguzar, North Turkestan, Mirzachul.

Genus 55. Baeothryon A. Dietr.

184. Baeothryon pumilum (Vahl) Á. & D. Löve

Perennial. Banks of rivers, swamps, alpine meadows, wet places, places near melting snow. Foothills, montane and alpine zones. Nuratau, Malguzar, North Turkestan.

Genus 56. Eleocharis R. Br.

185. Eleocharis argyrolepis Kier.

Perennial. Swamp meadows, banks of rivers, lakes and canals. Plain. Mirzachul. Weed.

186. Eleocharis quinqueflora (Hartmann) O. Schwarz. [*Eleocharis meridionalis* Zinserl.]
Perennial. Swamp meadows, banks of rivers, lakes and canals, saline depressions. Plain, foothills, montane and alpine zones. Nuratau, Nuratau Relic Mountains, Malguzar, North Turkestan, Mirzachul.

187. Eleocharis mitracarpa Steud.
Perennial. Banks of rivers, lakes and canals, swamps, wet meadows. Plain, foothills, montane zone. Malguzar, North Turkestan, Mirzachul.

188. Eleocharis uniglumis (Link.) Schult.
Perennial. Swamp meadows, saline lands, banks of rivers, lakes and canals. Plain, foothills, montane zone. Nuratau, Malguzar.

Genus 57. Fimbristylis Vahl.

189. Fimbristylis dichotoma (L.) Vahl
Annual. Wet places, banks of rivers, lakes and canals, pebbles. Plain. Mirzachul. Weed.

Genus 58. Cyperus L.

190. Cyperus glaber L.
Annual. Banks of rivers, lakes and canals, pebbles, swamp meadows, wet places near springs. Plain, foothills, montane zone. Nuratau. Fodder, weed.

191. Cyperus longus L.   Figure 84
Perennial. Banks of rivers, lakes and canals, pebbles, swamp meadows, wet places near springs. Plain, foothills, montane zone. Nuratau, Nuratau Relic Mountains, Malguzar, North Turkestan, Mirzachul. Medicinal, food (spice-aromatic), fodder, ornamental.

Figure 84 Cyperus longus (photography by Natalya Beshko)

192. Cyperus rotundus L.
Banks of rivers, lakes and canals, pebbles, swamp meadows, wet places near springs, fallow lands.

Plain, foothills, montane zone. Nuratau, Nuratau Relic Mountains, Malguzar, North Turkestan, Mirzachul. Medicinal, fodder, weed, ornamental.

Genus 59. Pycreus P. Beauv.

193. Pycreus globosus (All.) Reichenb
Annual. Banks of rivers and canals, pebbles, swamps, wet places near springs, pebbles, alkaline meadows. Plain, foothills, montane zone. Nuratau, Nuratau Relic Mountains. Fodder, weed.

Genus 60. Schoenus L.

194. Schoenus nigricans L.   Figure 85
Perennial. Swamp meadows, banks of rivers, wet places near springs. Montane zone. Nuratau.

Figure 85 Schoenus nigricans (photography by Natalya Beshko)

Genus 61. Kobresia Willd.

195. Kobresia pamiroalaica Ivanova
Perennial. Wet slopes, banks of rivers, swamps, highlands. Montane and alpine zones. North Turkestan. Fodder.

196. Kobresia persica Kük. & Bornm.
Perennial. Banks of rivers, swamps, highlands, wet stony slopes, screes. Montane and alpine zones. North Turkestan. Fodder.

197. Kobresia stenocarpa (Kar. & Kir.) Steud.
Perennial. Swamps, banks of rivers, wet places near springs. Montane and alpine zones. North Turkestan. Fodder.

Genus 62. Carex L.

198. Carex diluta M. Bieb.
Perennial. Swamp meadows, banks of rivers, wet places near springs. Plain, foothills, montane zone. Nuratau, Aktau, Nuratau Relic Mountains, Malguzar, North Turkestan. Fodder.

199. Carex divisa Huds.
Perennial. Swamp meadows, banks of rivers, lakes and canals, wet places near springs, depressions. Plain, foothills, montane zone. Nuratau, Aktau, Nuratau Relic Mountains, Malguzar, North Turkestan, Mirzachul. Fodder.

200. Carex enervis C. A. Mey.
Perennial. Swamp meadows, banks of rivers, wet places near springs. Montane and alpine zones. North Turkestan. Fodder.

201. Carex karoi (Freyn) Freyn
Perennial. Swamp meadows, banks of rivers, wet places near springs. Montane and alpine zones. North Turkestan. Fodder.

202. Carex melanantha C. A. Mey.   Figure 86
Perennial. Swamp meadows, banks of rivers, wet places near springs. Alpine zone. North Turkestan. Fodder.

Figure 86 Carex melanantha (photography by Natalya Beshko)

203. Carex melanostachya M. Bieb. ex Willd.

Perennial. Swamp meadows, fine-earth slopes, banks of rivers, wet places near springs. Plain, foothills, montane zone. Nuratau, Aktau, Malguzar, North Turkestan. Fodder.

204. Carex microglochin Wahlenb

Perennial. Swamp meadows, banks of rivers and canals, wet places near springs. Montane and alpine zones. North Turkestan. Fodder.

205. Carex orbicularis Boott.

Perennial. Swamp meadows, banks of rivers, lakes and canals, pebbles, wet places near springs. Plain, foothills, montane and alpine zones. Nuratau, Aktau, Malguzar, North Turkestan. Fodder.

206. Carex pachystilis J. Gay.

Perennial. Clay deserts, piedmont plain, fine-earth, gravelly and stony slopes, fallow lands. Plain, foothills, montane and alpine zones. Nuratau, Aktau, Nuratau Relic Mountains, Malguzar, North Turkestan, Mirzachul. Fodder.

207. Carex parva Nees.

Perennial. Swamps, banks of rivers, wet places near springs. Montane and alpine zones. North Turkestan. Fodder.

208. Carex physodes M. Bieb.   Figure 87

Perennial. Sandy deserts. Plain. Kyzylkum. Fodder.

Figure 87 Carex physodes (photography by Natalya Beshko)

209. Carex pseudofoetida Kük.

Perennial. Swamps, banks of rivers, wet places near springs, alpine meadows. Montane and alpine zones. North Turkestan. Fodder.

210. Carex serotina Mérat.

Perennial. Swamp meadows, banks of rivers, wet places near springs. Foothills, montane zone. Nuratau, Nuratau Relic Mountains, Malguzar, North Turkestan. Fodder.

211. Carex songorica Kar. & Kir.

Perennial. Swamp meadows, river valleys, ravines, pebbles, banks of canals. Plain, foothills, montane and alpine zones. Nuratau, Aktau, Nuratau Relic Mountains, Malguzar, North Turkestan. Fodder.

212. Carex stenophylla Wahlenb

Perennial. Piedmont plains, fine-earth slopes, meadows, river valleys, dry riverbeds, banks of canals. Plain, foothills, montane and alpine zones. Nuratau, Aktau, Nuratau Relic Mountains, Malguzar, North Turkestan. Fodder.

213. Carex subphysodes Popov ex V. I. Krecz.

Perennial. Sandy deserts. Plain. Kyzylkum. Fodder.

214. Carex stenocarpa Turcz. & Besser.

Perennial. Alpine and subalpine meadows, fine-earth slopes, rocks. Montane and alpine zones, Malguzar, North Turkestan. Fodder.

215. Carex turkestanica Regel   Figure 88

Perennial. Fine-earth and stony slopes, river valleys, rocks, screes. Montane zone. Malguzar, North Turkestan. Fodder.

Figure 88 Carex turkestanica (photography by Natalya Beshko)

Family 29. Poaceae [Gramineae]

Genus 63. Imperata Cyr.

216. Imperata cylindrica (L.) Raeusch.
Perennial. River valleys, banks of lakes and canals, fallow lands, fine-earth slopes. Plain, foothills, montane zone. North Turkestan. Fodder, industrial, ornamental.

Genus 64. Saccharum L.

217. Saccharum spontaneum L.
Perennial. Banks of rivers, lakes and canals, pebbles. Plain. Mirzachul, Kyzylkum. Fodder, food, industrial.

Genus 65. Erianthus Rich.

218. Erianthus ravennae (L.) P. Beauv.   Figure 89
Perennial. River valleys, banks of lakes and canals, wet places. Plain, foothills, montane zone. Nuratau, Malguzar, North Turkestan, Mirzachul. Fodder, industrial, ornamental.

Figure 89 Erianthus ravennae (photography by Natalya Beshko)

Genus 66. Botriochloa Kuntze

219. Botriochloa ischaemum (L.) H. Keng
Perennial. Fine-earth and stony slopes, banks of canals, roadsides. Plain, foothills, montane zone. Nuratau, Aktau, Nuratau Relic Mountains, Malguzar, North Turkestan, Mirzachul. Fodder, medicinal, weed.

Genus 67. *Sorghum Pers.

220. *Sorghum halepense (L.) Pers.
Perennial. Fields, fallow lands, gardens, river valleys, banks of canals. Plain, foothills, montane zone. Nuratau, Aktau, Malguzar, North Turkestan, Mirzachul. Fodder, food, medicinal, industrial, weed.

Genus 68. Digitaria Heist.

221. Digitaria ischaemum (Schreb.) Muehl.
Annual. Banks of rivers and canals, fields, fallow lands, gardens, wastelands. Plain, foothills, montane zone. Nuratau, Aktau, Malguzar, North Turkestan, Mirzachul. Fodder, weeds.

222. Digitaria sanguinalis (L.) Scop.
Annual. Banks of rivers and canals, fields, gardens, wastelands. Plain, foothills, montane zone. Nuratau, Aktau, Nuratau Relic Mountains, Malguzar, North Turkestan, Mirzachul. Fodder, weed.

Genus 69. *Echinochloa P. Beauv.

223. *Echinochloa crus-galli (L.) P. Beauv.
Annual. Fields, fallow lands, gardens, wastelands, banks of rivers and canals, pebbles. Plain, foothills, montane zone. Nuratau, Aktau, Nuratau Relic Mountains, Malguzar, North Turkestan, Mirzachul, Kyzylkum. Fodder, food, medicinal, weed.

Genus 70. Setaria P. Beauv.

224. Setaria lutescens (Weig.) F. T. Hubb.
Annual. Fields, fallow lands, gardens, wastelands, banks of canals. Plain, foothills, montane zone. Nuratau, Aktau, Malguzar, North Turkestan, Mirzachul. Fodder, weed.

225. *Setaria verticillata (L.) P. Beauv.
Annual. Fields, fallow lands, gardens, wastelands, banks of canals. Plain, foothills, montane zone. Nuratau, Aktau, Malguzar, North Turkestan, Mirzachul. Fodder, weed.

226. Setaria viridis (L.) P. Beauv.

Annual. Fields, fallow lands, gardens, wastelands, banks of canals, stony slopes. Plain, foothills, montane zone. Nuratau, Aktau, Nuratau Relic Mountains, Malguzar, North Turkestan, Mirzachul, Kyzylkum. Fodder, weed.

Genus 71. Phalaroides Wolf

227. Phalaroides arundinacea (L.) Rauschert

Perennial. Banks of rivers, lakes and canals, swamp meadows, wet places. Plain, foothills, montane zone. Nuratau, Malguzar, North Turkestan. Fodder, ornamental, weed.

Genus 72. Stipagrostis Nees

228. Stipagrostis karelinii (Trin. & Rupr.) H. Scholz [*Aristida karelinii* (Trin. & Rupr.) Roshev.]   Figure 90

Perennial. Sandy deserts. Plain. Mirzachul, Kyzylkum. Fodder.

Figure 90 Stipagrostis karelinii (photography by Natalya Beshko)

229. Stipagrostis pennata (Trin.) De Winter [*Aristida pennata* Trin.]
Perennial. Sandy deserts. Plain. Kyzylkum. Fodder.

Genus 73. Stipa L.

230. Stipa arabica Trin. & Rupr. [*Stipa szovitsiana* Trin.]
Perennial. Fine-earth, gravelly and stony slopes. Foothills, montane zone. Nuratau, Aktau, Nuratau Relic Mountains, Malguzar, North Turkestan. Fodder.

231. Stipa caragana Trin. [*Achnatherum caragana* (Trin.) Roshev.]
Perennial. Fine-earth, gravelly and stony slopes. Montane zone. Nuratau, Malguzar, North Turkestan. Fodder, industrial.

232. Stipa capillata L.   Figure 91
Perennial. Fine-earth, gravelly and stony slopes. Montane and alpine zones. Nuratau, Malguzar, North Turkestan. Fodder, ornamental.

Figure 91 Stipa capillata (photography by Natalya Beshko)

233. Stipa caucasica Schmalh.

Perennial. Stony slopes, screes. Foothills, montane and alpine zones. Nuratau, Malguzar, North Turkestan. Fodder.

234. Stipa hohenackeriana Trin. & Rupr.

Perennial. Fine-earth, gravelly and stony slopes. Foothills, montane zone. Nuratau, Aktau, Nuratau Relic Mountains, Malguzar, North Turkestan. Fodder.

235. Stipa kirghisorum P. A. Smirn.

Perennial. Fine-earth, gravelly and stony slopes, pebbles, rocks, screes. Montane and alpine zones. North Turkestan. Fodder.

236. Stipa kurdistanica Bor. [*Achnatherum turkomanicum* (Roshev.) Tzvelev]

Perennial. Fine-earth, gravelly and stony slopes. Montane zone. North Turkestan. Fodder, industrial.

237. Stipa lessingiana Trin. & Rupr.

Perennial. Fine-earth, gravelly and stony slopes, screes. Montane zone. North Turkestan. Fodder, ornamental.

238. Stipa lingua Junge

Perennial. Fine-earth, gravelly and stony slopes. Montane and alpine zones. North Turkestan. Fodder.

239. Stipa lipskyi Roshev.

Perennial. Fine-earth, gravelly and stony slopes. Montane zone. Nuratau, Malguzar, North Turkestan. Fodder.

240. Stipa margelanica P. A. Smirn.

Perennial. Stony slopes. Foothills, montane zone. North Turkestan. Fodder.

241. Stipa orientalis Trin. ex Ledeb.

Perennial. Stony and gravelly slopes, screes, rocks. Montane and alpine zones. Nuratau, Malguzar, North Turkestan. Fodder.

242. Stipa richteriana Kar. & Kir.

Perennial. Fine-earth, gravelly and stony slopes. Montane zone. North Turkestan. Fodder.

243. Stipa sareptana Beck.

Perennial. Fine-earth, gravelly and stony slopes. Montane zone. North Turkestan. Fodder.

244. Stipa splendens Trin. [*Achnatherum splendens* (Trin.) Nevski]

Perennial. Saline lands, river valleys. Plain, foothills, montane and alpine zones. North Turkestan. Fodder, industrial.

245. Stipa turkestanica Hack. [*Stipa trichoides* P. Smirn.]

Perennial. Stony slopes. Montane and alpine zones. North Turkestan. Fodder.

Genus 74. Piptatherum P. Beauv.

246. Piptatherum alpestre (Grig.) Roshev.

Perennial. Stony slopes, screes, moraines, swamp meadows. Alpine zone. North Turkestan. Fodder.

247. Piptatherum ferganense (Litv.) Roshev.
Perennial. Fine-earth, gravelly and stony slopes. Montane and alpine zones. Nuratau, Malguzar, North Turkestan. Fodder.

248. Piptatherum holciforme (M. Bieb.) Roem. & Schult.
Perennial. Stony slopes, screes. Montane zone. North Turkestan. Fodder.

249. Piptatherum laterale (Regel) Nevski
Perennial. Fine-earth, gravelly and stony slopes, screes, rocks. Montane and alpine zones. Malguzar, North Turkestan. Fodder.

250. Piptatherum latifolium (Roshev.) Nevski
Perennial. Fine-earth and stony slopes, rocks. Montane zone. Nuratau, Malguzar, North Turkestan. Fodder.

251. Piptatherum pamiralaicum (Grig.) Roshev.
Perennial. Fine-earth, gravelly and stony slopes, moraines. Montane and alpine zones. Nuratau, Malguzar, North Turkestan. Fodder.

252. Piptatherum platyanthum Nevski
Perennial. Stony slopes, moraines, pebbles, screes, rocks. Montane and alpine zones. North Turkestan. Fodder.

253. Piptatherum sogdianum (Grig.) Roshev.
Perennial. Fine-earth, gravelly and stony slopes, screes, rocks. Montane and alpine zones. Nuratau, Malguzar, North Turkestan.

254. Piptatherum songaricum (Trin. & Rupr.) Roshev.
Perennial. Fine-earth and stony slopes, rocks, screes. Foothills, montane and alpine zones. Nuratau, Aktau, Nuratau Relic Mountains, Malguzar, North Turkestan. Fodder.

255. Piptatherum microcarpum (Pilg.) Tzvelev [*Piptatherum vicarium* (Grig.) Roshev.]
Perennial. Stony slopes, rocks. Montane zone. North Turkestan. Fodder.

Genus 75. Milium L.

256. Milium vernale M. Bieb.
Annual. Fine-earth and stony slopes, wet places, banks of rivers and canals, gardens. Foothills, montane zone. Nuratau, North Turkestan. Fodder.

Genus 76. Crypsis Ait.

257. Crypsis schoeoides (L.) Lam.
Annual. River valleys, banks of canals, wastelands, saline depressions. Plain. Nuratau, Malguzar, North Turkestan, Mirzachul, Kyzylkum. Fodder.

Genus 77. Phleum L.

258. Phleum alpinum L.   Figure 92
Perennial. Banks of rivers, wet places, alpine meadows. Montane and alpine zone. North Turkestan. Fodder.

Figure 92 Phleum alpinum (photography by Natalya Beshko)

259. Phleum exaratum Griseb. [*Phleum graecum* Boiss. & Heldr.]
Annual. Stony slopes. Foothills, montane zone. North Turkestan. Fodder.
260. Phleum paniculatum Huds.   Figure 93
Annual. Fine-earth, gravelly and stony slopes, screes, pebbles, fallow lands. Foothills, montane zone. Nuratau, Aktau, Nuratau Relic Mountains, Malguzar, North Turkestan. Fodder.

Figure 93 Phleum paniculatum (photography by Natalya Beshko)

261. Phleum phleoides (L.) H. Karst.
Perennial. Fine-earth slopes. Montane zone. Nuratau, Aktau, Nuratau Relic Mountains, Malguzar, North Turkestan. Fodder.

262. *Phleum pratense L.
Perennial. Meadows, river valleys, banks of canals, swamps. Foothills, montane zone. Nuratau, Malguzar, North Turkestan. Fodder.

Genus 78. Alopecurus L.

263. Alopecurus arundinaceus Poir.
Perennial. Swamp meadows, banks of rivers, wet screes. Plain, foothills, montane and alpine zones. Nuratau, Malguzar, North Turkestan. Fodder.

264. Alopecurus himalaicus Hook. f.
Perennial. Fine-earth and gravelly slopes, screes, alpine meadows. Montane and alpine zones. North Turkestan. Fodder.

265. Alopecurus myosuroides Huds.
Annual. Banks of rivers and canals, meadows, pebbles, fields. Plain, foothills. Nuratau, Aktau, Nuratau Relic Mountains, Malguzar, North Turkestan. Fodder, weed.

266. Alopecurus pratensis L.
Perennial. Swamp meadows, banks of rivers, pebbles. Plain, foothills, montane and alpine zones. Nuratau, Nuratau Relic Mountains, Malguzar, North Turkestan. Fodder, ornamental.

Genus 79. Polypogon Desf.

267. Polypogon fugax Nees ex Steud.
Annual, perennial. Saline lands, banks of rivers and canals, fields, swamp meadows. Plain, foothills, montane zone. Nuratau, Malguzar, North Turkestan. Fodder.

268. Polypogon monspleniensis (L.) Desf.
Annual. Wet places, banks of rivers and canals, fields, swamp meadows. Plain, foothills, montane zone. Nuratau, Aktau, Nuratau Relic Mountains, Malguzar, North Turkestan.

269. Polypogon viridis (Gouan) Breistr. [*Polypogon semiverticillatus* (Forssk.) Hyl.]
Annual, perennial. Banks of rivers and canals, fields, swamp meadows. Plain, foothills, montane zone. Nuratau, Aktau, Nuratau Relic Mountains, Malguzar, North Turkestan. Fodder.

Genus 80. Agrostis L.

270. Agrostis gigantea Roth.
Perennial. Wet places, banks of rivers, swamp meadows. Foothills, montane and alpine zones.

Malguzar, North Turkestan. Fodder, ornamental.

271. Agrostis hissarica Roshev.
Perennial. River valleys. Alpine zone. Nuratau, Malguzar, North Turkestan. Fodder.

272. Agrostis olympica (Boiss.) Bor. [*Pentatherum agrostidiforme* (Roshev.) Nevski]
Perennial. Swamps, banks of rivers. Alpine zone. North Turkestan. Fodder.

273. Agrostis stolonifera L.
Perennial. Wet places, banks of rivers, swamp meadows. Foothills, montane and alpine zones. Nuratau, North Turkestan. Fodder.

Genus 81. Calamagrostis Adans.

274. Calamagrostis anthoxanthoides (Munro) Regel [*Calamagrostis laguroides* Regel]
Perennial. Stony slopes, screes. Alpine zone. North Turkestan. Fodder.

275. Calamagrostis dubia Bunge
Perennial. River valleys, saline lands, banks of canals, fields. Plain, foothills. Nuratau, Malguzar, North Turkestan. Fodder.

276. Calamagrostis epigeios (L.) Roth.
Perennial. Banks of rivers and canals, fallow lands, wet fine-earth slopes. Plain, foothills, montane and alpine zones. Malguzar, North Turkestan. Fodder, medicinal, industrial.

277. Calamagrostis holciformis Jaub. & Spach.
Perennial. Rocks, stony slopes. Alpine zone. North Turkestan. Fodder.

278. Calamagrostis pseudophragmites (Haller f.) Koeler.
Perennial. Banks of rivers, pebbles, meadows, wet stony slopes. Foothills, montane and alpine zones. Nuratau, North Turkestan. Fodder.

Genus 82. Apera Adans.

279. Apera interrupta (L.) P. Beauv.
Annual. Saline lands, river valleys. Plain, foothills, montane zone. Nuratau, Aktau, Nuratau Relic Mountains, Malguzar, North Turkestan, Mirzachul. Fodder.

Genus 83. Deschampsia P. Beauv.

280. Deschampsia cespitosa (L.) P. Beauv.
Perennial. River valleys, subalpine meadows. Alpine zone. North Turkestan. Fodder, industrial.

281. Deschampsia koelerioides Regel
Perennial. Alpine meadows, stony slopes. Alpine zone. North Turkestan. Fodder, industrial.

Genus 84. Trisetum Pers.

282. Trisetum spicatum (L.) K. Richt.
Perennial. Fine-earth and gravelly slopes, swamps, subalpine meadows, banks of rivers. Alpine zone. North Turkestan. Fodder, ornamental.

Genus 85. Avena L.

283. Avena barbata Pott ex Link.
Annual. Fine-earth, gravelly and stony slopes, fields, fallow lands. Foothills. North Turkestan. Fodder, weed.

284. *Avena fatua L.
Annual. Fine-earth, gravelly and stony slopes, fields, fallow lands. Plain, foothills. Nuratau, Aktau, Nuratau Relic Mountains, Malguzar, North Turkestan. Fodder, food, medicinal, weed.

285. *Avena sterilis subsp. **ludoviciana** (Durieu) Gillet & Magne [*Avena persica* Steud.; *Avena trichophylla* K. Koch]
Annual. River valleys, fine-earth slopes. Foothills. Nuratau, Aktau, Nuratau Relic Mountains, Malguzar, North Turkestan, Mirzachul, Kyzylkum. Fodder, weed.

Genus 86. Helictotrichon Bess.

286. Helictotrichon asiaticum (Roshev.) Grossh.
Perennial. Fine-earth, gravelly and stony slopes. Montane and alpine zones. Nuratau, Aktau, Nuratau Relic Mountains, Malguzar, North Turkestan. Fodder.

287. Helictotrichon desertorum (Less.) Pilg.
Perennial. Piedmont plain, fine-earth, gravelly and stony slopes. Plain, foothills, montane zone. Malguzar, North Turkestan. Fodder.

Genus 87. *Cynodon Rich.

288. *Cynodon dactylon (L.) Pers.   Figure 94
Perennial. River valleys, fields, gardens, fallow lands, banks of canals, roadsides, wastelands, fine-earth slopes. Plain, foothills, montane zone. Nuratau, Aktau, Nuratau Relic Mountains, Malguzar, North Turkestan, Mirzachul. Fodder, medicinal, industrial, weed.

Figure 94 Cynodon dactylon (photography by Natalya Beshko)

Genus 88. Enneapogon Desv.

289. Enneapogon persicus Boiss.
Perennial. Fine-earth and stony slopes, dry riverbeds, pebbles, rocks. Foothills, montane zone. Nuratau, Nuratau Relic Mountains, Malguzar, North Turkestan, Mirzachul, Kyzylkum.

Genus 89. Phragmites Adans.

290. Phragmites australis (Cav.) Trin. & Steud.   Figure 95
Perennial. Banks of rivers, lakes and canals, swamps, wet places, saline lands, fields. Plain, foothills, montane zone. Nuratau, Aktau, Nuratau Relic Mountains, Malguzar, North Turkestan, Mirzachul, Kyzylkum. Fodder, food, industrial, medicinal, ornamental.

Figure 95 **Phragmites australis** (photography by Natalya Beshko)

Genus 90. Eragrostis Host.

291. Eragrostis minor Host.
Annual. Fine-earth slopes, pebbles, banks of rivers and canals, fields, fallow lands. Plain, foothills. Nuratau, Aktau, Nuratau Relic Mountains, Malguzar, North Turkestan, Mirzachul, Kyzylkum. Fodder, weed.

292. Eragrostis pilosa (L.) Beauv.
Annual. River valleys, banks of canals, pebbles, fields, fallow lands, gardens, wastelands. Plain, foothills, montane zone. Nuratau, Aktau, Nuratau Relic Mountains, Malguzar, North Turkestan, Mirzachul. Fodder, weed.

Genus 91. Koeleria Pers.

293. Koeleria macrantha (Ledeb.) Schult. [*Koeleria cristata* (L.) Pers.]
Perennial. Fine-earth, gravelly and stony slopes. Montane and alpine zones. North Turkestan. Fodder, medicinal, ornamental.

Genus 92. Cutandia Willk.

294. Cutandia rigescens (Grossh.) Tzvelev
Annual. Wet places, saline lands, banks of rivers, lakes and canals, fields. Plain. Nuratau, Aktau, Nuratau Relic Mountains, Malguzar, North Turkestan, Mirzachul, Kyzylkum. Fodder, weed.

Genus 93. Catabrosa P. Beauv.

295. Catabrosa aquatica (L.) Beauv. [*Catabrosa capusii* Franch.]
Perennial. Wet places, banks of rivers, lakes and canals, swamps. Montane and alpine zones. Malguzar, North Turkestan. Fodder.

Genus 94. Melica L.

296. Melica altissima L.
Perennial. Stony slopes. Montane zone. Nuratau, North Turkestan. Poisonous.
297. Melica hohenackeri Boiss.
Perennial. Stony slopes, rocks, screes. Montane zone. Nuratau, Malguzar, North Turkestan.
298. Melica inaequiglumis Boiss.   Figure 96
Perennial. Stony slopes, screes, pebbles. Montane zone. Nuratau, Malguzar, North Turkestan.

Figure 96 Melica inaequiglumis (photography by Natalya Beshko)

Genus 95. Aeluropus Trin.

299. Aeluropus intermedius Regel
Perennial. Saline sandy soils, saline lands, depressions. Plain. Nuratau, Nuratau Relic Mountains, Malguzar, North Turkestan, Mirzachul, Kyzylkum. Fodder.

300. Aeluropus litoralis (Gouan.) Parl.
Perennial. Saline lands, swamp meadows, banks of rivers, lakes and canals. Plain, foothills. Nuratau Relic Mountains, Mirzachul, Kyzylkum. Fodder.

Genus 96. Dactylis L.

301. Dactylis glomerata L.
Perennial. Meadows, river valleys, fine-earth and stony slopes, gardens, wet places. Plain, foothills, montane zone. Nuratau, Malguzar, North Turkestan, Mirzachul. Fodder, medicinal, ornamental.

Genus 97. Sclerochloa P. Beauv.

302. Sclerochloa dura (L.) P. Beauv.
Annual. Banks of rivers and canals, roadsides. Plain, foothills, montane zone. Nuratau, Aktau, Malguzar, North Turkestan. Fodder, weed.

Genus 98. Schismus P. Beauv.

303. Schismus arabicus Nees.
Annual. Sandy and clay deserts, saline lands, piedmont plains, fine-earth and gravelly slopes. Plain, foothills, montane zone. Nuratau, Nuratau Relic Mountains, Malguzar, North Turkestan, Mirzachul, Kyzylkum. Fodder.

Genus 99. Poa L.

304. Poa alpina L.   Figure 97
Perennial. Alpine meadows. Alpine zone. North Turkestan. Fodder.

Figure 97 Poa alpina (photography by Natalya Beshko)

305. Poa angustifolia L.
Perennial. Fine-earth and stony slopes, screes, meadows. Montane and alpine zones, Malguzar, North Turkestan. Fodder.

306. Poa annua L.
Annual or perennial. Banks of rivers and canals, wet places near springs, wastelands, gardens. Plain, foothills, montane zone. Nuratau, Aktau, Nuratau Relic Mountains, Malguzar, North Turkestan, Mirzachul. Fodder, weed.

307. Poa bactriana Roshev.
Perennial. Fine-earth, gravelly and stony slopes, river valleys. Montane and alpine zones. Malguzar, North Turkestan. Fodder.

308. Poa bucharica Roshev.
Perennial. Fine-earth, gravelly and stony slopes, meadows. Montane and alpine zones. North Turkestan. Fodder.

309. Poa bulbosa L.   Figure 98
Perennial. This plant has a wide ecological range, and it is found almost everywhere in the Province. Plain, foothills, montane and alpine zones. Nuratau, Aktau, Nuratau Relic Mountains, Malguzar, North Turkestan, Mirzachul, Kyzylkum. Fodder.

Figure 98 Poa bulbosa (photography by Natalya Beshko)

310. Poa diaphora Trin.
Annual. Sandy, clayey, stony and saline soils, banks of rivers and lakes, wet places near springs. Plain, foothills, montane and alpine zones. Nuratau, Aktau, Nuratau Relic Mountains, Malguzar, North Turkestan, Mirzachul, Kyzylkum. Fodder.

311. Poa fragilis Ovcz.
Perennial. Fine-earth, gravelly and stony slopes. Montane and alpine zones, Nuratau, Malguzar, North Turkestan. Fodder.

312. Poa litvinoviana Ovcz.
Perennial. Stony slopes, rocks. Montane and alpine zones. Nuratau, Malguzar, North Turkestan. Fodder.

313. Poa nemoralis L.
Perennial. River valleys, ravines, meadows, stony slopes, rocks. Montane and alpine zones. Nuratau, Malguzar, North Turkestan. Fodder.

314. Poa palustris L.
Perennial. Wet places, river valleys, swamp meadows. Foothills, montane zone. Nuratau, Malguzar, North Turkestan. Fodder.

315. Poa pratensis L.
Perennial. River valleys, banks of canals, meadows. Plain, foothills, montane and alpine zones. Nuratau, Malguzar, North Turkestan, Mirzachul. Fodder, ornamental.

316. Poa relaxa Ovcz.
Perennial. Stony, Fine-earth and stony slopes, rocks, screes. Montane and alpine zones. Nuratau, Malguzar, North Turkestan. Fodder.

317. Poa supina Schrad.
Annual or perennial. Wet places, swamps, banks of rivers. Montane and alpine zones. Nuratau, Malguzar, North Turkestan. Fodder.

318. Poa trivialis L.
Perennial. River valleys, gardens, banks of canals, fallow lands. Plain, foothills, montane zone. Nuratau, Malguzar, North Turkestan, Mirzachul. Fodder.

Genus 100. Colpodium Trin.

319. Colpodium humile (M. Bieb.) Griseb.
Perennial. Sandy deserts, saline lands, wet places. Plain, foothills. Mirzachul, Kyzylkum. Fodder.

320. Colpodium parviflorum Boiss. & Buhse.
Perennial. Wet places. Plain. Mirzachul. Fodder.

Genus 101. Glyceria R. Br.

321. Glyceria plicata (Fr.) Fr.
Perennial. Banks of rivers, lakes and canals. Plain. Nuratau, Malguzar, North Turkestan. Fodder.

Genus 102. Puccinelia Parl.

322. Puccinelia distans (Jack.) Parl.
Perennial. Banks of rivers, lakes and canals, swamp meadows. Plain, foothills, montane and alpine zones. Nuratau, Nuratau Relic Mountains, Malguzar, North Turkestan, Mirzachul, Kyzylkum. Fodder, weed.

323. Puccinellia poecilantha (K. Koch) Grossh.
Perennial. Swamp meadows, saline lands, banks of rivers, lakes and canals. Plain, foothills. Nuratau, Malguzar, North Turkestan, Mirzachul. Fodder, weed.

Genus 103. Festuca L.

324. Festuca alaica Drobow.
Perennial. Stony slopes, screes, rocks, moraines. Montane and alpine zones. North Turkestan. Fodder.

325. Festuca amblyodes Krecz. & Bobrov
Perennial. Stony slopes, screes, rocks, moraines. Alpine zone. North Turkestan. Fodder.

326. Festuca arundinacea Schreb.
Perennial. Saline lands, banks of rivers, lakes and canals, gardens, meadows, fields, fallow lands, cliffs. Plain, foothills, montane zone. Nuratau, Malguzar, North Turkestan, Mirzachul. Fodder.

327. Festuca griffithiana (St.-Yves) Krivot. [*Leucopoa karatavica* (Bunge) V. Krecz. & Bobrov].
Perennial. Stony slopes, screes, rocks. Montane and alpine zones. North Turkestan. Fodder.

328. Festuca olgae (Regel) Krivot. [*Leucopoa olgae* (Regel) V. Krecz. & Bobrov]
Perennial. Stony slopes, screes, rocks. Montane and alpine zones. North Turkestan. Fodder.

329. Festuca pratensis Huds.
Perennial. Wet places, gardens, banks of rivers and canals, wet slopes. Plain, foothills, montane zone. North Turkestan. Fodder.

330. Festuca rubra L.
Perennial. River valleys, pebbles, meadows, swamps, fine-earth and stony slopes. Montane and alpine zones. Malguzar, North Turkestan. Fodder.

331. Festuca rupicola Heuff.
Perennial. Stony slopes, rocks. Montane and alpine zones. Malguzar, North Turkestan. Fodder.

332. Festuca valesiaca Schleich. & Gaudin.   Figure 99
Perennial. Fine-earth, gravelly and stony slopes, rocks, screes, pebbles, moraines. Montane and alpine zones. Nuratau, Aktau, Malguzar, North Turkestan. Fodder.

Figure 99 Festuca valesiaca (photography by Natalya Beshko)

Genus 104. Loliolum V. Krecz. & Bobrov

333. Loliolum subulatum (Banks & Sol.) Eig.
Annual. Piedmont plains, fine-earth, gravelly and stony slopes. Plain, foothills, montane zone. Nuratau, Aktau, Malguzar, North Turkestan.

Genus 105. Vulpia C. C. Gmel.

334. Vulpia ciliata Dumort.
Annual. Sandy, clayey, gravelly and saline soils, banks of rivers. Plain, foothills. Nuratau, Aktau, Nuratau Relic Mountains, Malguzar, North Turkestan, Mirzachul, Kyzylkum. Fodder, weed.

335. Vulpia myuros (L.) C. C. Gmel.
Annual. Fine-earth and stony slopes, pebbles, river valleys, saline lands. Plain, foothills, montane zone. Nuratau, Aktau, Nuratau Relic Mountains, Malguzar, North Turkestan, Mirzachul, Kyzylkum. Fodder, weed

336. Vulpia persica (Boiss. & Buhse) V. Krecz. & Bobrov
Annual. Clay deserts, fine-earth and gravelly slopes, pebbles, river valleys, saline depressions. Plain, foothills. Nuratau, Aktau, Nuratau Relic Mountains, Malguzar, North Turkestan, Mirzachul, Kyzylkum. Fodder, weed.

337. Vulpia unilateralis (L). Stace [*Nardurus krausei* (Regel) V. Krecz. & Bobrov]
Annual. Fine-earth and stony slopes, piedmont plains, sandy and clay deserts, pebbles, river valleys. Plain, foothills, montane zone. Nuratau, Aktau, Nuratau Relic Mountains, Malguzar, North Turkestan, Mirzachul, Kyzylkum. Fodder, weed.

Genus 106. Bromus L.

338. Bromus danthoniae Trin.   Figure 100
Annual. Clay deserts, piedmont plains, fine-earth, gravelly and stony slopes, fallow lands, pebbles. Plain, foothills, montane zone. Nuratau, Aktau, Nuratau Relic Mountains, Malguzar, North Turkestan, Mirzachul. Fodder, weed.

Figure 100 Bromus danthoniae (photography by Natalya Beshko)

339. Bromus inermis Leyss.
Perennial. Meadows, fine-earth and gravelly slopes, river valleys. Montane and alpine zones. Nuratau, Malguzar, North Turkestan. Fodder.

340. Bromus japonicus Thung.
Annual. Meadows, fine-earth slopes, roadsides, fallow lands, banks of rivers and canals. Foothills, montane and alpine zones. Nuratau, Aktau, Nuratau Relic Mountains, Malguzar, North Turkestan. Fodder, weed.

341. Bromus lanceolatus Roth.
Annual. Sandy and clay deserts, piedmont plains, fine-earth, gravelly and stony slopes, fields, fallow lands, banks of rivers, pebbles. Plain, foothills, montane zone. Nuratau, Aktau, Nuratau Relic Mountains, Malguzar, North Turkestan, Mirzachul, Kyzylkum. Fodder, weed.

342. Bromus oxyodon Schrenk
Annual. Fine-earth, gravelly and stony slopes, fields, fallow lands. Plain, foothills, montane zone. Nuratau, Malguzar, North Turkestan, Mirzachul. Fodder, weed.

343. Bromus paulsenii Hack.
Perennial. Stony slopes, banks of rivers, screes. Montane and alpine zones. Malguzar, North Turkestan. Fodder.

344. Bromus popovii Drobow
Annual. Saline lands, river valleys. Plain. Mirzachul. Fodder.

345. Bromus scoparius L.
Annual. Fine-earth, gravelly and stony slopes, fields, fallow lands, river valleys, roadsides, gardens. Plain, foothills, montane zone. Nuratau, Aktau, Nuratau Relic Mountains, Malguzar, North Turkestan. Fodder, weed.

346. Bromus sericeus Drobow
Annual. Sandy deserts, fine-earth and stony slopes, river valleys. Plain, foothills, montane and alpine zones. Nuratau, Aktau, Nuratau Relic Mountains, Malguzar, North Turkestan, Mirzachul, Kyzylkum. Fodder, weed.

347. Bromus sewerzowii Regel
Annual. River valleys, fields, fallow lands, fine-earth and stony slopes. Plain, foothills, montane zone. Nuratau, Aktau, Nuratau Relic Mountains, Malguzar, North Turkestan, Mirzachul, Kyzylkum. Fodder, weed.

348. Bromus scoparius L.
Annual. Sandy and clay deserts, piedmont plains, fine-earth, gravelly and stony slopes, meadows, pebbles, fields, fallow lands, gardens. Nuratau, Aktau, Nuratau Relic Mountains, Malguzar, North Turkestan, Mirzachul, Kyzylkum. Fodder, weed.

349. Bromus sterilis L.
Annual. River valleys, ravines, fine-earth slopes, meadows, gardens, roadsides, fields. Foothills, montane zone. Nuratau, Aktau, Malguzar, North Turkestan. Fodder, weed.

350. Bromus tectorum L.
Annual. Sandy and clay deserts, piedmont plains, fine-earth, gravelly and stony slopes, fields, fallow

lands, settlements, river valleys. Plain, foothills, montane and alpine zones. Nuratau, Aktau, Nuratau Relic Mountains, Malguzar, North Turkestan, Mirzachul, Kyzylkum. Fodder, weed.

Genus 107. Boissiera Hochsst & Steud.

351. Boissiera squarrosa (Sol.) Nevski.   Figure 101

Annual. Sandy and clay desert, piedmont plains, fine-earth, gravelly and stony slopes, fields, fallow lands, roadsides. Plain, foothills, montane and alpine zones. Nuratau, Aktau, Nuratau Relic Mountains, Malguzar, North Turkestan, Mirzachul, Kyzylkum. Fodder, weed.

Figure 101 Boissiera squarrosa (photography by Natalya Beshko)

Genus 108. Brachypodium P. Beauv.

352. Brachypodium sylvaticum P. Beauv.

Perennial. Fine-earth wet slopes, river valleys, ravines. Montane zone. Nuratau, Malguzar, North Turkestan. Fodder.

Genus 109. Lolium L.

353. *Lolium perenne L.
Perennial. Gardens, fields, fallow lands, banks of canals, dry slopes. Plain, foothills, montane zone. Malguzar, North Turkestan, Mirzachul. Fodder, medicinal.

354. Lolium persicum Boiss. & Hohen.
Annual. Fallow lands, fields, stony slopes, river valleys. Foothills, montane zone. Malguzar, North Turkestan. Fodder, weed.

355. *Lolium temulentum L.
Annual. Fallow lands, gardens, wet meadows, banks of canals, river valleys, fine-earth slopes. Foothills, montane zone. Malguzar, North Turkestan. Fodder, medicinal.

Genus 110. Henrardia Hubb.

356. Henrardia persica (Boiss.) C. E. Hubb.
Annual. Dry fine-earth, gravelly and stony slopes. Foothills. Nuratau, Malguzar, North Turkestan. Fodder, weed.

Genus 111. Psilurus Trin.

357. Psilurus incurvus (Gouan) Schinz & Thell.
Annual. Piedmont plains, fine-earth, gravelly and stony slopes. Plain, foothills. Nuratau, Aktau, Nuratau Relic Mountains, Malguzar, North Turkestan. Fodder.

Genus 112. Agropyron Gaertn.

358. Agropyron badamense Drobow
Perennial. Stony slopes. Montane and alpine zones. Nuratau, Malguzar, North Turkestan. Fodder.

359. Agropyron setuliferum (Nevski) Nevski
Perennial. Fine-earth and stony slopes. Montane zone. Malguzar, North Turkestan. Fodder.

Genus 113. Eremopyrum Jaub. & Spach.

360. Eremopyrum bonaepartis (Spreng.) Nevski   Figure 102
Annual. Sandy, clayey and gravelly soils, fallow lands. Plain, foothills. Nuratau, Aktau, Nuratau Relic Mountains, Malguzar, North Turkestan, Mirzachul, Kyzylkum. Fodder, weed.

Figure 102 **Eremopyrum bonaepartis** (photography by Natalya Beshko)

361. Eremopyrum distans (K. Koch) Nevski   Figure 103

Annual. Sandy, clayey, gravelly and saline soils, pebbles, fallow lands. Plain, foothills, montane zone. Nuratau, Aktau, Nuratau Relic Mountains, Malguzar, North Turkestan, Mirzachul, Kyzylkum. Fodder, weed.

Figure 103 **Eremopyrum distans** (photography by Natalya Beshko)

362. Eremopyrum orientale (L.) Jaub. & Spach.
Annual. Sandy, clayey, gravelly and saline soils, pebbles, fallow lands. Plain, foothills. Nuratau, Aktau, Nuratau Relic Mountains, Malguzar, North Turkestan, Mirzachul, Kyzylkum.

363. Eremopyrum triticeum (Gaertn.) Nevski
Annual. Sandy, clayey, gravelly and saline soils, river valleys, fallow lands. Plain, foothills. Nuratau, Aktau, Nuratau Relic Mountains, Malguzar, North Turkestan, Mirzachul, Kyzylkum. Fodder, weed.

Genus 114. Secale L.

364. *Secale cereale L.
Annual. Fields, fallow lands, river valleys, banks of canals, roadsides. Foothills, montane zone. Nuratau, Nuratau Relic Mountains, Malguzar, North Turkestan, Mirzachul, Kyzylkum. Fodder, food.

365. Secale segetale (Zhuk.) Roshev.
Annual. Fields, fallow lands, river valleys, banks of canals, roadsides. Plain, foothills. Nuratau, Nuratau Relic Mountains, Malguzar, North Turkestan, Mirzachul, Kyzylkum. Fodder, food.

366. Secale sylvestre Host.
Annual. Sandy and clayey soils. Plain, Mirzachul, Kyzylkum. Fodder, food.

Genus 115. Aegilops L.

367. Aegilops crassa Boiss.
Annual. Fine-earth and stony slopes, fields, fallow lands, river valleys. Plain, foothills, montane zone. Nuratau, Nuratau Relic Mountains, Malguzar, North Turkestan, Mirzachul, Kyzylkum. Fodder, weed.

368. Aegilops cylindrica Host.
Annual. Fine-earth and stony slopes, fields, fallow lands, river valleys. Plain, foothills. Nuratau, Aktau, Nuratau Relic Mountains, Malguzar, North Turkestan. Fodder, weed.

369. Aegilops juvenalis (Thell.) Eig.
Annual. Fine earth slopes. Foothills. Nuratau, Aktau, Nuratau Relic Mountains. Fodder, weed.

370. Aegilops triuncialis L. [*Aegilops squarrosa* L.]   Figure 104
Annual. Piedmont plains, fine-earth and stony slopes, fields, fallow lands, river valleys. Plain, foothills, montane zone. Nuratau, Aktau, Nuratau Relic Mountains, Malguzar, North Turkestan, Mirzachul, Kyzylkum. Fodder, weed.

Figure 104 Aegilops triuncialis (photography by Natalya Beshko)

Genus 116. Heteranthelium Hochst. ex Jaub. & Spach.

371. Heteranthelium piliferum (Sol.) Hochst. ex Jaub. & Spach.
Annual. Stony slopes, piedmont plains. Plain, foothills, montane zone. Nuratau, Aktau, Nuratau Relic Mountains, Malguzar, North Turkestan. Fodder.

Genus 117. Taeniatherum Nevski

372. Taeniatherum caput-medusae (L.) Nevski [*Taeniatherum asperum* (Simonk.) Nevski; *Taeniatherum crinitum* (Schreb.) Nevski]
Annual. Fallow lands, fields, roadsides, piedmont plain, fine-earth, gravelly and stony slopes. Plain, foothills, montane zone. Nuratau, Aktau, Nuratau Relic Mountains, Malguzar, North Turkestan, Mirzachul, Kyzylkum. Fodder, weed.

Genus 118. Hordeum L.

373. Hordeum bogdanii Wilensky
Perennial. Swamp meadows, saline lands, banks of rivers, lakes and canals. Plain, foothills, montane zone. Mirzachul, Kyzylkum. Fodder.

374. Hordeum brevisubulatum (Trin.) Link.
Perennial. Banks of rivers and lakes, swamp meadows, stony slopes. Plain, foothills, montane and alpine zones. Nuratau, Malguzar, North Turkestan, Mirzachul. Fodder.

375. Hordeum bulbosum L.   Figure 105
Perennial. Fine-earth, gravelly and stony slopes, river valleys, fields, gardens, banks of canals. Plain, foothills, montane zone. Nuratau, Aktau, Nuratau Relic Mountains, Malguzar, North Turkestan, Mirzachul. Fodder, food.

Figure 105 Hordeum bulbosum (photography by Natalya Beshko)

376. Hordeum marinum subsp. **gussoneanum** (Parl.) Thell. [*Hordeum geniculatum* All.]
Annual. River valleys. Plain, foothills. Nuratau, North Turkestan, Mirzachul. Fodder, weed.

377. Hordeum murinum subsp. **leporinum** (Link.) Arcang. [*Hordeum leporinum* Link.]
Annual. River valleys, gardens, fields, fallow lands, wastelands, settelments, sandy and clay deserts, fine-earth, gravelly and stony slopes. Plain, foothills, montane zone. Nuratau, Aktau, Nuratau Relic Mountains, Malguzar, North Turkestan, Mirzachul, Kyzylkum. Fodder, weed.

378. Hordeum spontaneum K. Koch   Figure 106
Annual. Fields, fallow lands, banks of canals, fine-earth, gravelly and stony slopes. Plain, foothills, montane zone. Nuratau, Aktau, Malguzar, North Turkestan, Mirzachul, Kyzylkum. Fodder, weed.

Figure 106 **Hordeum spontaneum** (photography by Natalya Beshko)

Genus 119. Elymus L.

379. Elymus bungeanus (Trin.) Melderis [*Agropyron ferganense* Drobow]
Perennial. Fine-earth and stony slopes. Montane zone. Malguzar, North Turkestan. Fodder.

380. Elymus dentatus (Hook. f.) Tzvelev [*Agropyron lachnophyllum* (Ovcz. & Sidorenko) Bondarenko; *Agropyron ugamicum* Drobow]
Perennial. Fine-earth and stony slopes. Montane and alpine zones. Malguzar, North Turkestan.

Fodder.

381. Elymus drobovii (Nevski) Tzvelev [*Agropyron drobovii* Nevski]
Perennial. Stony slopes. Montane zone. Nuratau. Fodder.

382. Elymus hispidus (Opiz) Melderis [*Agropyron intermedium* (Host) Beauv.; *Agropyron trichophorum* (Link.) K. Richt.]   Figure 107
Perennial. Fine-earth, gravelly and stony slopes. The species is dominant of so-called savannoids (dry herbaceous steppes), a vegetation type endemic to the Central Asian mountains. Foothills, montane zone. Nuratau, Aktau, Malguzar, North Turkestan. Fodder.

Figure 107 Elymus hispidus (photography by Natalya Beshko)

383. Elymus lolioides (P. Candargy) Melderis [*Agropyron lolioides* (Kar. & Kir.) Roshev.]
Perennial. Meadows, fine-earth slopes, banks of rivers and canals. Plain, foothills, montane zone. Nuratau, Malguzar, North Turkestan. Fodder.

384. Elymus repens (L.) Gould. [*Agropyron repens* (L.) P. Beauv.]
Perennial. Wet slopes, meadows, river valleys, gardens. Plain, foothills, montane zone. Nuratau, Aktau, Malguzar, North Turkestan, Mirzachul. Fodder, medicinal, weed.

385. Elymus uralensis (Nevski) Tzvelev [*Agropyron tianschanicum* Drobow]
Perennial. Stony slopes, screes. Montane zone. Malguzar, North Turkestan. Fodder.

Genus 120. Leymus Hochst.

386. Leymus alaicus (Korsch.) Tzvelev
Perennial. Stony slopes. Montane zone. North Turkestan. Fodder.

387. Leymus angustus (Trin.) Pilg.   Figure 108
Perennial. Fine-earth, gravelly and stony slopes. Montane zone. Nuratau. Fodder.

Figure 108 Leymus angustus (photography by Natalya Beshko)

388. Leymus multicaulis (Kar. & Kir.) Tzvelev

Perennial. River valleys, wet meadows, banks of canals. Plain, foothills, montane zone. North Turkestan. Fodder.

MAGNOLIOPSIDA [EUDICOTS]

Family 30. Ceratophyllaceae

Genus 121. Ceratophyllum L.

389. Ceratophyllum demersum L.
Perennial. Standing and slowly flowing water. Plain, foothills. Nuratau, Mirzachul, Kyzylkum. Fodder (for fish), ornamental.

Family 31. Papaveraceae

Genus 122. Papaver L.

390. Papaver dubium L. [*Papaver litwinowii* Fedde ex Bornm.]   Figure 109
Annual. Fine-earth, gravelly and stony slopes, rocks, screes, pebbles, dry riverbeds. Foothills, montane zone. Nuratau, Aktau, Nuratau Relic Mountains, Malguzar, North Turkestan. Alkaloid-bearing.

Figure 109 Papaver dubium (photography by Natalya Beshko)

391. Papaver nudicaule L. [*Papaver croceum* Ledeb.]   Figure 110
Perennial. Fine-earth, gravelly and stony slopes, rocks, screes, alpine meadows. Montane and alpine zones. North Turkestan. Ornamental, poisonous, alkaloid-bearing.

Figure 110 Papaver nudicaule (photography by Natalya Beshko)

392. Papaver pavoninum C. A. Mey.   Figure 111
Annual. Sandy and clay deserts, piedmont plains, fine-earth, gravelly and stony slopes, fallow lands, roadsides, wastelands, settlements, fields. Plain, foothills, montane zone. Nuratau, Aktau, Nuratau Relic Mountains, Malguzar, North Turkestan, Mirzachul, Kyzylkum. Medicinal, dye, ornamental, weed, poisonous, alkaloid-bearing.

Figure 111 Papaver pavoninum (photography by Natalya Beshko)

Genus 123. Roemeria Medik.

393. Roemeria hybrida (L.) DC.   Figure 112
Annual. Sandy and clay deserts, saline depressions, pebbles, fine-earth, gravelly and stony slopes. Plain, foothills, montane zone. Nuratau, Aktau, Nuratau Relic Mountains, Malguzar, North Turkestan, Mirzachul, Kyzylkum. Weed.

Figure 112 Roemeria hybrida (photography by Natalya Beshko)

394. Roemeria refracta DC.   Figure 113

Annual. Sandy and clay deserts, fine-earth, gravelly and stony slopes, fallow lands, roadsides, gardens, settlements, fields. Plain, foothills, montane zone. Nuratau, Aktau, Nuratau Relic Mountains, Malguzar, North Turkestan, Mirzachul, Kyzylkum. Medicinal, dye, ornamental, weed, poisonous, alkaloid-bearing.

Figure 113 **Roemeria refracta** (photography by Natalya Beshko)

Genus 124. Glaucium Adans.

395. Glaucium elegans Fisch. & C. A. Mey.   Figure 114

Annual. Fine-earth, gravelly and stony slopes, screes, pebbles, fallow lands. Plain, foothills, montane zone. Nuratau, Aktau, Nuratau Relic Mountains, Malguzar, North Turkestan. Medicinal, ornamental, poisonous, alkaloid-bearing.

Figure 114 **Glaucium elegans** (photography by Natalya Beshko)

396. Glaucium fimbrilligerum (Trautv.) Boiss. Figure 115

Biennial. Fine-earth, gravelly and stony slopes, dry riverbeds, pebbles, screes, cliffs, fallow lands. Foothills, montane zone. Nuratau, Aktau, Nuratau Relic Mountains, Malguzar, North Turkestan. Medicinal, ornamental, poisonous, alkaloid-bearing.

Figure 115 **Glaucium fimbrilligerum** (photography by Natalya Beshko)

397. Glaucium squamigerum Kar. & Kir.

Perennial. Stony slopes, rocks, screes, pebbles, ravines, banks of rivers. Montane and alpine zones. Malguzar, North Turkestan. Poisonous, alkaloid-bearing.

Genus 125. Hypecoum L.

398. Hypecoum pendulum L. [*Hypecoum parviflorum* Kar. & Kir.] Figure 116

Annual. Sandy and clay deserts, fine-earth, gravelly and stony slopes, fallow lands, fields. Plain, foothills, montane zone. Nuratau, Aktau, Nuratau Relic Mountains, Malguzar, North Turkestan, Mirzachul, Kyzylkum. Weed, alkaloid-bearing.

Figure 116 **Hypecoum pendulum** (photography by Natalya Beshko)

399. Hypecoum trilobum Trautv.
Annual. Fine-earth and stony slopes, sandy and clay deserts, saline depressions, wastelands, fallow lands. Plain, foothills, montane zone. Nuratau, Aktau, Nuratau Relic Mountains, Malguzar, North Turkestan, Mirzachul, Kyzylkum. Weed, alkaloid-bearing.

Genus 126. Corydalis DC.

400. Corydalis aitchisonii Popov [*Corydalis sewerzowii* Regel]   Figure 117
Perennial. Fine-earth, gravelly and stony slopes, rocks, screes. Foothills, montane zone. Nuratau, Aktau, Nuratau Relic Mountains, Malguzar, North Turkestan. Ornamental, alkaloid-bearing.

Figure 117 Corydalis aitchisonii (photography by Natalya Beshko)

401. Corydalis glaucescens Regel   Figure 118
Perennial. Fine-earth and stony slopes, shady wet places. Montane zone. Malguzar, North Turkestan. Ornamental, medicinal, alkaloid-bearing.

Figure 118 **Corydalis glaucescens** (photography by Natalya Beshko)

402. Corydalis gortschakovii Schrenk
Perennial. Stony slopes, rocks, screes, subalpine and alpine meadows. Alpine zone. North Turkestan. Ornamental, alkaloid-bearing.

403. Corydalis ledebouriana Kar. & Kir.   Figure 119
Perennial. Fine-earth, gravelly and stony slopes, places near melting snow. Montane and alpine zones. Nuratau, Malguzar, North Turkestan. Ornamental, medicinal, alkaloid-bearing.

Figure 119 **Corydalis ledebouriana** (photography by Natalya Beshko)

404. Corydalis nudicaulis Regel

Perennial. Fine-earth, gravelly slopes, rocks, places near melting snow. Montane and alpine zones. North Turkestan.

405. Corydalis schelesnowiana Regel & Schmalh.

Perennial. Stony slopes, rocks, screes. Montane zone. North Turkestan.

Genus 127. Fumaria L.

406. Fumaria vaillantii Loisel

Annual. Fine-earth, gravelly and stony slopes, screes, fallow lands, wastelands, roadsides, settlements, gardens, fields. Plain, foothills, montane zone. Nuratau, Aktau, Nuratau Relic Mountains, Malguzar, North Turkestan, Mirzachul, Kyzylkum. Medicinal weed, alkaloid-bearing, poisonous.

Family 32. Berberidaceae

Genus 128. Berberis L.

407. Berberis integerrima Bunge

Shrub. Fine-earth, gravelly and stony slopes, river valleys. Montane zone. Malguzar, North Turkestan. Medicinal, food, dye, ornamental, meliferous, afforestation.

408. Berberis nummularia Bunge   Figure 120

Shrub. Stony and gravelly slopes, rocks, screes, river valleys. Foothills, montane zone. Malguzar, North Turkestan. Medicinal, food, dye, ornamental, melliferous, afforestation.

Figure 120 Berberis nummularia (photography by Natalya Beshko)

409. Berberis oblonga C. K. Schneid   Figure 121
Shrub. Stony slopes, screees, river valleys. Montane zone. Malguzar, North Turkestan. Medicinal, food, dye, ornamental, melliferous, afforestation.

Figure 121 Berberis oblonga (photography by Natalya Beshko)

Genus 129. Gymnospermium Spach

410. Gymnospermium albertii (Regel) Takht.
Perennial. Fine-earth, gravelly and stony slopes, screes. Montane and alpine zones. Nuratau, Malguzar, North Turkestan. Ornamental, alkaloid-bearing, saponin-bearing.

Genus 130. Leontice L.

411. Leontice eversmannii Bunge
Perennial. Fine-earth, gravelly and stony slopes, piedmont plains, river valleys, fallow lands. Plain, foothills, montane zone. Nuratau, Aktau, Nuratau Relic Mountains, Malguzar, North Turkestan. Ornamental, alkaloid-bearing, saponin-bearing, poisonous.

Genus 131. Bongardia C. A. Mey.

412. Bongardia chrysogonum (L.) Boiss.   Figure 122

Perennial. Fine-earth, gravelly and stony slopes, piedmont plains, fallow lands. Plain, foothills, montane zone. Nuratau, Aktau, Nuratau Relic Mountains, Malguzar, North Turkestan. Alkaloid-bearing, saponin-bearing, ornamental.

Figure 122 Bongardia chrysogonum (photography by Natalya Beshko)

Family 33. Ranunculaceae

Genus 132. Aquilegia L.

413. Aquilegia lactiflora Kar. & Kir.

Perennial. Fine-earth and stony slopes, rocks, shady wet places, swamps, river valleys, places near melting snow. Montane and alpine zones. North Turkestan. Ornamental, poisonous.

414. Aquilegia vicaria Nevski   Figure 123

Perennial. Shady wet places, rocks, banks of rivers, swamps, wet places near springs. Montane and alpine zones. Malguzar, North Turkestan. Ornamental, poisonous.

Figure 123 Aquilegia vicaria (photography by Natalya Beshko)

Genus 133. Isopyrum L.

415. Isopyrum anemonoides Kar. & Kir.   Figure 124
Perennial. Shady wet places, fine-earth and stony slopes, rocks, alpine meadows, swamps, banks of rivers, places near melting snow. Montane and alpine zones. North Turkestan.

Figure 124 Isopyrum anemonoides (photography by Natalya Beshko)

Genus 134. Paraquilegia Drumm. & Hutch.

416. Paraquilegia caespitosa J. R. Drumm. & Hutch.
Perennial. Stony slopes, rocks, screes. Alpine zone. North Turkestan.

Genus 135. Thalictrum L.

417. Thalictrum isopyroides C. A. Mey.
Perennial. Fine-earth, gravelly and stony slopes, rocks, screes. Foothills, montane zone. Nuratau, Aktau, Nuratau Relic Mountains, Malguzar, North Turkestan. Medicinal, alkaloid-bearing.

418. Thalictrum minus L.   Figure 125
Perennial. River valleys, fine-earth, gravelly and stony slopes. Montane zone. Nuratau, Malguzar, North Turkestan. Medicinal, alkaloid-bearing.

Figure 125 Thalictrum minus (photography by Natalya Beshko)

419. Thalictrum sultanabadense Stapf.
Perennial. Fine-earth, gravelly and stony slopes, rocks, screes. Foothills, montane zone. Nuratau, Aktau, Nuratau Relic Mountains, Malguzar, North Turkestan. Alkaloid-bearing.

Genus 136. Adonis L.

420. *Adonis aestivalis L.   Figure 126
Annual. Fine-earth, gravelly and stony slopes, wastelands, fields, fallow lands, roadsides. Plain, foothills, montane zone. Nuratau, Aktau, Nuratau Relic Mountains, Malguzar, North Turkestan. Medicinal, weed, poisonous.

Figure 126 Adonis aestivalis (photography by Natalya Beshko)

421. Adonis turkestanica (Korsh.) Adolf.   Figure 127

Perennial. Fine-earth, gravelly and stony slopes, subalpine meadows. Montane and alpine zones. North Turkestan. Medicinal, weed, poisonous.

Figure 127 Adonis turkestanica (photography by Natalya Beshko)

Genus 137. Trollius L.

422. Trollius komarovii Pachom.   Figure 128

Perennial. Stream banks, swamps, places near melting snow. Alpine zone. North Turkestan.

Figure 128 Trollius komarovii (photography by Natalya Beshko)

Genus 138. Aconitum L.

423. Aconitum zeravschanicum Steinb

Perennial. Fine-earth and stony slopes, screes, subalpine and alpine meadows, places near melting snow. Montane and alpine zones. Malguzar, North Turkestan. Alkaloid-bearing, medicinal, poisonous, ornamental.

424. Aconitum talassicum Popov   Figure 129

Perennial. Banks of rivers, ravines, wet places. Montane and alpine zones. Malguzar, North

Turkestan. Ornamental, alkaloid-bearing, medicinal, poisonous. UzbRDB 3.

Figure 129 **Aconitum talassicum** (photography by Natalya Beshko)

Genus 139. Consolida S. F. Gray.

425. Consolida camptocarpa (Fisch. & C. A. Mey. ex Ledeb.) Nevski [*Delphinium camptocarpum* Fisch. & C. A. Mey.]   Figure 130
Annual. Sandy and clay deserts. Plain. Kyzylkum. Poisonous, alkaloid-bearing.

Figure 130 Consolida camptocarpa (photography by Natalya Beshko)

426. Consolida leptocarpa Nevski [*Delphinium leptocarpum* Nevski]   Figure 131
Annual. Sandy and clay deserts, river valleys, fine-earth and stony slopes. Plain, foothills, montane zone. Nuratau, Nuratau Relic Mountains, Malguzar, North Turkestan, Kyzylkum. Poisonous,

alkaloid-bearing.

Figure 131 Consolida leptocarpa (photography by Natalya Beshko)

427. Consolida rugulosa (Boiss.) Schrödinger [*Delphinium rugulosum* Boiss.]
Annual. Sandy and clay deserts, saline lands, fine-earth and stony slopes, fallow lands. Plain, foothills. Nuratau Relic Mountains, Kyzylkum. Alkaloid-bearing, poisonous.

Genus 140. Delphinium L.

428. Delphinium barbatum Bunge   Figure 132
Annual. Fine-earth and stony slopes, screes, river valleys, fallow lands, fields, roadsides. Foothills, montane zone. Nuratau, Aktau, Nuratau Relic Mountains, Malguzar, North Turkestan. Dye, alkaloid-bearing.

Figure 132 Delphinium barbatum (photography by Natalya Beshko)

429. Delphinium batalinii Huth Figure 133

Perennial. Fine-earth and stony slopes, screes, banks of rivers, pebbles. Montane and alpine zones. Nuratau, Malguzar, North Turkestan. Alkaloid-bearing.

Figure 133 Delphinium batalinii (photography by Natalya Beshko)

430. Delphinium biternatum Huth   Figure 134
Perennial. Fine-earth, gravelly and stony slopes, rocks, screes, banks of rivers, subalpine meadows. Montane and alpine zones. Nuratau, Malguzar. Dye, poisonous, alkaloid-bearing.

Figure 134 Delphinium biternatum (photography by Natalya Beshko)

431. Delphinium confusum Popov
Perennial. Fine-earth, gravelly and stony slopes, alpine meadows, banks of rivers. Montane and alpine zones. North Turkestan. Dye, alkaloid-bearing, poisonous.

432. Delphinium longipedunculatum Regel & Schmalh.   Figure 135
Perennial. Fine-earth, gravelly and stony slopes. Foothills, montane zone. Nuratau, Aktau, Nuratau Relic Mountains, Malguzar, North Turkestan. Dye, alkaloid-bearing, poisonous.

Figure 135 Delphinium longipedunculatum (photography by Natalya Beshko)

433. Delphinium oreophilum Huth

Perennial. Fine-earth and stony slopes, rocks, screes, pebbles, moraines, banks of rivers, swamps, subalpine and alpine meadows. Montane and alpine zones. Malguzar, North Turkestan. Alkaloid-bearing.

434. Delphinium poltoratzkii Rupr.

Perennial. Fine-earth and stony slopes, rocks, screes, ravines, banks of rivers. Montane and alpine zones. Malguzar, North Turkestan. Alkaloid-bearing.

435. Delphinium propinquum Nevski

Perennial. Fine-earth and stony slopes, places near melting snow. Alpine zone. North Turkestan. Alkaloid-bearing.

436. Delphinium semibarbatum Bien. ex Boiss.   Figure 136

Perennial. Clay deserts, relic mountains, piedmont plains, foothills, fine-earth and stony slopes, screes, ravines, fallow lands, gardens. Plain, foothills, montane zone. Nuratau, Aktau, Nuratau Relic Mountains, Malguzar, North Turkestan. Dye, medicinal, ornamental, poisonous, alkaloid-bearing.

Figure 136 **Delphinium semibarbatum** (photography by Natalya Beshko)

Genus 141. Nigella L.

437. Nigella integrifolia Regel   Figure 137
Annual. Sandy and clay deserts, relic mountains, foothills, fine-earth and stony slopes, screes, fallow lands, fields. Plain, foothills, montane zone. Nuratau, Aktau, Nuratau Relic Mountains, Malguzar, North Turkestan, Mirzachul, Kyzylkum.

Figure 137 *Nigella integrifolia* (photography by Natalya Beshko)

Genus 142. Eranthis Salisb.

438. Eranthis longistipitata Regel   Figure 138
Perennial. Fine-earth, gravelly and stony slopes, river valleys, places near melting snows. Foothills, montane zone. Nuratau, Aktau, Nuratau Relic Mountains, Malguzar, North Turkestan.

Figure 138 *Eranthis longistipitata* (photography by Natalya Beshko)

Genus 143. Callianthemum C. A. Mey.

439. Callianthemum alatavicum Freyn.
Perennial. Fine-earth and stony slopes, screes, rocks, moraines, pebbles, alpine meadows, banks of rivers, wet places, places near melting snow. Alpine zone. North Turkestan. Alkaloid-bearing.

Genus 144. Anemone L.

440. Anemone biflora DC. var. **petiolulosa** (Juz.) S. Ziman [*Anemone petiolulosa* Juz.]   Figure 139
Perennial. Fine-earth, gravelly and stony slopes. Foothills, montane zone. Nuratau, Aktau, Nuratau Relic Mountains, Malguzar, North Turkestan.

Figure 139 Anemone biflora var. **petiolulosa** (photography by Natalya Beshko)

441. Anemone narcissiflora subsp. **protracta** (Ulbr.) Ziman & Fedor. [*Anemone protracta* (Ulbr.) Juz.]   Figure 140
Perennial. Fine-earth and stony slopes, alpine meadows, places near melting snow. Montane and alpine zones. Malguzar, North Turkestan. Ornamental, poisonous.

Figure 140 Anemone narcissiflora subsp. **protracta** (photography by Natalya Beshko)

442. Anemone tschernaewii (Czern.) Regel
Perennial. Fine-earth and stony slopes, shady wet places. Montane zone. Malguzar, North Turkestan.

Genus 145. Clematis L.

443. Clematis orientalis L.   Figure 141
Shrub or liana. Banks of rivers, canals and lakes, fine-earth and stony slopes. Plain, foothills, montane zone. Nuratau, Aktau, Nuratau Relic Mountains, Malguzar, North Turkestan, Mirzachul, Kyzylkum. Ornamental, alkaloid-bearing, poisonous.

Figure 141 Clematis orientalis (photography by Natalya Beshko)

Genus 146. Pulsatilla Adans.

444. Pulsatilla campanella Fisch. ex Krylov.
Perennial. Fine-earth and stony slopes, moraines. Alpine zone. North Turkestan. Ornamental, alkaloid-bearing.

Genus 147. Ceratocephala Moench

445. Ceratocephala falcata (L.) Pers.   Figure 142

Annual. Sandy and clay desert, pebbles, fine-earth and stony slopes, screes, wastelands. Plain, foothills, montane zone. Nuratau, Aktau, Nuratau Relic Mountains, Malguzar, North Turkestan, Mirzachul, Kyzylkum. Weed, poisonous.

Figure 142 Ceratocephala falcata (photography by Natalya Beshko)

446. Ceratocephala testiculata (Crantz) Besser

Annual. Sandy and clay deserts, pebbles, fine-earth and stony slopes, screes, wastelands. Plain, foothills, montane zone. Nuratau, Aktau, Nuratau Relic Mountains, Malguzar, North Turkestan, Mirzachul, Kyzylkum. Weed, poisonous.

Genus 148. Halerpestes Green

447. Halerpestes sarmentosa (Adams) Kom.

Perennial. Swamps, banks of rivers. Montane and alpine zones. North Turkestan.

Genus 149. Ranunculus L.

448. *Ranunculus arvensis L.

Annual. Clayey soils, banks of rivers and canals, wastelands, gardens, fields, fallow lands. Plain, foothills, montane zone. Nuratau, Aktau, Nuratau Relic Mountains, Malguzar, North Turkestan. Weed, poisonous.

449. Ranunculus baldshuanicus Regel & Kom.   Figure 143

Perennial. Wet places, meadows, banks of rivers and canals. Plain, foothills, montane zone. Nuratau, Aktau, Malguzar, North Turkestan. Alkaloid-bearing.

Figure 143 Ranunculus baldshuanicus (photography by Natalya Beshko)

450. Ranunculus brevirostris Edgew

Perennial. Banks of rivers, wet places. Plain, foothills, montane zone. Nuratau, Malguzar, North Turkestan. Alkaloid-bearing.

451. Ranunculus komarovii Freyn

Perennial. Fine-earth, gravelly and stony slopes. Foothills, montane zone. Nuratau. Alkaloid-bearing.

452. Ranunculus linearilobus Bunge

Perennial. Sandy, clayey and skeleton soils. Plain, foothills. Nuratau, Aktau, Nuratau Relic Mountains, Malguzar, Kyzylkum. Medicinal, poisonous.

453. Ranunculus mindshelkensis B. Fedtsch.   Figure 144

Perennial. Fine-earth and stony slopes. Montane and alpine zones. Nuratau, Malguzar, North Turkestan. Alkaloid-bearing.

Figure 144 Ranunculus mindshelkensis (photography by Natalya Beshko)

454. Ranunculus natans C. A. Mey.

Perennial. Standing and slowly flowing water. Montane and alpine zones. North Turkestan.

455. Ranunculus olgae Regel

Perennial. Fine-earth and stony slopes, places near melting snow. Montane and alpine zones. Malguzar, North Turkestan. Alkaloid-bearing.

456. Ranunculus oxyspermus Willd.

Perennial. River valleys, banks of canals, fields, gardens, roadsides, fallow lands, fine-earth and gravelly slopes. Plain, foothills, montane zone. Nuratau, Aktau, Malguzar, North Turkestan. Weed, alkaloid-bearing.

457. Ranunculus paucidentatus Schrenk   Figure 145

Perennial. Fine-earth and stony slopes, places near melting snow. Montane and alpine zones. Nuratau, Aktau, Malguzar, North Turkestan. Alkaloid-bearing.

Figure 145 Ranunculus paucidentatus (photography by Natalya Beshko)

458. Ranunculus pinnatisectus Popov   Figure 146

Perennial. Sandy and clay deserts, piedmont plains, fine-earth, gravelly and stony slopes, banks of canals. Plain, foothills, montane zone. Nuratau, Aktau, Nuratau Relic Mountains, Malguzar, North Turkestan, Mirzachul, Kyzylkum. Weed, alkaloid-bearing.

Figure 146 **Ranunculus pinnatisectus** (photography by Natalya Beshko)

459. Ranunculus pulchellus C. A. Mey.
Perennial. Wet places, swamps, alpine meadows, banks of rivers. Alpine zone. North Turkestan. Alkaloid-bearing.

460. Ranunculus repens L.
Perennial. Wet places, banks of rivers and canals, swamps, gardens. Plain, foothills, montane and alpine zones. Nuratau, Aktau, Nuratau Relic Mountains, Malguzar, North Turkestan. Medicinal, alkaloid-bearing, poisonous.

461. Ranunculus rionii Lagger [*Batrachium rionii* (Lagger) Nym.]
Perennial. Standing and slowly flowing water. Plain, foothills, montane and alpine zones. Nuratau, North Turkestan, Mirzachul, Kyzylkum.

462. Ranunculus rubrocalyx Regel ex Kom.
Perennial. Fine-earth and stony slopes, rocks, alpine meadows, banks of rivers, swamps, places near melting snow. Alpine zone. North Turkestan.

463. Ranunculus rufosepalus Franch.   Figure 147
Perennial. Fine-earth and stony slopes, screes, moraines, alpine meadows, places near melting snow. Montane and alpine zones. Malguzar, North Turkestan.

Figure 147 **Ranunculus rufosepalus** (photography by Natalya Beshko)

464. Ranunculus sceleratus L.

Annual or biennial. Wet places, swamps, banks of rivers and canals. Plain, foothills, montane zone. Nuratau, Nuratau Relic Mountains, Malguzar, North Turkestan. Weed, poisonous, alkaloid-bearing, weed.

465. Ranunculus sewerzowii Regel   Figure 148

Perennial. Fine-earth and stony slopes, piedmont plains, sandy and clay deserts. Plain, foothills, montane zone. Nuratau, Aktau, Nuratau Relic Mountains, Malguzar, North Turkestan, Mirzachul, Kyzylkum. Alkaloid-bearing.

Figure 148 **Ranunculus sewerzowii** (photography by Natalya Beshko)

466. Ranunculus songaricus Schrenk
Perennial. Fine-earth, gravelly and stony slopes, subalpine meadows. Montane and alpine zones. Malguzar, North Turkestan.

467. Ranunculus sphaerospermus Boiss. & C. I. Blanche [*Batrachium pachycaulon* Nevski]
Perennial. Standing and slowly flowing water. Plain, foothills, montane and alpine zones. Nuratau, North Turkestan.

468. Ranunculus tenuilobus Regel & Kom.
Perennial. Fine-earth, gravelly and stony slopes. Montane zone. Nuratau, Malguzar, North Turkestan.

469. Ranunculus trichophyllus Chaix [*Batrachium divaricatum* (Schrank) Schur; *Batrachium trichophyllum* (Chaix) Bosch]
Perennial. Standing and slowly flowing water. Plain, foothills, montane and alpine zones. Nuratau, North Turkestan, Mirzachul, Kyzylkum.

Family 34. Platanaceae

Genus 150. Platanus L.

470. Platanus orientalis L.
Tree. River valleys. Montane zone. Nuratau, Malguzar, North Turkestan, Mirzachul (planted). Industrial, ornamental, medicinal, afforestation.

Family 35. Paeoniaceae

Genus 151. Paeonia L.

471. Paeonia tenuifolia L. [*Paeonia hybrida* Pall.]   Figure 149
Perennial. Fine-earth, gravelly and stony slopes, subalpine and alpine meadows, places near melting snow. Montane and alpine zones. North Turkestan. Ornamental. UzbRDB 3.

Figure 149 Paeonia tenuifolia (photography by Natalya Beshko)

Family 36. Grossulariaceae

Genus 152. Ribes L.

472. Ribes meyeri Maxim.   Figure 150
Shrub. Wet slopes, banks of rivers. Montane zone. Malguzar, North Turkestan. Food, ornamental.

Figure 150 Ribes meyeri (photography by Natalya Beshko)

Family 37. Saxifragaceae

Genus 153. Saxifraga L.

473. Saxifraga hirculus L.
Perennial. Banks of rivers, wetstony slopes, screes, moraines, alpine meadows, places near melting

snow. Alpine zone. North Turkestan. UzbRDB 2.

474. Saxifraga sibirica L.

Perennial. Stony slopes, screes, rocks, moraines, alpine meadows. Alpine zone. North Turkestan. Meliferous.

Family 38. Crassulaceae

Genus 154. Clemensia Rose

475. Clementsia semenovii (Regel & Herder) Boriss.   Figure 151

Perennial. Wet stony slopes, pebbles, swamps, banks of rivers. Alpine zone. North Turkestan. Medicinal, ornamental, dye.

Figure 151 Clementsia semenovii (photography by Natalya Beshko)

Genus 155. Rhodiola L.

476. Rhodiola heterodonta (Hook f. & Thomson) Boriss.   Figure 152

Perennial. Stony slopes, rocks, alpine meadows. Alpine zone. North Turkestan. Medicinal, ornamental.

Figure 152 **Rhodiola heterodonta** (photography by Natalya Beshko)

Genus 156. Sedum L.

477. Sedum hispanicum L. [*Sedum pentapetalum* Boriss.]   Figure 153
Annual. Stony slopes. Montane zone. Malguzar.

Figure 153 **Sedum hispanicum** (photography by Natalya Beshko)

478. Sedum tetramerum Trautv.   Figure 154
Annual. Stony, slopes, clay deserts, piedmont plain, pebbles. Plain, foothills, montane zone. Nuratau, Aktau, Nuratau Relic Mountains, Malguzar.

Figure 154 Sedum tetramerum (photography by Natalya Beshko)

Genus 157. Pseudosedum (Boiss.) A. Berger

479. Pseudosedum bucharicum Boriss.
Perennial. Stony slopes, rocks. Montane zone. Malguzar, North Turkestan. Ornamental.

480. Pseudosedum campanuliflorum Boriss.   Figure 155
Perennial. Stony slopes, rocks. Montane zone. Malguzar, North Turkestan. Ornamental, threatened species. UzbRDB 1.

Figure 155 Pseudosedum campanuliflorum (photography by Natalya Beshko)

481. Pseudosedum lievenii (Ledeb.) A. Berger.
Perennial. Stony slopes, rocks. Foothills, montane zone. Nuratau, Malguzar. Ornamental.
482. Pseudosedum longidentatum Boriss.   Figure 156
Perennial. Stony slopes, rocks, screes. Foothills, montane and alpine zones. Nuratau, Aktau, Nuratau Relic Mountains, Malguzar, North Turkestan. Ornamental.

Figure 156 Pseudosedum longidentatum (photography by Natalya Beshko)

Genus 158. Rosularia (DC.) Stapf.

483. Rosularia radicosa (Boiss. & Hohen.) Eggli [*Rosularia paniculata* (Regel & Schmalh.) A. Berger]
Perennial. Stony slopes. Montane and alpine zones. Nuratau, Malguzar, North Turkestan.
484. Rosularia subspicata (Freyn & Sint.) Boriss.
Perennial. Stony slopes. Montane zone. Nuratau.

Family 39. Haloragaceae

Genus 159. Myriophyllum L.

485. Myriophyllum spicatum L.
Perennial. Standing and slowly flowing water. Plain, foothills, montane zone. Nuratau, Mirzachul. Medicinal, fodder, dye, ornamental, industrial.

Family 40. Vitaceae

Genus 160. Vitis L.

486. Vitis vinifera L.   Figure 157

Shrub. Ravines, mountain valleys. Montane zone, Nuratau. Food, medicinal, ornamental, melliferous. UzbRDB 2.

Figure 157 Vitis vinifera (photography by Natalya Beshko)

Family 41. Zygophyllaceae

Genus 161. Zygophyllum L.

487. Zygophyllum jaxarticum Popov

Perennial. Saline lands. Plain. Mirzachul.

488. Zygophyllum macrophyllum Regel & Schmalh.   Figure 158

Perennial. Saline lands, banks of canals, fields, fallow lands, wastelands. Plain. Mirzachul.

Figure 158 Zygophyllum macrophyllum (photography by Natalya Beshko)

489. Zygophyllum miniatum Cham. Figure 159

Perennial. Clay deserts, relic mountains. Plain, foothills. Nuratau Relic Mountains, Mirzachul.

Figure 159 **Zygophyllum miniatum** (photography by Natalya Beshko)

490. Zygophyllum oxianum Boriss. Figure 160

Perennial. Saline lands, sandy soils, river valleys, banks of canals, fields, fallow lands, wastelands. Plain, foothills. Nuratau, Nuratau Relic Mountains, Mirzachul, Kyzylkum. Weed, medicinal, poisonous.

Figure 160 **Zygophyllum oxianum** (photography by Natalya Beshko)

Genus 162. Tribulus L.

491. *Tribulus terrestris L.

Annual. Sandy, clayey and saline soils, pebbles, fields, fallow lands, wastelands, roadsides. Plain, foothills, montane zone. Nuratau, Aktau, Nuratau Relic Mountains, Malguzar, North Turkestan,

Mirzachul, Kyzylkum. Weed, medicinal, poisonous.

Family 42. Fabaceae [Leguminosae]

Genus 163. *Gleditsia L.

492. *Gleditsia caspia Desf.
Tree. River valleys, fine-earth slopes (cultivated and naturalized). Foothills, montane zone. Nuratau. Ornamental, afforestation, industrial, fodder.

Genus 164. Sophora L.

493. Sophora alopecuroides L. [*Vexibia alopecuroides* (L.) Yakovlev]   Figure 161
Perennial. Fine-earth and gravelly slopes, river valleys, meadows, swamps, wet places, fields, fallow lands, roadsides, banks of canals. Plain, foothills, montane zone. Nuratau, Aktau, Malguzar, North Turkestan. Weed, poisonous.

Figure 161 Sophora alopecuroides (photography by Natalya Beshko)

494. Sophora lehmanni (Bunge) Yakovlev [*Ammothamnus lehmanni* Bunge]
Semishrub. Sandy deserts. Plain. Kyzylkum. Poisonous, ornamental, afforestation, melliferous, insecticide.

495. Sophora pachycarpa C. A. Mey. [*Vexibia pachycarpa* (C. A. Mey.) Yakovlev]   Figure 162
Perennial. Sandy, clayey and saline soils, fine-earth and gravelly slopes, banks of canals, fields, fallow lands, roadsides. Plain, foothills, montane zone. Nuratau, Aktau, Nuratau Relic Mountains, Malguzar, Mirzachul, Kyzylkum. Weed, poisonous.

Figure 162 Sophora pachycarpa (photography by Natalya Beshko)

Genus 165. Ammodendron Fisch.

496. Ammodendron conollyi Bunge ex Boiss.   Figure 163
Tree or shrub. Sandy deserts, saline lands. Plain. Kyzylkum. Dye, ornamental, industrial, afforestation, melliferous.

Figure 163 Ammodendron conollyi (photography by Natalya Beshko)

497. Ammodendron lehmannii Bunge ex Boiss.
Tree or shrub. Sandy deserts. Plain. Kyzylkum, Mirzachul. Ornamental, afforestation, melliferous.

Genus 166. Ononis L.

498. Ononis spinosa subsp. **antiquorum** (L.) Briq.
Perennial. Fine-earth and stony slopes, river valleys, banks of canals, roadsides, fields. Foothills, montane zone. Nuratau. Fodder.

Genus 167. Trigonella L.

499. Trigonella adscendens (Nevski) Afan. & Gontsch. [*Melissitus adscendens* (Nevski) Ikonn.]
Perennial. Fine-earth and stony slopes, pebbles. Montane and alpine zones, Malguzar, North Turkestan. Fodder.

500. Trigonella aristata (Vassilcz.) Sojak [*Melissitus aristatus* (Vass.) Latsch.]
Perennial. Stony slopes, dry riverbeds, river valleys. Montane zone. Malguzar, North Turkestan. Fodder.

501. Trigonella cancellata Desf.
Annual. Fine-earth and stony slopes, pebbles, fallow lands, dry riverbeds. Foothills, montane zone. North Turkestan. Fodder.

502. Trigonella geminiflora Bunge   Figure 164

Annual. Fine-earth and stony slopes, sandy and clay deserts, fallow lands, roadsides, banks of rivers, dry riverbeds. Plain, foothills, montane zone. Nuratau, Aktau, Nuratau Relic Mountains, Malguzar, North Turkestan, Mirzachul, Kyzylkum. Fodder.

Figure 164 **Trigonella geminiflora** (photography by Natalya Beshko)

503. Trigonella gontscharovii Vassilcz. [*Melissitus gontscharovii* (Vassilcz.) Latsch.]

Perennial. Stony slopes. Montane and alpine zones. Malguzar, North Turkestan. Fodder.

504. Trigonella grandiflora Bunge   Figure 165

Annual. Fine-earth and stony slopes, banks of canals, roadsides, dry riverbeds, fallow lands. Plain, foothills, montane zone. Nuratau, Aktau, Nuratau Relic Mountains, Malguzar, North Turkestan, Mirzachul, Kyzylkum. Fodder.

Figure 165 **Trigonella grandiflora** (photography by Natalya Beshko)

505. Trigonella popovii Korovin [*Melissitus popovii* (Korovin) Golosk.]
Perennial. Fine-earth, gravelly slopes, river valleys. Montane and alpine zones. Nuratau, Malguzar, North Turkestan. Fodder.

506. Trigonella verae Sirj.
Annual. Fine-earth, gravelly and stony slopes. Montane zone. North Turkestan. Fodder.

Genus 168. Melissitus Medik.

507. Melissitus pamiricus (Boriss.) Golosk
Perennial. Stony slopes, rocks, screes, pebbles. Montane and alpine zones. Malguzar, North Turkestan. Fodder.

Genus 169. Medicago L.

508. Medicago lupulina L.
Annual or biennial. Fine-earth, gravelly and stony slopes, river valleys, banks of canals, pebbles, meadows, gardens, fields. Foothills, montane zone. Nuratau, Aktau, Nuratau Relic Mountains, Malguzar, North Turkestan, Mirzachul, Kyzylkum. Fodder, medicinal, melliferous.

509. Medicago medicaginoides (Retz.) E. Small [*Trigonella arcuata* C. A. Mey.]
Annual. Fine-earth and stony slopes, fallow lands, banks of rivers, dry riverbeds. Foothills, montane zone. Malguzar, North Turkestan.

510. Medicago minima (L.) L.
Annual or biennial. Clay deserts, fine-earth, gravelly and stony slopes, dry riverbeds, river valleys, fallow lands. Plain, foothills, montane zone. Nuratau, Aktau, Nuratau Relic Mountains, Malguzar, North Turkestan, Mirzachul, Kyzylkum. Fodder, melliferous, weed.

511. Medicago monantha (C. A. Mey.) Trautv. [*Trigonella noeana* Boiss.]
Annual. Fine-earth and stony slopes, fields, fallow lands, dry riverbeds. Foothills, montane zone. Malguzar. Fodder, melliferous, weed.

512. Medicago orbicularis (L.) Bartal
Annual. Fine-earth, gravelly and stony slopes, river valleys, pebbles, fields, fallow lands. Foothills, montane zone. Nuratau, Aktau, Nuratau Relic Mountains, Malguzar, North Turkestan, Mirzachul, Kyzylkum. Fodder, melliferous, weed.

513. Medicago orthoceras (Kar. & Kir.) Trautv. [*Trigonella orthoceras* Kar. & Kir.]
Annual. Fine-earth and stony slopes, fields, fallow lands, dry riverbeds, roadsides. Plain, foothills, montane zone. Nuratau, Malguzar. Fodder, medicinal, dye.

514. Medicago polymorpha L. [*Medicago denticulata* Willd.]
Annual. Piedmont plains, fine-earth and gravelly slopes, gardens, fields. Plain, foothills, montane zone. North Turkestan. Fodder, medicinal, melliferous.

515. Medicago rigidula (L.) All.

Annual. Fine-earth, gravelly and stony slopes, piedmont plains, clay deserts, river valleys, pebbles. Plain, foothills, montane zone. Nuratau, Aktau, Nuratau Relic Mountains, Malguzar, North Turkestan, Mirzachul, Kyzylkum. Fodder, melliferous, weed.

516. *Medicago sativa L.

Perennial. Fine-earth, gravelly and stony slopes, meadows, screes, river valleys, pebbles, gardens, pastures, fields, fallow lands, roadsides (naturalized). Plain, foothills, montane zone. Nuratau, Aktau, Nuratau Relic Mountains, Malguzar, North Turkestan, Mirzachul, Kyzylkum. Fodder, medicinal, melliferous.

Genus 170. Melilotus Mill.

517. Melilotus albus Medik.

Biennial. Meadows, river valleys, banks of canals, fine-earth slopes, pebbles, wet places, fallow lands, fields, gardens. Plain, foothills, montane zone. Nuratau, Aktau, Nuratau Relic Mountains, Malguzar, North Turkestan, Mirzachul, Kyzylkum. Fodder, medicinal, melliferous, insecticide.

518. Melilotus officinalis (L.) Pall.

Biennial. Fine-earth slopes, meadows, river valleys, roadsides, gardens, fields, fallow lands. Plain, foothills. Nuratau, Nuratau Relic Mountains, Malguzar, North Turkestan, Mirzachul. Fodder, medicinal, melliferous, insecticide.

Genus 171. Trifolium L.

519. Trifolium campestre Schreb. [*Chrysaspis campestris* (Schreb.) Desv.]

Annual. River valleys, meadows, fine-earth slopes, fields. Foothills, montane zone. Malguzar. Fodder, melliferous.

520. Trifolium fragiferum L.[*Trifolium neglectum* C. A. Mey.; *Amoria bonannii* (C. Presl.) Roskov; *Amoria fragifera* (L.) Roskov]

Perennial. Wet places, meadows, banks of rivers and canals, swamps, pebbles, fields, fallow lands. Plain, foothills, montane zone. Nuratau, Aktau, Malguzar, North Turkestan. Fodder, melliferous.

521. Trifolium pratense L.   Figure 166

Perennial. River valleys, banks of canals, fine-earth slopes, wet places, meadows, fields, fallow lands. Plain, foothills, montane zone. Nuratau, Aktau, Nuratau Relic Mountains, Malguzar, North Turkestan, Mirzachul. Fodder, medicinal, essential oil, dye, melliferous.

Figure 166 **Trifolium pratense** (photography by Natalya Beshko)

522. Trifolium repens L.
Perennial. Wet places, meadows, banks of rivers and canals, swamps, pebbles, fields, fallow lands. Plain, foothills, montane zone. Nuratau, Aktau, Nuratau Relic Mountains, Malguzar, North Turkestan, Mirzachul. Fodder, medicinal, melliferous.

Genus 172. Lotus L.

523. Lotus krylovii Schischkin & Serg. [*L. sergievskiae* Kamelin & Kovalevsk.]
Perennial. River valleys, banks of rivers, lakes and canals, fallow lands, meadows, fields, gardens, saline lands. Plain, foothills, montane zone. Nuratau, Malguzar, North Turkestan. Fodder, weed.

Genus 173. Cullen Medik.

524. Cullen drupaceum (Bunge) C. H. Stirt. [*Psoralea drupacea* Bunge]   Figure 167
Perennial. Fine-earth slopes, piedmont plain, sandy and clay deserts, banks of rivers, lakes and canals, fallow lands, fields. Plain, foothills, montane zone. Nuratau, Aktau, Nuratau Relic Mountains, Malguzar, North Turkestan, Mirzachul, Kyzylkum. Medicinal, essential oil,

meliferous, weed.

Figure 167 Cullen drupaceum (photography by Natalya Beshko)

Genus 174. Smirnowia Bunge

525. Smirnowia turkestana Bunge
Shrub. Sandy deserts. Plain. Kyzylkum. Medicinal.

Genus 175. Colutea L.

526. Colutea paulsenii Freyn   Figure 168
Shrub. Fine-earth and stony slopes. Montane zone. Malguzar, North Turkestan. Ornamental, afforestation, melliferous.

Figure 168 Colutea paulsenii (photography by Natalya Beshko)

Genus 176. Halimodendron Fisch. ex DC.

527. Halimodendron halodendron (Pall.) Voss.   Figure 169

Shrub. Saline lands, banks of rivers, lakes and canals, pebbles. Plain, foothills, montane zone. Nuratau, Nuratau Relic Mountains, Malguzar, North Turkestan, Mirzachul, Kyzylkum. Dye, fodder, ornamental, afforestation, melliferous.

Figure 169 Halimodendron halodendron (photography by Natalya Beshko)

Genus 177. Caragana Fabr.

528. Caragana alaica Pojark.
Shrub. Stony slopes, ravines, river valleys. Montane zone. Malguzar, North Turkestan. Melliferous.

529. Caragana turkestanica Kom.
Shrub. Fine-earth, gravelly and stony slopes. Montane zone. Malguzar, North Turkestan. Melliferous.

Genus 178. Chesneya Lindl.

530. Chesneya ternata (Korsh.) Popov   Figure 170
Perennial. Fine-earth, gravelly and stony slopes, rocks, screes, pebbles, banks of rivers. Foothills, montane zone. Nuratau, Malguzar, North Turkestan.

Figure 170 Chesneya ternata (photography by Natalya Beshko)

531. Chesneya turkestanica Franch.
Perennial. Fine-earth, gravelly and stony slopes. Montane zone. Malguzar, North Turkestan.

Genus 179. Astragalus L.

532. Astragalus adpressipilosus Gontsch.
Perennial. Stony slopes, pebbles. Montane and alpine zones. Malguzar, North Turkestan.

533. Astragalus alabugensis B.
Fedtsch. Perennial. Stony slopes. Montane zone. Nuratau.

534. Astragalus alopecias Pall.   Figure 171
Perennial. Fine-earth slopes, piedmont plains, sandy and clay deserts, screes, fallow lands, banks of rivers and canals, roadsides. Plain, foothills, montane zone. Nuratau, Aktau, Malguzar, North Turkestan, Mirzachul. Fodder, ornamental.

Family 171 Astragalus alopecias (photography by Natalya Beshko)

535. Astragalus alpinus L.
Perennial. Fine-earth and gravelly slopes, banks of rivers, subalpine and alpine meadows. Montane and alpine zones. North Turkestan. Fodder, ornamental, melliferous.

536. Astragalus ammotrophus Bunge
Perennial. Stony and gravelly slopes, clay deserts. Plain, foothills. Nuratau Relic Mountains.

537. Astragalus angreni Lipsky

Perennial. Fine-earth slopes. Montane zone. Malguzar, North Turkestan.

538. Astragalus aphanassjievii Gontsch.

Perennial. Fine-earth and gravelly slopes. Montane and alpine zones. Malguzar, North Turkestan.

539. Astragalus aschuturi B. Fedtsch.

Perennial. Fine-earth, gravelly and stony slopes, screes, dry riverbeds. Montane and alpine zones, Malguzar, North Turkestan.

540. Astragalus bactrianus Bunge   Figure 172

Shrub. Fine-earth, gravelly and stony slopes, pebbles, fallow lands. Foothills, montane zone. Nuratau, Aktau, Malguzar, North Turkestan.

Figure 172 Astragalus bactrianus (photography by Natalya Beshko)

541. Astragalus bakaliensis Bunge

Annual. Sandy deserts, stony slopes, fields, fallow lands. Plain, foothills, montane zone. Nuratau, Nuratau Relic Mountains, Malguzar, North Turkestan. Fodder, melliferous.

542. Astragalus belolipovii Kamelin ex F. O. Khass. & N. Sulajm.

Perennial. Fine-earth and stony slopes, among juniper forests. Montane zone. Malguzar, North Turkestan. UzbRDB 1.

543. Astragalus camptoceras Bunge

Annual. Fine-earth, gravelly and stony slopes, piedmont plains. Plain, foothills, montane zone.

Nuratau, Aktau, Nuratau Relic Mountains, Malguzar, North Turkestan.

544. Astragalus campylorrhynchus Fisch. & C. A. Mey.   Figure 173

Annual. Fine-earth, gravelly and stony slopes, sandy and clay deserts. Plain, foothills, montane zone. Nuratau, Aktau, Nuratau Relic Mountains, Malguzar, North Turkestan, Mirzachul, Kyzylkum. Fodder.

Figure 173 **Astragalus campylorrhynchus** (photography by Natalya Beshko)

545. Astragalus campylotrichus Bunge

Annual. Fine-earth, gravelly and stony slopes, sandy and clay deserts. Plain, foothills, montane zone. Nuratau, Aktau, Nuratau Relic Mountains, Malguzar, North Turkestan, Mirzachul. Fodder.

546. Astragalus chodshenticus B. Fedtsch.

Semishrub. River valleys, saline lands. Plain, foothills. Malguzar.

547. Astragalus commixtus Bunge

Annual. Sandy, clayey and sceleton soils. Plain, foothills. Nuratau, Aktau, Nuratau Relic Mountains, Malguzar, North Turkestan, Mirzachul, Kyzylkum. Fodder.

548. Astragalus compositus Pavlov

Annual. Fine-earth, gravelly and stony slopes. Foothills, montane zone. Nuratau, Malguzar, North Turkestan. Fodder, weed.

549. Astragalus contortuplicatus L.

Annual. Saline lands, banks of rivers, lakes and canals, fields. Plain. Mirzachul. Fodder, weed.

550. Astragalus cornu-bovis Lipsky
Annual. Sandy and clay deserts. Plain. Kyzylkum. Fodder.

551. Astragalus dendroides Kar. & Kir.
Shrub. Fine-earth, gravelly and stony slopes. Montane zone. Malguzar, North Turkestan.

552. Astragalus dictamnoides Gontsch.
Perennial. Fine-earth, gravelly and stony slopes. Montane zone. North Turkestan.

553. Astragalus dipelta Bunge
Annual. Fine-earth and stony slopes, fields, fallow lands. Foothills, montane zone. Malguzar, North Turkestan. Fodder.

554. Astragalus dolichocarpus Popov
Semishrub. Fine-earth and gravelly slopes. Foothills, montane zone. Nuratau, Malguzar.

555. Astragalus eximius Bunge   Figure 174
Perennial. Fine-earth, gravelly and stony slopes. Foothills, montane zone. Nuratau, Malguzar, North Turkestan. Ornamental.

Figure 174 Astragalus eximius (photography by Natalya Beshko)

556. Astragalus falcigerus Popov
Semishrub. Gravelly and stony slopes. Montane zone. Nuratau.

557. Astragalus farctissimus Lipsky
Perennial. Fine-earth, gravelly and stony slopes. Montane zone. Nuratau, Aktau, Malguzar, North Turkestan.

558. Astragalus ferganensis (Popov) B. Fedtsch. ex A. S. Korol.
Perennial. Stony slopes, pebbles. Foothills. North Turkestan.

559. Astragalus filicaulis Fisch. & C. A. Mey.
Annual. Sandy, clayey and skeleton soils, fields, fallow lands. Plain, foothills, montane zone. Nuratau, Aktau, Nuratau Relic Mountains, Malguzar, North Turkestan, Mirzachul, Kyzylkum. Fodder.

560. Astragalus flexus Fisch.   Figure 175

Perennial. Sandy deserts. Plain. Kyzylkum. Fodder.

Figure 175 Astragalus flexus (photography by Natalya Beshko)

561. Astragalus floccosifolius Sumnev.   Figure 176

Perennial. Fine-earth and stony slopes, river valleys, fallow lands. Montane zone. Malguzar, North Turkestan.

Figure 176 Astragalus floccosifolius (photography by Natalya Beshko)

562. Astragalus globiceps Bunge   Figure 177

Perennial. Fine-earth, gravelly and stony slopes, fields, fallow lands. Foothills, montane zone. Nuratau, Aktau, Malguzar, North Turkestan. Fodder, ornamental.

Figure 177 **Astragalus globiceps** (photography by Natalya Beshko)

563. Astragalus guttatus Bans & Solander [*Astragalus striatellus* M. Bieb.]

Annual. Fine-earth, gravelly and stony slopes. Foothills. Nuratau. Fodder.

564. Astragalus harpilobus Kar. & Kir.

Annual. Sandy deserts. Plain. Kyzylkum. Fodder, weed.

565. Astragalus inaequalifolius Basil.

Perennial. Stony slopes, rocks, screes. Montane zone. Nuratau.

566. Astragalus intarrensis Franch.

Perennial. Stony slopes. Montane zone. Malguzar, North Turkestan.

567. Astragalus iskanderi Lipsky

Shrub. Fine-earth and stony slopes, river valleys. Montane zone. Nuratau, Malguzar, North Turkestan.

568. Astragalus isfahanicus Boiss.

Perennial. Fine-earth, gravelly and stony slopes. Montane zone. Malguzar, North Turkestan.

569. Astragalus jagnobicus Lipsky

Perennial. Fine-earth, gravelly and stony slopes, pebbles. Montane and alpine zones. Malguzar, North Turkestan.

570. Astragalus juratzkanus Freyn & Sint. [*Astragalus maverranagri* Popov]

Semishrub. Fine-earth, gravelly and stony slopes, roadsides. Foothills, montane zone. Nuratau, Aktau, Nuratau Relic Mountains, Malguzar, North Turkestan.

571. Astragalus kelleri Popov   Figure 178

Perennial. Fine-earth and stony slopes. Foothills, montane zone. Nuratau, Aktau, Nuratau Relic Mountains, Malguzar. UzbRDB 2.

Figure 178 **Astragalus kelleri** (photography by Natalya Beshko)

572. Astragalus knorringianus Boriss.   Figure 179
Perennial. Fine-earth, gravelly and stony slopes. Foothills, montane zone. Nuratau, Aktau, Nuratau Relic Mountains, Malguzar, North Turkestan. UzbRDB 2.

Figure 179 **Astragalus knorringianus** (photography by Natalya Beshko)

573. Astragalus lasiosemius Boiss.
Semishrub. Fine-earth, gravelly and stony slopes, screes. Montane and alpine zones. Malguzar, North Turkestan.

574. Astragalus lasiostylus Fisch. [*Astragalus vladimiri* Sirj.]
Shrub. Fine-earth and stony slopes, rocks, screes. Montane and alpine zones. Nuratau, Malguzar, North Turkestan.

575. Astragalus leiophysa Bunge
Perennial. Sandy deserts. Plain. Kyzylkum.

576. Astragalus lentilobus Kamelin & Kovalevsk.
Perennial. Fine-earth and gravelly slopes, screes, moraines, subalpine and alpine meadows. Montane and alpine zones. Malguzar, North Turkestan. Fodder.

577. Astragalus leptophysus Vved.   Figure 180
Perennial. Stony slopes, screes. Montane zone. Nuratau, North Turkestan, Nuratau Reserve. UzbRDB 2.

Figure 180 **Astragalus leptophysus** (photography by Natalya Beshko)

578. Astragalus leptostachys Pall.　Figure 181
Perennial. Fine-earth, gravelly and stony slopes, pebbles. Montane and alpine zones. Nuratau, Malguzar, North Turkestan. Fodder.

Figure 181 Astragalus leptostachys (photography by Natalya Beshko)

579. Astragalus lipskyi Popov　Figure 182
Perennial. Fine-earth and gravelly slopes. Montane zone. Nuratau, Malguzar, North Turkestan.

Figure 182 Astragalus lipskyi (photography by Natalya Beshko)

580. Astragalus lithophilus Kar. & Kir.
Perennial. Fine-earth, gravelly and stony slopes, pebbles, moraines, screes, alpine meadows. Montane and alpine zones, North Turkestan.

581. Astragalus macrocladus Bunge　Figure 183
Semishrub. Sandy deserts. Plain. Kyzylkum. Fodder, melliferous.

Figure 183 Astragalus macrocladus (photography by Natalya Beshko)

582. Astragalus macronyx Bunge   Figure 184

Perennial. Fine-earth, gravelly and stony slopes. Plain, foothills, montane zone. Nuratau, Aktau, Nuratau Relic Mountains, Malguzar, North Turkestan.

Figure 184 Astragalus macronyx (photography by Natalya Beshko)

583. Astragalus macropetalus Schrenk

Perennial. Fine-earth, gravelly slopes, Clay deserts. Plain, foothills. Nuratau.

584. Astragalus macrotropis Bunge

Perennial. Fine-earth, gravelly and stony slopes, fallow lands, rocks, banks of rivers, pebbles. Foothills, montane zone. Nuratau, Aktau, Malguzar, North Turkestan.

585. Astragalus marguzaricus Lipsky

Perennial. Fine-earth and stony slopes, banks of rivers, pebbles. Montane zone. Nuratau, Malguzar, North Turkestan.

586. Astragalus masenderanus Bunge [*Astragalus kurdaicus* Saposhn. ex Sumn.; *Astragalus skorniakovii* B. Fedtsch.]   Figure 185

Perennial. Fine-earth, gravelly and stony slopes. Montane and alpine zones. Nuratau, Malguzar, North Turkestan. Fodder.

Figure 185 Astragalus masenderanus (photography by Natalya Beshko)

587. Astragalus mucidus Bunge ex Boiss.   Figure 186

Perennial. Fine-earth, gravelly and stony slopes, piedmont plains. Plain, foothills, montane zone. Nuratau, Aktau, Nuratau Relic Mountains, Malguzar, North Turkestan.

Figure 186 Astragalus mucidus (photography by Natalya Beshko)

588. Astragalus nematodes Bunge

Perennial. Fine-earth, gravelly and stony slopes, rocks. Foothills. Nuratau.

589. Astragalus nivalis Kar. & Kir.

Perennial. Gravelly and stony slopes, pebbles, banks of rivers, moraines. Montane and alpine zones. North Turkestan.

590. Astragalus nobilis Bunge ex B. Fedtsch.   Figure 187

Perennial. Fine-earth, gravelly and stony slopes, rocks. Montane zone. Nuratau, Aktau, Malguzar, North Turkestan.

Figure 187 Astragalus nobilis (photography by Natalya Beshko)

591. Astragalus nuciferus Bunge   Figure 188
Perennial. Fine-earth and stony slopes, screes. Montane zone. Malguzar, North Turkestan.

Figure 188 Astragalus nuciferus (photography by Natalya Beshko)

592. Astragalus ophiocarpus Benth. ex Boiss.
Annual. Sandy and clay deserts, banks of rivers. Plain. Mirzachul. Fodder.

593. Astragalus orbiculatus Ledeb.
Perennial. Sandy deserts, saline lands, wet places, banks of rivers. Plain. Malguzar, Mirzachul, Kyzylkum.

594. Astragalus oxyglottis Stev. ex M. Bieb.
Annual. Fine-earth, skeleton, rarely sandy soils. Plain, foothills, montane zone. Nuratau, Nuratau Relic Mountains, Malguzar, Mirzachul. Fodder.

595. Astragalus patentipilosus Kitam. [*Astragalus korovinianus* Barneby]
Perennial. Clayey soils, among sagebrush vegetation. Plain, foothills. Nuratau, Mirzachul. Fodder.

596. Astragalus patentivillosus Gontsch.
Perennial. Fine-earth and gravelly slopes, moraines. Montane and alpine zone. North Turkestan.

597. Astragalus paucijugus Schrenk
Shrub. Sandy deserts. Plain. Kyzylkum. Fodder, meliferous.

598. Astragalus pauper Bunge
Perennial. Fine-earth and gravelly slopes, alpine meadows. Alpine zone. North Turkestan. Fodder.

599. Astragalus peduncularis Benth. [*Astragalus corydalinus* Bunge]
Perennial. Stony slopes, pebbles, screes, wet meadows. Montane and alpine zone. Malguzar, North Turkestan.

600. Astragalus persipolitanus Boiss. [*Astragalus ammophilus* Kar. & Kir.]
Annual. Sandy, clayey and skeleton soils, screes. Plain, foothills, montane zone. Nuratau, Nuratau Relic Mountains, Mirzachul, Kyzylkum. Fodder.

601. Astragalus petunnikovii Litv.
Perennial. Sandy deserts. Plain. Kyzylkum.

602. Astragalus platyphyllus Kar. & Kir.

Perennial. Stony slopes. Montane zone. Malguzar, North Turkestan.

603. Astragalus pterocephalus Bunge [*Astragalus stipulosus* Boriss.]

Shrub. Fine-earth, gravelly and stony slopes, screes, rocks. Montane and alpine zones. Nuratau, Malguzar, North Turkestan.

604. Astragalus pulcher Korovin   Figure 189

Perennial. Stony slopes, rocks. Montane zone. Nuratau.

Figure 189 Astragalus pulcher (photography by Natalya Beshko)

605. Astragalus quisqualis Bunge

Perennial. Fine-earth, gravelly and stony slopes, screes. Montane zone. Malguzar, North Turkestan. Fodder.

606. Astragalus retamocarpus Boiss. & Hohen.

Perennial. Fine-earth, gravelly and stony slopes. Foothills, montane zone. Malguzar. Fodder.

607. Astragalus rubromarginatus Czerniak.

Perennial. Sandy deserts. Plain. Kyzylkum.

608. Astragalus russanovii F. O. Khass., Sarybaeva & Esankulov

Perennial. Fine-earth, gravelly and stony slopes. Montane zone. Malguzar, North Turkestan.

609. Astragalus saratagius Bunge

Perennial. Stony slopes, screes. Montane and alpine zones. North Turkestan.

610. Astragalus sarytavicus Popov

Perennial. Fine-earth, gravelly and stony slopes. Montane zone. Malguzar, North Turkestan.

611. Astragalus schachimardanus Basil.

Perennial. Stony slopes, rocks, screes. Montane zone. Malguzar, North Turkestan.

612. Astragalus schmalhausenii Bunge

Annual. Fine-earth, gravelly and stony slopes, fields, fallow lands. Foothills, montane zone. Nuratau, Aktau, Nuratau Relic Mountains, Malguzar, North Turkestan. Fodder, weed.

613. Astragalus schrenkianus Fisch. & C. A. Mey.

Perennial. Stony slopes, rocks, screes. Montane zone. Nuratau.

614. Astragalus schugnanicus B. Fedtsch.
Perennial. Fine-earth, gravelly and stony slopes, alpine meadows, stream banks. Alpine zone. North Turkestan. Fodder.

615. Astragalus sesamoides Boiss.
Annual. Fine-earth, gravelly and stony slopes, pebbles, river valleys, fields, fallow lands. Foothills, montane zone. Nuratau, Aktau, Nuratau Relic Mountains, Malguzar, North Turkestan, Mirzachul. Fodder.

616. Astragalus sewerzowii Bunge [*Astragalus subbarbellatus* Bunge]
Perennial. Fine-earth slopes, subalpine meadows. Montane and alpine zones. Nuratau, Aktau, Malguzar, North Turkestan. Fodder.

617. Astragalus sieversianus Pall.   Figure 190
Perennial. Fine-earth, gravelly and stony slopes, fallow lands. Foothills, montane zone. Nuratau, Aktau, Malguzar, North Turkestan. Fodder, medicinal, ornamental.

Figure 190 Astragalus sieversianus (photography by Natalya Beshko)

618. Astragalus sogdianus Bunge   Figure 191

Perennial. Fine-earth, gravelly and stony slopes, piedmont plain. Plain, foothills, montane zone. Nuratau, Aktau, Nuratau Relic Mountains, Malguzar, North Turkestan, Mirzachul.

Figure 191 **Astragalus sogdianus** (photography by Natalya Beshko)

619. Astragalus stalinskyi Širj.

Annual. Fine-earth, gravelly and stony slopes, piedmont plain. Plain, foothills, montane zone. Nuratau, Aktau, Nuratau Relic Mountains, Malguzar, North Turkestan, Mirzachul. Fodder.

620. Astragalus stenanthus Bunge

Perennial. Fine-earth, gravelly and stony slopes, river valleys. Foothills, montane zone. Nuratau, Malguzar.

621. Astragalus stenocystis Bunge

Perennial. Fine-earth, gravelly and stony slopes. Foothills, montane zone. Nuratau, Aktau, Nuratau Relic Mountains, Malguzar, North Turkestan.

622. Astragalus subinduratus Gontsch.

Perennial. Stony slopes, banks of rivers. Montane and alpine zones. North Turkestan.

623. Astragalus subscaposus Popov ex Boriss.

Perennial. Fine-earth, gravelly and stony slopes, wet places near springs, meadows. Montane and alpine zones. North Turkestan. Fodder.

624. Astragalus substipitatus Gontsch.

Perennial. Fine-earth, gravelly and stony slopes. Foothills, montane zone. Nuratau, Malguzar, North

Turkestan.

625. Astragalus subverticillatus Gontsch.

Perennial. Fine-earth, gravelly slopes. Montane zone. Malguzar, North Turkestan.

626. Astragalus tibetanus Benth. ex Bunge

Perennial. Fine-earth, gravelly and stony slopes, river valleys. Montane and alpine zones. Malguzar, North Turkestan. Fodder.

627. Astragalus titovii Gontsch.

Perennial. Stony slopes, swamps, places near melting snow. Montane and alpine zones. Malguzar, North Turkestan. Fodder.

628. Astragalus transoxanus Fisch. ex Bunge

Shrub. Fine-earth, gravelly and stony slopes. Montane zone. Nuratau.

629. Astragalus turbinatus Bunge

Perennial. Sandy deserts. Plain. Kyzylkum.

630. Astragalus turczaninovii Kar. & Kir.   Figure 192

Perennial. Sandy deserts. Plain. Kyzylkum.

Figure 192 **Astragalus turczaninovii** (photography by Natalya Beshko)

631. Astragalus turkestanus Bunge   Figure 193

Perennial. Fine-earth and gravelly slopes. Foothills, montane zone. Nuratau, Aktau, Nuratau Relic Mountains, Malguzar, North Turkestan, Mirzachul. Fodder.

Figure 193 Astragalus turkestanus (photography by Natalya Beshko)

632. Astragalus unifoliolatus Bunge

Semishrub. Sandy deserts. Plain. Kyzylkum. Fodder, melliferous.

633. Astragalus uninodus Popov & Vved.

Annual. River valleys, stony slopes. Montane zone. Malguzar, North Turkestan.

634. Astragalus urgutinus Lipsky

Shrub. Fine-earth, gravelly and stony slopes, rocks, cliffs. Foothills, montane and alpine zones. Nuratau, Aktau, Malguzar, North Turkestan.

635. Astragalus variegatus Franch.   Figure 194

Shrub. Fine-earth, gravelly and stony slopes, banks of rivers, fallow lands. Montane and alpine zones. Nuratau, Malguzar, North Turkestan.

Figure 194 Astragalus variegatus (photography by Natalya Beshko)

636. Astragalus vicarius Lipsky

Annual. Fine-earth, gravelly and stony slopes, fallow lands, fields. Foothills, montane zone. Nuratau, Nuratau Relic Mountains, Malguzar, North Turkestan, Mirzachul. Fodder.

637. Astragalus villosissimus Bunge

Semishrub. Sandy and clay deserts, saline lands. Plain, foothills. Nuratau Relic Mountains, Mirzachul, Kyzylkum. Fodder, melliferous.

638. Astragalus xanthomeloides Korovin & Popov   Figure 195

Perennial. Stony slopes, screes, pebbles. Montane zone. Nuratau, Malguzar, North Turkestan.

Figure 195 Astragalus xanthomeloides (photography by Natalya Beshko)

Genus 180. Oxytropis DC.

639. Oxytropis aspera Gontsch. Fedtsch.
Perennial. Stony slopes, places near melting snow. Montane and alpine zones. North Turkestan.

640. Oxytropis capusii Franch.   Figure 196
Perennial. Fine-earth, gravelly and stony slopes. Montane and alpine zones. Malguzar, North Turkestan.

Figure 196 Oxytropis capusii (photography by Natalya Beshko)

641. Oxytropis chesneyoides Gontsch.
Perennial. Rocks, stony slopes. Foothills, montane zone. Nuratau, Malguzar.

642. Oxytropis humifusa Kar. & Kir.
Perennial. Fine-earth, gravelly and stony slopes, moraines, alpine meadows, places near melting snow. Alpine zone. North Turkestan.

643. Oxytropis immersa (Baker) B. Fedtsch.
Perennial. Fine-earth, gravelly and stony slopes, rocks, moraines, alpine meadows. Alpine zone. North Turkestan.

644. Oxytropis integripetala Bunge
Perennial. Fine-earth, gravelly and stony slopes. Foothills, montane zone. North Turkestan.

645. Oxytropis kamelinii Vassilcz.
Perennial. Stony slopes. Montane zone. Malguzar. Endemic.

646. Oxytropis lehmanni Bunge
Perennial. Fine-earth and stony slopes, rocks, alpine meadows. Montane and alpine zones. Malguzar, North Turkestan.

647. Oxytropis leucocyanea Bunge
Semishrub. Stony slopes, screes. Montane and alpine zones. North Turkestan.

648. Oxytropis lipskyi Gontsch.
Perennial. Fine-earth, gravelly and stony slopes. Montane zone. North Turkestan.

649. Oxytropis macrocarpa Kar. & Kir.   Figure 197
Perennial. Fine-earth, gravelly and stony slopes. Montane zone. Nuratau, Malguzar, North Turkestan.

Figure 197 Oxytropis macrocarpa (photography by Natalya Beshko)

650. Oxytropis michelsonii B. Fedtsch.
Perennial. Stony slopes, places near melting snow. Montane and alpine zones. North Turkestan.

651. Oxytropis microsphaera Bunge
Perennial. Stony slopes, rocks, places near melting snow. Montane and alpine zones. Malguzar, North Turkestan.

652. Oxytropis platonychia Bunge
Perennial. Stony slopes, screes, pebbles, alpine meadows. Alpine zone. North Turkestan.

653. Oxytropis pseudorosea Filim.   Figure 198
Perennial. Stony slopes. Montane zone. Nuratau, UzbRDB 2.

Figure 198 Oxytropis pseudorosea (photography by N. Yu. Beshko)

654. Oxytropis riparia Litv.

Perennial. Meadows, river valleys, wet slopes. Montane zone. North Turkestan. Endemic to the Nuratau Mountains.

655. Oxytropis rosea Bunge   Figure 199

Perennial. Dry riverbeds, pebbles, gravelly and stony slopes, rocks. Foothills, montane zone. Malguzar, North Turkestan.

Figure 199 Oxytropis rosea (photography by Natalya Beshko)

656. Oxytropis savellanica Bunge

Semishrub. Stony slopes, screes, rocks, alpine meadows, stream banks. Montane and alpine zones. North Turkestan.

657. Oxytropis seravschanica Gontsch.

Perennial. Stony slopes, dry riverbeds. Montane and alpine zones. Malguzar, North Turkestan.

658. Oxytropis tachtensis Franch.

Perennial. Fine-earth and stony slopes. Montane zone. Nuratau, Malguzar, North Turkestan.

659. Oxytropis trichocalycina Bunge   Figure 200

Perennial. Stony slopes, rocks, screes. Montane and alpine zones. Nuratau, Malguzar, North Turkestan.

Figure 200 **Oxytropis trichocalycina** (photography by Natalya Beshko)

Genus 181. Glycyrrhiza L.

660. Glycyrrhiza aspera Pall.
Perennial. Banks of rivers, saline lands, fine-earth slopes, fallow lands, fields. Plain, foothills, montane zone. Nuratau, Nuratau Relic Mountains, Malguzar, Mirzachul. Medicinal, fodder, melliferous.

661. Glycyrrhiza glabra L.
Perennial. Fine-earth slopes, meadows, fields, fallow lands, river valleys, banks of canals. Plain, foothills, montane zone. Nuratau, Aktau, Nuratau Relic Mountains, Malguzar, North Turkestan, Mirzachul. Medicinal, food, fodder, dye, industrial, melliferous, weed.

662. Glycyrrhiza triphylla Fisch. & C. A. Mey. [*Meristotropis triphylla* Fisch. & C. A. Mey.]
Perennial. Fine-earth, gravelly and stony slopes, river valleys, pebbles, fallow lands. Foothills, montane zone. Nuratau. Medicinal, fodder, melliferous.

Genus 182. Hedysarum L.

663. Hedysarum flavescens Regel & Schmalh.
Perennial. Fine-earth, gravelly and stony slopes, screes, rocks, river valleys, alpine meadows. Montane and alpine zones. North Turkestan. Fodder, ornamental, meliferous.

664. Hedysarum jaxarticum Popov & Kir.
Perennial. Clayey soils. Plain, foothills. North Turkestan. Fodder, ornamental, meliferous.

665. Hedysarum mogianicum (B. Fedtsch.) B. Fedtsch.
Perennial. Fine-earth, gravelly and stony slopes. Montane zone. Nuratau. Fodder, ornamental, melliferous.

666. Hedysarum montanum (B. Fedtsch.) B. Fedtsch.
Perennial. Pebbles, fine-earth and stony slopes, fallow lands, roadsides. Foothills, montane zone. Malguzar, North Turkestan. Fodder, ornamental, melliferous.

667. **Hedysarum nuratense** Popov   Figure 201

Perennial. Fine-earth, gravelly and stony slopes. Montane zone. Nuratau, Malguzar, North Turkestan. Fodder, ornamental, melliferous.

Figure 201 Hedysarum nuratense (photography by Natalya Beshko)

668. Hedysarum plumosum Boiss. & Hausskn.   Figure 202
Perennial. Stony slopes, screes, rocks, pebbles. Montane and alpine zones. Malguzar, North Turkestan. Fodder, ornamental, melliferous.

Figure 202 **Hedysarum plumosum** (photography by Natalya Beshko)

Genus 183. Onobrychis Hill

669. Onobrychis chorassanica Bunge   Figure 203
Perennial. Fine-earth, gravelly slopes, fallow lands. Foothills, montane zone. Nuratau, Aktau, Malguzar, North Turkestan. Fodder, melliferous.

Figure 203 Onobrychis chorassanica (photography by Natalya Beshko)

670. Onobrychis echidna Lipsky

Shrub. Stony slopes, screes. Montane and alpine zones. Malguzar, North Turkestan. Fodder, melliferous.

671. Onobrychis gontscharovii Vassilcz.

Perennial. Rocks, stony slopes. Montane zone. Nuratau. Fodder, melliferous.

672. Onobrychis grandis Lipsky   Figure 204

Perennial. Fine-earth, gravelly and stony slopes. Montane zone. Malguzar, North Turkestan. Fodder, melliferous.

Figure 204 Onobrychis grandis (photography by Natalya Beshko)

673. Onobrychis micrantha Schrenk

Annual. Sandy and clay deserts, fine-earth, gravelly and stony slopes, fields, fallow lands. Plain, foothills, montane zone. Nuratau, Aktau, Nuratau Relic Mountains, Malguzar, North Turkestan, Mirzachul, Kyzylkum. Fodder, meliferous, weed.

674. Onobrychis pulchella Schrenk   Figure 205

Annual. Fine-earth and stony slopes, pebbles, fields, fallow lands, piedmont plains, clay deserts. Foothills, montane zone. Nuratau, Aktau, Nuratau Relic Mountains, Malguzar, North Turkestan, Mirzachul. Fodder, meliferous, weed.

Figure 205 Onobrychis pulchella (photography by Natalya Beshko)

675. Onobrychis saravschanica B. Fedtsch.
Perennial. Fine-earth and stony slopes, screes. Montane and alpine zones. Nuratau, Malguzar, North Turkestan. Fodder, melliferous.

Genus 184. Alhagi Hill

676. Alhagi canescens (Regel) B. Keller & Shap.
Perennial. Sandy and clay deserts, roadsides, fallow and saline lands, banks of canals. Plain, foothills. Nuratau, Nuratau Relic Mountains, Mirzachul, Kyzylkum. Fodder, medicinal, melliferous.

677. Alhagi kirghisorum Schrenk   Figure 206
Perennial. Sandy and clay deserts, fine-earth, gravelly and stony slopes, roadsides, fallow lands, fields, banks of canals. Plain, foothills, montane zone. Nuratau, Aktau, Nuratau Relic Mountains, Malguzar, North Turkestan, Mirzachul, Kyzylkum. Fodder, medicinal, melliferous.

Figure 206 Alhagi kirghisorum (photography by Natalya Beshko)

678. Alhagi pseudalhagi (M. Bieb.) Desv. ex B. Keller & Shap.
Perennial. Saline wet places, sandy deserts, river valleys, banks of lakes and canals, fallow lands, fields. Plain, foothills. Nuratau, Aktau, Nuratau Relic Mountains, Malguzar, North Turkestan, Mirzachul, Kyzylkum. Fodder, medicinal, melliferous.

Genus 185. Cicer L.

679. Cicer flexuosum Lipsky
Perennial. Stony slopes, screes, pebbles. Montane zone. Nuratau, Malguzar, North Turkestan. Fodder.

680. Cicer grande (Popov) Korotkova   Figure 207
Perennial. Fine-earth and stony slopes. Montane zone. Nuratau, Nuratau Reserve. UzbRDB 2. Fodder, food, threatened species.

Figure 207 Cicer grande (photography by N. Yu. Beshko)

681. Cicer macracanthum Popov

Perennial. Fine-earth and stony slopes, pebbles. Montane and alpine zones. Malguzar, North Turkestan.

682. Cicer paucijugum (Popov) Nevski.

Perennial. Fine-earth and stony slopes. Montane and alpine zones. North Turkestan. Fodder.

683. Cicer pungens Boiss.

Perennial. Fine-earth and stony slopes, pebbles. Montane and alpine zones. Nuratau, Malguzar, North Turkestan.

684. Cicer songaricum DC.   Figure 208

Perennial. Fine-earth and stony slopes. Montane and alpine zones. Malguzar, North Turkestan. Fodder, food.

Figure 208 Cicer songaricum (photography by Natalya Beshko)

Genus 186. Vicia L.

685. Vicia anatolica Turrill [*Vicia hajastana* Grossh.]
Annual. Stony slopes. Foothills, montane zone. Nuratau, Nuratau Relic Mountains, Malguzar, North Turkestan. Fodder.

686.*Vicia ervilia (L.) Willd.
Annual. Fine-earth, gravelly and stony slopes, fallow lands, fields. Foothills, montane zone, Nuratau, Aktau, Malguzar, North Turkestan. Fodder, food, weed.

687. Vicia hyrcanica Fisch. & C. A. Mey.
Annual. Fine-earth slopes, meadows, river valleys, wastelands, gardens, fields, fallow lands. Plain, foothills, montane zone. Nuratau, Aktau, Nuratau Relic Mountains, Malguzar, North Turkestan, Mirzachul. Fodder, weed.

688. Vicia kokanica Regel & Schmalh.
Perennial. Fine-earth and stony slopes, screes, river valleys. Montane zone. Nuratau, Malguzar, North Turkestan. Fodder.

689. Vicia michauxii Spreng.   Figure 209
Annual. Fine-earth, gravelly and stony slopes, river valleys, meadows, fallow lands, fields, gardens. Plain, foothills, montane zone. Nuratau, Aktau, Malguzar, North Turkestan. Fodder, weed.

Figure 209 Vicia michauxii (photography by Natalya Beshko)

690. *Vicia peregrina L. [*Vicia gracilior* (Popov) Popov ex B. Fedtsch.]

Annual. Fine-earth, gravelly and stony slopes, river valleys, meadows, screes, fallow lands, fields, roadsides. Plain, foothills, montane zone. Nuratau, Aktau, Nuratau Relic Mountains, Malguzar. Fodder, weed.

691. *Vicia sativa subsp. **nigra** (L). Ehrh. [*Vicia angustifolia* L.]

Annual. Fine-earth slopes, meadows, river valleys, wastelands, gardens, fields, fallow lands. Plain, foothills, montane zone. Nuratau, Aktau, Nuratau Relic Mountains, Malguzar, North Turkestan, Mirzachul. Fodder, melliferous.

692. Vicia subvillosa (Ledeb.) Boiss.   Figure 210

Perennial. Fine-earth, gravelly and stony slopes. Foothills, montane zone. Nuratau, Aktau, Nuratau Relic Mountains, Malguzar, North Turkestan. Fodder.

Figure 210 Vicia subvillosa (photography by Natalya Beshko)

693. Vicia tenuifolia Roth.

Perennial. Fine-earth, gravelly slopes, river valleys, meadows, fallow lands. Foothills, montane zone. Malguzar, North Turkestan. Fodder, melliferous.

694. *Vicia villosa Roth.

Annual. Meadows, gardens, fallow lands, river valleys, fine-earth slopes. Plain, foothills, montane zone. Nuratau, Malguzar, North Turkestan. Fodder, weed, melliferous.

Genus 187. Lens Mill.

695. Lens culinaris subsp. **orientalis** (Boiss.) Ponert [*Lens orientalis* (Boiss.) Schmalh.]
Annual. Fine-earth, gravelly slopes, fallow lands. Foothills, montane zone. Nuratau, Aktau, Nuratau Relic Mountains, Malguzar, North Turkestan, Mirzachul. Fodder, food, weed.

Genus 188. Lathyrus L.

696. Lathyrus aphaca L.
Annual. Fine-earth, gravelly and stony slopes, river valleys, banks of canals, pebbles, fallow lands, gardens, fields, meadows. Plain, foothills, montane zone. Nuratau, Aktau, Nuratau Relic Mountains, Malguzar, North Turkestan, Mirzachul. Fodder, medicinal, dye, melliferous.

697.*Lathyrus cicera L.
Annual. Fine-earth and stony slopes, river valleys, meadows, pebbles, banks of streams and canals, fields. Plain, foothills, montane zone. Nuratau, Malguzar. Fodder, medicinal, melliferous.

698. Lathyrus inconspicuus L.
Annual. Fine-earth, gravelly slopes, ravines, fallow lands, fields, gardens, banks of rivers, dry riverbeds. Foothills, montane zone. Nuratau, Aktau, Nuratau Relic Mountains, Malguzar, North Turkestan, Mirzachul. Fodder.

699. Lathyrus pratensis L.   Figure 211
Perennial. Fine-earth and stony slopes, ravines, banks of rivers, lakes and canals, pebbles, wet meadows. Plain, foothills, montane and alpine zones. Nuratau, Malguzar, North Turkestan, Mirzachul. Fodder, medicinal, melliferous.

Figure 211 **Lathyrus pratensis** (photography by Natalya Beshko)

700. *Lathyrus sativus L. [*Lathyrus asiaticus* (Zalkind) Kudr.]
Annual. River valleys, fields, fallow lands, wastelands, ravines. Plain, foothills, montane and alpine zones. Nuratau, Nuratau Relic Mountains. Fodder, food, weed.

701. Lathyrus tuberosus L.
Perennial. Fine-earth and stony slopes, river valleys, wet meadows, banks of canals, fallow lands, fields. Foothills, montane zone. North Turkestan. Fodder, food, medicinal.

Family 43. Polygalaceae

Genus 189. Polygala L.

702. Polygala hybrida DC.
Perennial. Fine-earth and stony slopes, banks of streams, wet places, subalpine meadows, gardens. Foothills, montane and alpine zones. Malguzar, North Turkestan. Medicinal, ornamental, melliferous.

Family 44. Rosaceae

Genus 190. Spiraea L.

703. Spiraea hypericifolia L.   Figure 212
Shrub. River valleys, fine-earth, gravelly and stony slopes, screes, rocks, pebbles. Montane zone. Malguzar, North Turkestan. Medicinal, afforestation, ornamental, melliferous.

Figure 212 Spiraea hypericifolia (photography by Natalya Beshko)

704. Spiraea pilosa Franch.   Figure 213
Shrub. Stony slopes, banks of rivers, rocks, ravines. Montane zone. North Turkestan. Ornamental, melliferous.

Figure 213 Spiraea pilosa (photography by Natalya Beshko)

Genus 191. Cotoneaster Medik.

705. Cotoneaster goloskokovii Pojark.
Shrub. Stony slopes, screes, ravines. Montane zone. North Turkestan. Industrial, afforestation, ornamental, melliferous.

706. Cotoneaster multiflorus Bunge   Figure 214
Shrub. Rocks, stony slopes, ravines. Montane zone. Nuratau, North Turkestan. Industrial, afforestation, ornamental, melliferous.

Figure 214 Cotoneaster multiflorus (photography by Natalya Beshko)

707. Cotoneaster nummularioides Pojark.   Figure 215

Shrub. Fine-earth, gravelly and stony slopes, rocks, screes, pebbles, banks of rivers, ravines. Montane zone. Nuratau, Malguzar, North Turkestan. Industrial, afforestation, ornamental, melliferous.

Figure 215 Cotoneaster nummularioides (photography by Natalya Beshko)

708. Cotoneaster nummularius Fisch. & C.A. Mey.

Shrub. Fine-earth, gravelly and stony slopes, rocks, subalpine meadows. Montane zone. Nuratau, Aktau, Malguzar, North Turkestan. Industrial, afforestation, ornamental, melliferous.

709. Cotoneaster oliganthus Pojark.   Figure 216

Shrub. Fine-earth, gravelly and stony slopes, rocks. Montane zone. Malguzar, North Turkestan. Industrial, afforestation, ornamental, melliferous.

Figure 216 Cotoneaster oliganthus (photography by Natalya Beshko)

710. Cotoneaster songaricus (Regel & Herder) Popov   Figure 217
Shrub. Fine-earth, gravelly and stony slopes, rocks. Montane and alpine zones. Nuratau, North Turkestan. Industrial, afforestation, ornamental, melliferous.

Figure 217 Cotoneaster songaricus (photography by Natalya Beshko)

711. Cotoneaster suavis Pojark.

Shrub. Fine-earth, gravelly and stony slopes, river valleys. Montane zone. Malguzar, North Turkestan. Industrial, afforestation, ornamental, melliferous.

Genus 192. Pyrus L.

712. *Pyrus communis L.

Tree River valleys (planted and naturalized). Montane zone. Nuratau. Food, ornamental, industrial, dye, melliferous.

713. Pyrus korshinskyi Litv.

Tree. Fine-earth and stony slopes, river valleys. Montane zone. Malguzar. Ornamental, melliferous. IUCN CR.

714. Pyrus regelii Rehd.   Figure 218

Shrub. Fine-earth, gravelly and stony slopes, river valleys. Montane zone. Nuratau, Malguzar, North Turkestan. Ornamental, melliferous.

Figure 218 Pyrus regelii (photography by Natalya Beshko)

715. Pyrus turcomanica Maleev.
Tree. Fine-earth, gravelly and stony slopes, river valleys. Foothills, montane zone. Nuratau. Food, ornamental, industrial, dye, melliferous.

Genus 193. Malus Mill.

716. Malus sieversii (Ledeb.) M. Roem   Figure 219
Tree. Fine-earth, gravelly and stony slopes, ravines, river valleys. Foothills, montane zone. Nuratau, Aktau, Malguzar, North Turkestan. Food, fodder, industrial, dye, ornamental, melliferous. IUCN VU.

Figure 219 Malus sieversii (photography by Natalya Beshko)

Genus 194. Sorbus L.

717. Sorbus persica Hedl.   Figure 220
Tree. Fine-earth, gravelly and stony slopes. Montane zone. Nuratau, Malguzar, North Turkestan. Afforestation, industrial, ornamental, melliferous.

Figure 220 Sorbus persica (photography by Natalya Beshko)

718. Sorbus tianschanica Rupr.   Figure 221

Tree. Fine-earth, gravelly and stony slopes. Montane zone. Malguzar, North Turkestan. Medicinal, food, afforestation, industrial, ornamental, melliferous.

Figure 221 **Sorbus tianschanica** (photography by Natalya Beshko)

Genus 195. Crataegus L.

719. Crataegus pontica C. Koch   Figure 222

Tree. Fine-earth slopes, stream banks. Foothills, montane zone. Nuratau, Malguzar, North Turkestan. Food, afforestation, industrial, ornamental, melliferous.

Figure 222 **Crataegus pontica** (photography by Natalya Beshko)

720. Crataegus songarica C. Koch   Figure 223
Tree. Fine-earth, gravelly and stony slopes, screes, banks of rivers. Foothills, montane zone. Nuratau, Malguzar, North Turkestan. Food, medicinal, afforestation, industrial, ornamental, melliferous.

Figure 223 Crataegus songarica (photography by Natalya Beshko)

721. Crataegus turkestanica Pojark.   Figure 224
Tree. Fine-earth, gravelly and stony slopes, rocks, screes, river valleys. Foothills, montane zone. Nuratau, Malguzar, North Turkestan. Food, medicinal, afforestation, industrial, ornamental, melliferous.

Figure 224 Crataegus turkestanica (photography by Natalya Beshko)

Genus 196. Rubus L.

722. Rubus caesius L.   Figure 225
Shrub. Woodlands, wet stony slopes, screes, pebbles, fallow lands, river valleys, banks of canals, gardens. Plain, foothills, montane and alpine zones. Nuratau, Aktau, Malguzar, North Turkestan. Medicinal, food, dye, ornamental, fodder, meliferous.

Figure 225 Rubus caesius (photography by Natalya Beshko)

Genus 197. Potentilla L.

723. Potentilla asiatica (Th. Wolf) Juz.
Perennial. Fine-earth, gravelly and stony slopes, river valleys. Montane and alpine zones. Malguzar, North Turkestan. Medicinal, melliferous.

724. Potentilla asiae-mediae Ovcz. & Koczk.
Perennial. Fine-earth, gravelly and stony slopes. Montane and alpine zones. Malguzar, North Turkestan. Melliferous.

725. Potentilla desertorum Bunge
Perennial. Fine-earth, gravelly and stony slopes, rocks. Montane and alpine zones. Nuratau, Malguzar. melliferous.

726. Potentilla flabellata Regel & Schmalh.
Perennial. Wet fine-earth and stony slopes. Alpine zone. North Turkestan. Melliferous.

727. Potentilla gelida C. A. Mey.   Figure 226
Perennial. Fine-earth and stony slopes, rocks, moraines, alpine meadows. Montane and alpine zones. North Turkestan. Medicinal, fodder, melliferous.

Figure 226 **Potentilla gelida** (photography by Natalya Beshko)

728. Potentilla hololeuca Boiss. ex Lehm.
Perennial. Fine-earth, gravelly and stony slopes, river valleys, alpine meadows. Alpine zone. North Turkestan. melliferous.

729. Potentilla inclinata Vill. [*Potentilla impolita* Wahlenb.]
Perennial. Fine-earth and stony slopes, fallow lands. Montane and alpine zones. North Turkestan. melliferous.

730. Potentilla orientalis Juz.   Figure 227
Perennial. Fine-earth, gravelly and stony slopes, river valleys. Foothills, montane and alpine zones. Nuratau, Malguzar, North Turkestan. Fodder, melliferous.

Figure 227 **Potentilla orientalis** (photography by Natalya Beshko)

731. Potentilla pamiroalaica Juz.
Perennial. Fine-earth and stony slopes, rocks, pebbles, alpine meadows. Alpine zone. Malguzar, North Turkestan. Melliferous.

732. Potentilla pedata Willd. ex Hornem
Perennial. Fine-earth and stony slopes. Foothills, montane zone. Nuratau, Aktau, Nuratau Relic Mountains, Malguzar, North Turkestan. Medicinal, melliferous.

733. Potentilla reptans L.   Figure 228

Perennial. Meadows, banks of rivers, lakes and canals, wet places. Plain, foothills, montane zone. Nuratau, Aktau, Malguzar, North Turkestan. Medicinal, melliferous.

Figure 228 Potentilla reptans (photography by Natalya Beshko)

734. Potentilla soongorica Bunge

Perennial. Fine-earth, gravelly slopes. Foothills, montane zone. Nuratau, Malguzar, North Turkestan. Melliferous.

735. Potentilla supina L.

Perennial. River valleys, swamps, wet places, banks of canals, roadsides, wastelands, fields, fine-earth and stony slopes. Plain, foothills, montane zone. Nuratau, Malguzar, North Turkestan. Medicinal, fodder, melliferous.

736. Potentilla vvedenskyi Botsch.

Perennial. Rocks, alpine meadows, moraines, stony slopes. Alpine zone. North Turkestan. Melliferous.

Genus 198. Pentaphylloides Duham.

737. Pentaphylloides parvifolia (Fisch. ex Lehm.) Soják

Shrub. Fine-earth and stony slopes, rocks, pebbles, river valleys, alpine meadows. Montane and alpine zones. North Turkestan. Medicinal, fodder, ornamental, meliferous.

Genus 199. Sibbaldia L.

738. Sibbaldia tetrandra Bunge

Shrub. Moraines, fine-earth and stony slopes, screes, rocks, alpine meadows, places near melting snow. Alpine zone. North Turkestan.

Genus 200. Geum L.

739. Geum heterocarpum Boiss. [*Orthurus heterocarpus* (Boiss.) Juz.]   Figure 229
Perennial. Fine-earth, gravelly and stony slopes, stream banks, shady wet places. Montane zone. Nuratau, Malguzar, North Turkestan. Medicinal, meliferous.

Figure 229 Geum heterocarpum (photography by Natalya Beshko)

Genus 201. Orthurus Juz.

740. Orthurus kokanicus (Regel & Schmalh. ex Regel) Juz.
Perennial. Fine-earth and stony slopes, rocks. Montane and alpine zones. Nuratau, Malguzar, North Turkestan. Medicinal, meliferous.

Genus 202. Alchemilla L.

741. Alchemilla fontinalis Juz.
Perennial. Stream banks, alpine meadows. Alpine zone. North Turkestan.
742. Alchemilla krylovii Juz.
Perennial. Alpine meadows, stream banks, swamps. Alpine zone. North Turkestan.
743. Alchemilla sibirica Zamelis
Perennial. Wet places, stream banks. Montane and alpine zones. North Turkestan.

Genus 203. Agrimonia L.

744. Agrimonia asiatica Juz.
Perennial. Fine-earth and stony slopes, river valleys, meadows, gardens, banks of canals, roadsides. Foothills, montane zone. Nuratau, Malguzar, North Turkestan.

Genus 204. Sanguisorba L.

745. *Sanguisorba minor Scop. [*Poterium lasiocarpum* Boiss. & Hausskn.]
Perennial. Fine-earth and stony slopes, river valleys, fallow lands, fields, meadows, gardens, roadsides. Foothills, montane zone. Nuratau, Aktau, Nuratau Relic Mountains, Malguzar, North Turkestan, Mirzachul. Fodder, meliferous.

746. *Sanguisorba minor subsp. **balearica** (Bourg. & Nyman) Muoz Garm. & C. Navarro [*Poterium polygamum* Waldst. & Kitag.]
Perennial. Fine-earth and stony slopes, river valleys, fallow lands, fields, meadows, gardens, roadsides. Foothills, montane zone. Nuratau, Aktau, Nuratau Relic Mountains, Malguzar, North Turkestan, Mirzachul. Fodder, meliferous.

Genus 205. Rosa L.

747. Rosa beggeriana Schrenk & Fisch. ex C. A. Mey.   Figure 230
Shrub. Fine-earth slopes, river valleys, banks of canals, roadsides. Plain, foothills, montane zone. Nuratau, Nuratau Relic Mountains, Malguzar, North Turkestan, Mirzachul. Medicinal, ornamental.

Figure 230 Rosa beggeriana (photography by Natalya Beshko)

748. Rosa canina L.   Figure 231

Shrub. River valleys, fine-earth, gravelly and stony slopes. Plain, foothills, montane zone. Nuratau, Aktau, Nuratau Relic Mountains, Malguzar, North Turkestan, Mirzachul. Medicinal, food, tanniferous, ornamental, afforestation, meliferous.

Figure 231 Rosa canina (photography by Natalya Beshko)

749. Rosa corymbifera Borkh

Shrub. River valleys, fine-earth and stony slopes. Montane zone. Nuratau. Medicinal, food, ornamental, afforestation, meliferous.

750. Rosa ecae Aitch.   Figure 232

Shrub. Fine-earth, gravelly and stony slopes. Montane and alpine zones. Malguzar, North Turkestan. Medicinal, food, ornamental, meliferous, weed.

Figure 232 Rosa ecae (photography by Natalya Beshko)

751. Rosa fedtschenkoana Regel   Figure 233

Shrub. Fine-earth, gravelly and stony slopes, ravines, river valleys. Montane and alpine zones. Malguzar, North Turkestan. Medicinal, food, ornamental, afforestation, meliferous.

Figure 233 Rosa fedtschenkoana (photography by Natalya Beshko)

752. Rosa hissarica Slobodova

Shrub. Stony slopes, screes, rocks. Montane and alpine zones. North Turkestan. Medicinal, ornamental, meliferous.

753. Rosa lehmanniana Bunge

Shrub. Ravines, river valleys, wet fine-earth slopes. Montane zone. Nuratau. Medicinal, ornamental, meliferous.

754. Rosa kokanica (Regel) Regel ex Juz.   Figure 234

Shrub. Fine-earth, gravelly and stony slopes, ravines, river valleys. Montane and alpine zones. Nuratau, Malguzar, North Turkestan. Medicinal, ornamental, afforestation, meliferous.

Figure 234 Rosa kokanica (photography by Natalya Beshko)

755. **Rosa maracandica** Bunge   Figure 235

Shrub. Fine-earth, gravelly and stony slopes, rocks, pebbles, river valleys. Foothills, montane zone. Nuratau, Aktau, Malguzar, North Turkestan. Medicinal, ornamental, meliferous.

Figure 235 Rosa maracandica (photography by Natalya Beshko)

756. Rosa nanothamnus Boulenger

Shrub. Fine-earth, gravelly and stony slopes. Montane and alpine zones. Nuratau, Malguzar, North Turkestan. Medicinal, ornamental, meliferous.

757. Rosa transturkestanica N. F. Russanov.

Shrub. Fine-earth, gravelly and stony slopes. Montane zone. North Turkestan. Ornamental, meliferous.

Genus 206. Hulthemia Dumort.

758. Hulthemia persica (Michx. & Juss.) Bornm.

Shrub. Wastelands, pebbles, sandy soils, piedmont plains, gravelly and stony slopes, fallow lands, fields. Plain, foothills, montane zone. Nuratau, Aktau, Nuratau Relic Mountains, Malguzar, North Turkestan, Mirzachul. Ornamental, weed, meliferous.

Genus 207. Prunus L.

759. Prunus armeniaca L. [*Armeniaca vulgaris* Lam.]

Tree. Fine-earth slopes, river valleys. Foothills, montane zone. Nuratau, Malguzar, North Turkestan. Food, oil, medicinal, tanniferous, dye, afforestation, industrial, ornamental, meliferous. IUCN EN.

760. Prunus bucharica (Korsh.) B. Fedtsch. ex Rehder [*Amygdalus bucharica* Korsh.]   Figure 236
Tree or shrub. Fine-earth and stony slopes, rocks, screes, river valleys. Foothills, montane zone. Nuratau, Aktau, Malguzar, North Turkestan. Medicinal, food, oil, ornamental, dye, afforestation, industrial, meliferous. IUCN VU.

Figure 236 **Prunus bucharica** (photography by Natalya Beshko)

761. *Prunus cerasus L. [*Cerasus vulgaris* Mill.]

Tree. River valleys (naturalized). Nuratau. Food, medicinal, ornamental, meliferous.

762. Prunus divaricata Ledeb.   Figure 237

Tree. Fine-earth, gravelly and stony slopes, screes, river valleys, ravines. Montane zone. Nuratau, Malguzar, North Turkestan. Medicinal, food, oil, ornamental, afforestation, meliferous.

Figure 237 Prunus divaricata (photography by Natalya Beshko)

763. *Prunus domestica L.

Tree. Fine-earth slopes, river valleys (naturalized). Montane zone. Nuratau. Medicinal, food, oil, ornamental, afforestation, meliferous.

764.*Prunus dulcis (Mill.) D.A. Webb [*Amygdalus communis* L.]
Tree. Fine-earth and stony slopes, screes, banks of rivers. Foothills, montane zone. Nuratau, Malguzar, North Turkestan. Medicinal, food, oil, ornamental, afforestation, industrial, meliferous.

765. Prunus erythrocarpa (Nevski) Gilli [*Cerasus erythrocarpa* Nevski]   Figure 238
Shrub. Fine-earth and stony slopes, rocks, screes, ravines, banks of rivers. Foothills, montane zone. Nuratau, Malguzar, North Turkestan. Ornamental, industrial, afforestation, meliferous.

Figure 238 Prunus erythrocarpa (photography by Natalya Beshko)

766. Prunus mahaleb L. [*Cerasus mahaleb* (L.) Mill.]   Figure 239
Tree. Fine-earth and stony slopes, rocks, screes, ravines, banks of rivers. Montane zone. Nuratau, Malguzar, North Turkestan. Food, ornamental, afforestation, meliferous.

Figure 239 Prunus mahaleb (photography by Natalya Beshko)

767. Prunus spinosissima (Bunge) Franch. [*Amygdalus spinosissima* Bunge]   Figure 240
Shrub. Fine-earth, gravelly and stony slopes, rocks. Foothills, montane zone. Nuratau, Aktau, Nuratau Relic Mountains, Malguzar, North Turkestan, Mirzachul. Oil, dye, ornamental, afforestation, meliferous.

Figure 240 Prunus spinosissima (photography by Natalya Beshko)

768. Prunus verrucosa Franch.[*Cerasus amygdaliflora* Nevski, *Cerasus verrucosa* (Franch.) Nevski]   Figure 241

Shrub. Stony slopes, rocks, screes. Montane and alpine zones. Nuratau, Malguzar, North Turkestan. Ornamental, afforestation, meliferous.

Figure 241 Prunus verrucosa (photography by Natalya Beshko)

Family 45. Elaeagnaceae

Genus 208. Hippophae L.

769. Hippophae rhamnoides L.   Figure 242

Shrub. River valleys, ravines. Montane zone. North Turkestan. Medicinal, dye, food, ornamental, meliferous, afforestation.

Figure 242 Hippophae rhamnoides (photography by Natalya Beshko)

Genus 209. Elaeagnus L.

770. Elaeagnus angustifolia L.   Figure 243

Tree. Banks of lakes and canals, river valleys. Plain, foothills, montane zone. Nuratau, Aktau, Nuratau Relic Mountains, Malguzar, North Turkestan, Mirzachul, Kyzylkum. Food, medicinal, afforestation, industrial, tanniferous, dye, ornamental, meliferous.

Figure 243 Elaeagnus angustifolia (photography by Natalya Beshko)

Family 46. Rhamnaceae

Genus 210. Sageretia Brongn.

771. Sageretia thea (Osbeck) M. S. Johnst. [*Sageratia laetevirens* (Kom). Gontsch.]
Shrub. Fine-earth and stony slopes, rocks. Montane zone. Nuratau, Malguzar, North Turkestan. Ornamental, afforestation.

Genus 211. Rhamnus L.

772. Rhamnus cathartica L.
Tree or shrub. Stony and gravelly slopes, river valleys, ravines. Montane zone. Nuratau, Malguzar, North Turkestan. Medicinal, ornamental, afforestation, industrial, dye, oil, meliferous.

773. Rhamnus coriacea (Regel) Kom.   Figure 244
Shrub. Fine-earth, gravelly and stony slopes, screes, cliffs, ravines. Foothills, montane zone. Nuratau, Malguzar, North Turkestan. Ornamental, afforestation.

Figure 244 **Rhamnus coriacea** (photography by Natalya Beshko)

Family 47. Ulmaceae

Genus 212. Ulmus L.

774. *Ulmus glabra Huds.
Tree. Fine-earth and stony slopes, river valleys (naturalized), settlements, forest belts, cemeteries (planted). Foothills, montane zone. Nuratau, Malguzar, North Turkestan. Industrial, ornamental, afforestation.

775. *Ulmus laevis Pall.
Tree. Fine-earth and stony slopes, river valleys (naturalized), settlements, forest belts, cemeteries (planted). Plain, foothills, montane zone. Nuratau, North Turkestan. Industrial, ornamental, afforestation.

776. *Ulmus minor Mill
Tree. Fine-earth and stony slopes, river valleys (naturalized), settlements, forest belts, cemeteries (planted). Plain, foothills, montane zone. Nuratau, Malguzar. Industrial, ornamental, afforestation.

777. Ulmus pumila L.
Tree or shrub. Fine-earth and stony slopes, river valleys, ravines, rocks. Plain, foothills, montane zone. Nuratau, Aktau, Nuratau Relic Mountains, Mirzachul, Kyzylkum. Industrial, ornamental, afforestation.

Genus 213. Celtis L.

778. Celtis australis subsp. **caucasica** (Willd.) C. C. Towns. [*Celtis caucasica* Willd.]
Tree. Fine-earth and stony slopes, screes, rocks, river valleys. Foothills, montane zone. Nuratau, Malguzar, North Turkestan. Industrial, ornamental, afforestation, medicinal, tanniferous.

Family 48. Cannabaceae

Genus 214. Cannabis L.

779. *Cannabis sativa L.
Annual. Roadsides, fallow lands, wastelands, banks of canals. Plain, Mirzachul (naturalized). Narcotic, industrial, oil, weed.

Family 49. *Moraceae

Genus 215. *Morus L.

780. *Morus alba L.
Tree. Fine-earth and stony slopes, river valleys (naturalized), settlements, gardens, forest belts, burials (planted). Plain, foothills, montane zone. Nuratau, Nuratau Relic Mountains, Malguzar, North Turkestan. Food, fodder, medicinal, ornamental, industrial, dye, tanniferous.

781. *Morus nigra L.
Tree. Fine-earth and stony slopes, river valleys (naturalized), settlements, gardens (planted). Plain, foothills, montane zone. Nuratau, Malguzar, North Turkestan. Food, medicinal, ornamental, industrial, dye, tanniferous.

Family 50. Urticaceae

Genus 216. Urtica L.

782. Urtica dioica L.
Perennial. River valleys, roadsides, wastelands, gardens, riparian forests, banks of canals. Plain, foothills, montane and alpine zones. Nuratau, Nuratau Relic Mountains, Malguzar, North Turkestan, Mirzachul. Medicinal, food, dye.

Genus 217. Parietaria L.

783. Parietaria debilis G. Forst [*Parietaria micrantha* Ledeb.]
Annual. Rocks, ravines, shady wet places. Montane and alpine zones, Nuratau, Malguzar. Alkaloid-bearing.

784. Parietaria lusitanica subsp. **serbica** (Pančić) P. W. Ball [*Parietaria serbica* Pančić]
Annual. Ravines, screes, rocks. Montane zone. Malguzar, North Turkestan. Alkaloid-bearing.

Family 51. Juglandaceae

Genus 218. Juglans L.

785. Juglans regia L.
Tree. River valleys, wet slopes. Montane zone. Nuratau, Malguzar, North Turkestan. Food, industrial,

oil, afforestation, ornamental, medicinal, dye. IUCN NT.

Family 52. Betulaceae

Genus 219. Betula L.

786. Betula pendula Roth.   Figure 245
Tree. River valleys, wet slopes. Montane zone. North Turkestan. Industrial, ornamental, afforestation.

Figure 245 Betula pendula (photography by Natalya Beshko)

Family 53. Cucurbitaceae

Genus 220. Bryonia L.

787. *Bryonia cretica subsp. **dioica** (Jacq.) Tutin [*Bryonia dioica* Jacq.]   Figure 246
Perennial. Stony slopes, ravines, shady places. Foothills, montane zone. Nuratau. Medicinal,

essential oil, ornamental, poisonous.

Figure 246 Bryonia cretica subsp. **dioica** (photography by Natalya Beshko)

788. Bryonia melanocarpa Nabiev   Figure 247

Perennial. Sandy deserts. Plain. Kyzylkum. Medicinal, poisonous. Threatened species, UzbRDB 2. Endemic to the South-eastern Kyzylkum.

Figure 247 Bryonia melanocarpa (photography by N. Yu. Beshko)

Family 54. Datiscaceae

Genus 221. Datisca L.

789. Datisca cannabina L.   Figure 248

Perennial. Banks of rivers, swamp meadows, pebbles, wet slopes. Foothills, montane zone. Nuratau, Malguzar, North Turkestan. Medicinal, dye, ornamental.

Figure 248 Datisca cannabina (photography by Natalya Beshko)

Family 55. Celastraceae

Genus 222. Parnassia L.

790. Parnassia laxmannii Pall. ex Schult.
Perennial. Meadows, stony slopes, rocks, ravines, banks of rivers, swamps. Montane and alpine zones. North Turkestan.

Family 56. Hypericaceae

Genus 223. Hypericum L.

791. Hypericum elongatum Ledeb. ex Rchb.   Figure 249
Perennial. Fine-earth, gravelly and stony slopes, rocks, screes. Montane and alpine zones. Malguzar,

North Turkestan. Medicinal.

Figure 249 **Hypericum elongatum** (photography by Natalya Beshko)

792. Hypericum perforatum L.   Figure 250
Perennial. Fine-earth, gravelly and stony slopes, river valleys, dry riverbeds. Foothills, montane zone. Nuratau, Aktau, Malguzar, North Turkestan. Medicinal, dye, essential oil, meliferous.

Figure 250 **Hypericum perforatum** (photography by Natalya Beshko)

793. Hypericum scabrum L.   Figure 251
Perennial. Fine-earth, gravelly and stony slopes, screes, rocks, dry riverbeds. Foothills, montane and alpine zones. Nuratau, Aktau, Nuratau Relic Mountains, Malguzar, North Turkestan. Medicinal, dye,

essential oil, meliferous.

Figure 251 Hypericum scabrum (photography by Natalya Beshko)

Family 57. Violaceae

Genus 224. Viola L.

794. Viola alaica Vved.
Perennial. Screes, wet stony slopes, rocks, shady places. Montane zone. Malguzar, North Turkestan.

795. Viola occulta Lehm.   Figure 252
Annual. Fine-earth, gravelly and stony slopes, meadows, fallow lands. Foothills, montane zone. Nuratau, Aktau, Malguzar, North Turkestan.

Figure 252 Viola occulta (photography by Natalya Beshko)

796. Viola suavis M. Bieb.   Figure 253
Perennial. Fine-earth, gravelly and stony slopes, stream banks, shady places, gardens. Foothills, montane zone. Nuratau, North Turkestan. Medicinal, ornamental, meliferous.

Figure 253 Viola suavis (photography by Natalya Beshko)

797. Viola turkestanica Regel & Schmalh.
Perennial. Fine-earth and stony slopes, rocks, screes, shady wet places, meadows. Montane and alpine zones. Malguzar, North Turkestan.

Family 58. Salicaceae

Genus 225. Populus L.

798. Populus afghanica (Aitch. & Hemsl.) C. K. Schneid.   Figure 254
Tree. River valleys. Foothills, montane zone. Nuratau, Aktau, Malguzar, North Turkestan. Industrial, ornamental, afforestation, tanniferous.

Figure 254 **Populus afghanica** (photography by Natalya Beshko)

799. Populus alba L.   Figure 255
Tree. River valleys, banks of canals. Plain, foothills, montane zone. Nuratau, Aktau, Malguzar, North Turkestan, Mirzachul. Industrial, ornamental, afforestation, tanniferous, meliferous.

Figure 255 **Populus alba** (photography by Natalya Beshko)

800. Populus euphratica Oliv.   Figure 256

Tree. River valleys, banks of lakes and canals, riparian forests. Plain, foothills. Kyzylkum, Mirzachul. Industrial, ornamental, afforestation, tanniferous, meliferous.

Figure 256 Populus euphratica (photography by Natalya Beshko)

801. *Populus nigra L.

Tree. River valleys. Plain, foothills, montane zone (naturalized). Nuratau. Industrial, ornamental, afforestation, tanniferous, medicinal, meliferous.

802. Populus pruinosa Schrenk   Figure 257

Tree. River valleys, banks of lakes and canals, riparian forests. Plain. Kyzylkum, Mirzachul. Industrial, ornamental, afforestation, tanniferous, dye.

Figure 257 Populus pruinosa (photography by Natalya Beshko)

803. Populus talassica Kom.

Tree. River valleys, fine-earth and stony slopes. Montane and alpine zones. Malguzar, North

Turkestan. Industrial, ornamental, afforestation.

Genus 226. Salix L.

804. Salix alba L.
Tree. Banks of rivers. Plain, foothills, montane zone. Nuratau, Aktau, Malguzar, North Turkestan, Mirzachul, Kyzylkum. Industrial, ornamental, afforestation, tanniferous, medicinal, meliferous.

805. Salix blakii Goerz.
Tree or shrub. Banks of rivers. Foothills, montane zone. Nuratau, Malguzar, North Turkestan, Kyzylkum.

806. Salix olgae Regel
Shrub. Banks of rivers, pebbles. Plain, foothills, montane zone. Malguzar, North Turkestan.

807. Salix pycnostachya Andersson.
Tree or Shrub. Banks of rivers. Montane and alpine zones. Nuratau, Malguzar, North Turkestan.

808. Salix rosmarinifolia L.
Shrub. Wet places, river valleys. Montane zone. Nuratau.

809. Salix songarica Andersson
Tree or shrub. River valleys. Plain. Mirzachul. Industrial, ornamental, afforestation, tanniferous, dye, meliferous.

810. Salix wilhelmsiana M. Bieb.
Shrub. River valleys. Plain, foothills, montane and alpine zones. North Turkestan, Mirzachul. Industrial, ornamental, afforestation, tanniferous, meliferous.

Family 59. Euphorbiaceae

Genus 227. Andrachne L.

811. Andrachne fedtschenkoi Kossinsky
Semishrub. Stony slopes, rocks. Montane zone, Nuratau, Aktau, Malguzar.

812. Andrachne telephioides L.
Semishrub. Stony slopes, fields. Foothills, montane zone. Nuratau, Aktau, Nuratau Relic Mountains, Malguzar, North Turkestan. Medicinal, alkaloid-bearing, dye, fodder.

Genus 228. Chrozophora Neck.

813. Chrozophora tinctoria (L.) A. Juss. [*Chrozophora hierosolymitana* Spreng.; *Chrozophora obliqua* (Vahl.) A. Juss. ex. Spreng.]
Annual. Sandy deserts, wastelands, fine-earth, gravelly and stony slopes, fields, fallow lands,

wastelands. Plain, foothills, montane zone. Nuratau, Aktau, Nuratau Relic Mountains, Malguzar, North Turkestan, Mirzachul, Kyzylkum. Dye, poisonous.

814. Chrozophora sabulosa Kar. & Kir. [*Chrozophora. gracilis* Fisch. & C. A. Mey. ex Ledeb.]
Annual. Sandy deserts, banks of rivers, lakes and canals. Plain. Mirzachul, Kyzylkum. Dye, fodder.

Genus 229. Euphorbia L.

815. Euphorbia chamaesyce L. [*Chamaesyce canescens* (L.) Prokh.]   Figure 258
Annual. Pebbles, river valleys, dry riverbeds, fine-earth and stony slopes, fields, wastelands. Foothills, montane zone. Nuratau, Aktau, Malguzar, North Turkestan. Weed.

Figure 258 **Euphorbia chamaesyce** (photography by Natalya Beshko)

816. Euphorbia cheirolepis Fisch. & C. A. Mey. ex Karelin [*Cystidospermum cheirolepis* (Fisch. & C. A. Mey. ex Ledeb.) Prokh.]
Annual. Sandy deserts. Plain. Mirzachul, Kyzylkum. Fodder.

817. Euphorbia densa Schrenk
Annual. Sandy deserts, stony slopes. Plain, foothills. Nuratau, Nuratau Relic Mountains, Mirzachul, Kyzylkum.

818. Euphorbia esula L. [*Euphorbia jaxartica* Prokh.; *Euphorbia glomerulans* (Prokh.) Prokh.]   Figure 259

Perennial. River valleys, banks of canals, wet stony slopes, screes, pebbles, fallow lands. Plain, foothills, montane and alpine zones. Nuratau, Malguzar, North Turkestan. Dye, poisonous.

Figure 259 Euphorbia esula (photography by Natalya Beshko)

819. Euphorbia falcata L.   Figure 260

Annual. Stony, fine-earth slopes, pebbles, fields. Foothills, montane and alpine zones. Nuratau, Nuratau Relic Mountains, Malguzar, North Turkestan, Mirzachul. Medicinal, weed.

Figure 260 Euphorbia falcata (photography by Natalya Beshko)

820. Euphorbia franchetii B. Fedtsch.

Annual. Stony slopes, clay deserts, fields, fallow lands. Foothills, montane and alpine zones. Nuratau, Nuratau Relic Mountains, Malguzar, North Turkestan, Mirzachul.

821. Euphorbia granulata Forssk. [*Chamaesyce turcomanica* (Boiss.) Prokh.]
Annual. Gravelly slopes, pebbles, sandy and clay deserts, saline lands. Plains, foothills, montane zone. Nuratau, Malguzar, North Turkestan, Mirzachul.

822. *Euphorbia helioscopia L.   Figure 261
Annual. Stony slopes, roadsides, banks of canals, gardens, fallow lands, wastelands. Plain, foothills, montane zone. Nuratau, Aktau, Nuratau Relic Mountains, Malguzar, North Turkestan, Mirzachul, Kyzylkum. Poisonous, dye, weed.

Figure 261 **Euphorbia helioscopia** (photography by Natalya Beshko)

823. Euphorbia humilis C. A. Mey. ex Ledeb.
Perennial. Stony slopes, rocks, screes. Montane and alpine zones. Malguzar, North Turkestan.

824. Euphorbia inderiensis Less. ex Kar. & Kir.   Figure 262
Annual. Fine-earth and stony slopes, sandy soils, saline lands. Plain, foothills, montane zone. Nuratau, Malguzar, Mirzachul, Kyzylkum.

Figure 262 **Euphorbia inderiensis** (photography by Natalya Beshko)

825. Euphorbia kanaorica Boiss. [*Euphorbia polytimetica* (Prokh.) Prokh.]
Perennial. Rocks, screes, fine-earth and stony slopes. Alpine zone. North Turkestan.

826. Euphorbia rapulum Kar. & Kir.   Figure 263
Perennial. Fine-earth, gravelly and stony slopes. Plain, foothills, montane zone. Nuratau, Malguzar. Medicinal, poisonous, rubber-bearing.

Figure 263 **Euphorbia rapulum** (photography by Natalya Beshko)

827. Euphorbia rosularis Fed.
Perennial. Stony slopes, screes. Montane and alpine zones, Malguzar, North Turkestan.

828. Euphorbia sarawschanica Regel   Figure 264
Perennial. Stony slopes, rocks. Montane and alpine zones. Malguzar, North Turkestan. Medicinal, poisonous.

Figure 264 **Euphorbia sarawschanica** (photography by Natalya Beshko)

829. Euphorbia sororia Schrenk
Annual. Clayey hills. Foothills. Malguzar.

830. Euphorbia szowitsii Fisch. & C. A. Mey.
Annual. Fine-earth, gravelly slopes, screes, pebbles. Montane zone. Nuratau, Malguzar, North Turkestan.

831. Euphorbia turczaninowii Kar. & Kir.   Figure 265
Annual. Sandy deserts. Plain. Kyzylkum.

Figure 265 Euphorbia turczaninowii (photography by Natalya Beshko)

Family 60. Linaceae

Genus 230. Linum L.

832. Linum corymbulosum Rchb.
Annual. gardens, fallow lands, river valleys, meadows, fine-earth, gravelly and stony slopes. Plain, foothills, montane zone. Nuratau, Malguzar, North Turkestan.

833. Linum macrorhizum Juz.
Perennial. Fine-earth and stony slopes, screes, fallow lands. Foothills, montane zone. Nuratau, Malguzar, North Turkestan.

834. Linum olgae Juz.
Perennial. Fine-earth and stony slopes, rocks. Montane and alpine zones. North Turkestan. Ornamental, oil.

835. *Linum usitatissimum L. [*Linum humile* Mill.]
Annual. Roadsides, gardens, fallow lands, fields. Foothills, montane zone. Nuratau, Malguzar, North Turkestan. Food, medicinal, industrial, oil.

Family 61. Geraniaceae

Genus 231. Geranium L.

836. Geranium collinum Stephan ex Willd. [*Geranium regelii* Nevski]   Figure 266
Perennial. Meadows, river valleys, stream banks, wet places near springs, fine-earth and stony slopes, rocks. Foothills, montane and alpine zones. Nuratau, Malguzar, North Turkestan. Medicinal, dye, tanniferous, fodder, ornamental.

Figure 266 Geranium collinum (photography by Natalya Beshko)

837. Geranium divaricatum Ehrh.   Figure 267
Annual. Shady wet places, fine-earth and stony slopes, screes, wastelands, gardens. Foothills, montane zone. Nuratau, Aktau, Nuratau Relic Mountains, Malguzar, North Turkestan. Tanniferous, ornamental, weed.

Figure 267 Geranium divaricatum (photography by Natalya Beshko)

838. Geranium kotschyi subsp. **charlesii** (Aitch. & Hemsl.) P. H. Davis [*Geranium charlesii* (Aitch. & Hemsl.) Vved. ex Nevski]   Figure 268
Perennial. Fine-earth and stony slopes. Montane and alpine zones. Nuratau, Aktau, Nuratau Relic Mountains, Malguzar, North Turkestan.

Figure 268 Geranium kotschyi subsp. **charlesii** (photography by Natalya Beshko)

839. Geranium linearilobum DC. [*Geranium transversale* (Kar. & Kir.) Vved. ex Pavlov]
Perennial. Fine-earth, gravelly and stony slopes, piedmont plains. Plain, foothills, montane zone. Nuratau, Aktau, Nuratau Relic Mountains, Malguzar, North Turkestan. Ornamental.

840. *Geranium pusillum L.
Annual. Fields, fallow lands, roadsides, gardens, wastelands, fine-earth and stony slopes. Plain, foothills, montane zone. Nuratau, Aktau, Nuratau Relic Mountains, Malguzar, North Turkestan.

841. Geranium robertianum L.
Annual. Ravines, banks of rivers, shady wet places. Montane zone. Nuratau. Medicinal, ornamental.

842. Geranium rotundifolium L.   Figure 269
Annual. Pebbles, fine-earth, gravelly slopes, shady places, gardens. Foothills, montane zone.

Nuratau, Aktau, Nuratau Relic Mountains, Malguzar, North Turkestan. Medicinal, tanniferous.

Figure 269 **Geranium rotundifolium** (photography by Natalya Beshko)

843. Geranium saxatile Kar. & Kir.   Figure 270

Perennial. Wet places, stony slopes, rocks. Montane and alpine zones. Malguzar, North Turkestan. Ornamental.

Figure 270 **Geranium saxatile** (photography by Natalya Beshko)

Genus 232. Erodium L'Her.

844. Erodium ciconium (Jusl.) L'Her.
Annual. Sandy and clay deserts, piedmont plains, clayey hills. Plain, foothills. Nuratau, Aktau, Nuratau Relic Mountains, Malguzar, North Turkestan, Mirzachul, Kyzylkum. Fodder, weed.

845. Erodium cicutarium (L.) L'Her.　Figure 271
Annual. Fine-earth, gravelly and stony slopes, river valleys, gardens, banks of canals, fallow lands. Plain, foothills, montane zone. Nuratau, Aktau, Nuratau Relic Mountains, Malguzar, North Turkestan, Mirzachul. Medicinal, fodder, weed, meliferous.

Figure 271 Erodium cicutarium (photography by Natalya Beshko)

846. Erodium hoefftianum C. A. Mey.
Annual. Clayey and skeleton soils. Plain, foothills. Nuratau, Malguzar.

847. Erodium oxyrrhynchum M. Bieb.
Annual or biennial. Stony slopes, sandy soils, pebbles. Plain, foothills, montane zone. Nuratau, Aktau, Nuratau Relic Mountains, Malguzar, Mirzachul, Kyzylkum.

Family 62. Lythraceae

Genus 233. Lythrum L.

848. Lythrum hyssopifolia L.
Annual. Saline lands, banks of rivers, lakes and canals. Plain, foothills, montane zone. Nuratau. Medicinal, ornamental, meliferous.

849. Lythrum nanum Kar. & Kir.
Annual. Saline lands. Plain. Mirzachul.

Family 63. Onagraceae

Genus 234. Epilobium L.

850. Epilobium angustifolium L. [*Chamaenerion angustifolium* (L.) Scop.]   Figure 272
Perennial. Wet fine-earth and stony slopes, river valleys, places near melting snow and springs. Montane and alpine zones. Malguzar, North Turkestan. Medicinal, ornamental, meliferous.

Figure 272 Epilobium angustifolium (photography by Natalya Beshko)

851. Epilobium hirsutum L. (*Epilobium velutinum* Nevski)   Figure 273
Perennial. Banks of rivers and canals, wet places, swamps, meadows. Plain, foothills, montane zone. Nuratau, Aktau, Nuratau Relic Mountains, Malguzar, North Turkestan, Mirzachul. Medicinal, ornamental, fodder, meliferous.

Figure 273 **Epilobium hirsutum** (photography by Natalya Beshko)

852. Epilobium komarovii Ovcz.
Perennial. Banks of rivers, swamps, wet places near springs. Montane and alpine zones. Nuratau, Malguzar, North Turkestan.

853. Epilobium minutiflorum Hausskn.   Figure 274
Perennial. Banks of rivers and canals, pebbles, wet places near springs, swamps, alpine meadows. Plain, foothills, montane and alpine zones. Nuratau, Aktau, Nuratau Relic Mountains, Malguzar, North Turkestan, Mirzachul.

Figure 274 **Epilobium minutiflorum** (photography by Natalya Beshko)

854. Epilobium palustre L.

Perennial. Meadows, banks of rivers, swamps. Plain, foothills, montane and alpine zones. North Turkestan. Medicinal, fodder, meliferous.

855. Epilobium roseum subsp. **subsessile** (Boiss.) P. H. Raven [*Epilobium nervosum* Boiss.]

Perennial. Wet slopes, banks of rivers. Montane and alpine zones. Malguzar, North Turkestan.

856. Epilobium tetragonum L.

Perennial. Wet meadows, banks of rivers and canals, wet screes, swamps, saline lands, gardens, fields, fallow lands. Plain, foothills, montane zone. Nuratau, North Turkestan. Medicinal, fodder, meliferous.

857. Epilobium tianschanicum Pavlov

Perennial. Banks of rivers, wet places near springs, swamps, wet rocks and screes, alpine meadows. Montane and alpine zones. Nuratau, Malguzar, North Turkestan.

858. Epilobium turkestanicum Pazij & Vved.

Perennial. Wet stony slopes and rocks, wet places near melting snow, swamps, alpine meadows, pebbles, banks of rivers. Montane and alpine zones. Nuratau, Malguzar, North Turkestan.

Family 64. Biebersteiniaceae

Genus 235. Biebersteinia Steph.

859. Biebersteinia multifida DC.   Figure 275

Perennial. Fine-earth, gravelly and stony slopes. Foothills, montane zone. Nuratau, Aktau, Nuratau Relic Mountains, Malguzar, North Turkestan. Medicinal.

Figure 275 Biebersteinia multifida (photography by Natalya Beshko)

Family 65. Nitrariaceae

Genus 236. Peganum L.

860. Peganum harmala L.   Figure 276
Perennial. Saline lands, sandy and clay deserts, roadsides, surroundings of settlements, wells and sheepyards, wastelands. Plain, foothills, montane zone. Nuratau, Nuratau Relic Mountains, Malguzar, North Turkestan, Mirzachul, Kyzylkum. Medicinal, weed, poisonous.

Figure 276 Peganum harmala (photography by Natalya Beshko)

Family 66. Anacaridaceae

Genus 237. Pistacia L.

861. Pistacia vera L.   Figure 277

Tree or shrub. Fine-earth, gravelly and stony slopes, rocks, screes. Foothills, montane zone. Nuratau, Nuratau Relic Mountains, Malguzar, North Turkestan. Food, medicinal, ornamental, industrial, oil, dye, afforestation.

Figure 277 Pistacia vera (photography by Natalya Beshko)

Family 67. Sapindaceae

Genus 238. Acer L.

862. Acer pubescens Franch   Figure 278

Tree. Fine-earth, gravelly and stony slopes, river valleys. Montane zone. Nuratau, Malguzar, North

Turkestan. Ornamental, industrial, afforestation.

Figure 278 Acer pubescens (photography by Natalya Beshko)

863. Acer semenovii Regel & Herder   Figure 279
Tree. Fine-earth, gravelly and stony slopes, river valleys. Montane zone. Nuratau, Aktau, Malguzar, North Turkestan. Ornamental, industrial, afforestation.

Figure 279 Acer semenovii (photography by Natalya Beshko)

864. Acer turkestanicum Pax
Tree. Fine-earth, gravelly and stony slopes, river valleys. Montane zone. Malguzar, North Turkestan. Ornamental, industrial, afforestation.

Family 68. Rutaceae

Genus 239. Haplophyllum Juss.

865. Haplophyllum acutifolium (DC.) G. Don. [*Haplophyllum perforatum* (M. Bieb.) Vved.]
Perennial. Sandy and clay deserts, fine-earth, gravelly and stony slopes, pebbles, rocks, fields, fallow

lands. Plain, foothills, montane zone. Nuratau, Aktau, Nuratau Relic Mountains, Malguzar, North Turkestan, Mirzachul, Kyzylkum. Medicinal, dye.

866. Haplophyllum bungei Trautv.

Perennial. Sandy and clay deserts. Plain. Kyzylkum. Medicinal, dye.

867. Haplophyllum lasianthum Bunge [*Haplophyllum versicolor* Fisch. & C. A. Mey.]

Perennial. Sandy and clay deserts, fine-earth, gravelly and stony slopes, pebbles, fallow lands, fields. Plain, foothills. Nuratau, Nuratau Relic Mountains, Malguzar, North Turkestan, Mirzachul, Kyzylkum. Medicinal, weed.

868. Haplophyllum latifolium Kar. & Kir.   Figure 280

Perennial. Fine-earth, gravelly and stony slopes, fallow lands. Foothills, montane zone. Nuratau, Nuratau Relic Mountains, Malguzar, North Turkestan, Mirzachul. Medicinal.

Figure 280 **Haplophyllum latifolium** (photography by Natalya Beshko)

869. Haplophyllum pedicellatum Bunge ex Boiss.   Figure 281

Perennial. Sandy and clay deserts, saline lands, fallow lands, fine-earth slopes. Plain, foothills. Nuratau, Nuratau Relic Mountains, Malguzar, North Turkestan, Mirzachul, Kyzylkum. Medicinal, poisonous.

Figure 281 Haplophyllum pedicellatum (photography by Natalya Beshko)

870. Haplophyllum ramosissimum (Paulsen) Vved.
Semishrub. Sandy deserts. Plain. Kyzylkum.

871. Haplophyllum robustum Bunge
Perennial. Sandy and clay deserts. Plain. Mirzachul, Kyzylkum. Medicinal.

Genus 240. Dictamnus L.

872. Dictamnus angustifolius G. Don. f. & Sweet   Figure 282
Perennial. Fine-earth, gravelly and stony slopes, river valleys, stream banks, screes, meadows. Montane and alpine zones. Malguzar, North Turkestan. Poisonous, medicinal, ornamental, meliferous.

Figure 282 Dictamnus angustifolius (photography by Natalya Beshko)

Family 69. *Simaroubaceae

Genus 241. *Ailanthus Desf.

873. *Ailantus altissima (Mill.) Swingle
Tree. Introduced alien species. Settlements, forest belts (planted), river valleys, fine-earth slopes

(naturalized). Plain, foothills, montane zone. Nuratau, Malguzar, North Turkestan, Mirzachul. Ornamental, industrial, afforestation.

Family 70. Malvaceae

Genus 242. Malva L.

874. *Malva neglecta Wallr

Perennial. Fine-earth and skeleton soils, pebbles, wastelands, dry riverbeds. Plain, foothills, montane and alpine zones. Nuratau, Aktau, Nuratau Relic Mountains, Malguzar, North Turkestan, Mirzachul, Kyzylkum. Medicinal, food, fodder, meliferous, weed.

Genus 243. Alcea L.

875. Alcea litvinovii (Iljin) Iljin   Figure 283

Perennial. Fine-earth, gravelly slopes, river valleys. Plain, foothills, montane zone. Nuratau. Medicinal, ornamental, meliferous.

Figure 283 Alcea litvinovii (photography by Natalya Beshko)

876. Alcea nudiflora (Lindl.) Boiss.   Figure 284

Perennial. Fine-earth, gravelly and stony slopes, screes, fallow lands, roadsides. Plain, foothills, montane zone. Nuratau, Nuratau Relic Mountains, Malguzar, North Turkestan. Medicinal, ornamental, meliferous.

Figure 284 Alcea nudiflora (photography by Natalya Beshko)

877. Alcea rhyticarpa (Trautv.) Iljin

Perennial. Sandy and clayey soils, fine-earth, gravelly and stony slopes. Plain, foothills, montane zone. Nuratau, Nuratau Relic Mountains, Mirzachul, Kyzylkum. Medicinal, ornamental, meliferous.

Genus 244. Althaea L.

878. Althaea armeniaca Ten.

Perennial. River valleys, banks of canals, fields. Plain, foothills. Nuratau, Mirzachul. Medicinal, ornamental, meliferous.

879. Althaea officinalis L.

Perennial. River valleys, fields, fallow lands. Plain, foothills, montane zone. North Turkestan. Medicinal, ornamental, meliferous.

Genus 245. *Abutilon Mill.

880. *Abutilon therophrasti Medik

Annual. River valleys, fallow lands, banks of canals, roadsides, fields, gardens. Plain, foothills, montane zone. Nuratau, Nuratau Relic Mountains, Malguzar, North Turkestan, Mirzachul. Medicinal, oil, fiber, weed.

Genus 246. *Hibiscus L.

881. *Hibiscus trionum L.
Annual. Banks of canals, fallow lands, fields. Plain. Mirzachul. Medicinal, oil, fiber, weed, meliferous.

Family 71. Thymelaceae

Genus 247. Thymelaea Mill.

882. Thymelaea passerina (L.) Coss. & Germ.
Annual. Fine-earth and stony slopes, banks of rivers and canals, fields, fallow lands, wastelands. Montane zone. Nuratau, Malguzar, North Turkestan. Weed, poisonous.

Genus 248. Diarthron Turcz.

883. Diarthron arenaria (Pobed.) Kit Tan [*Dendrostellera arenaria* Pobed.]
Shrub. Sandy deserts. Plain. Kyzylkum. Fodder, industrial, afforestation.
884. Diarthron vesiculosum (Fisch. & C. A. Mey.) C. A. Mey.
Annual. Stony slopes, fallow lands, fields, piedmont plains, sandy and clay deserts. Plain, foothills, montane and alpine zones. Nuratau, Aktau, Nuratau Relic Mountains, Malguzar, North Turkestan, Mirzachul, Kyzylkum. Weed, poisonous.

Family 72. Cistaceae

Genus 249. Helianthemum Adans.

885. Helianthemum songaricum Schrenk ex Fisch. & C. A. Mey.
Semishrub. Stony slopes, rocks, screes, pebbles, banks of rivers. Foothills, montane and alpine zones. North Turkestan. Dye, meliferous.

Family 73. *Resedaceae

Genus 250. *Reseda L.

886. *Reseda lutea L.
Biennial or perennial. Fine-earth, gravelly and stony slopes, fields, fallow lands, dry riverbeds, banks of rivers, roadsides, wastelands. Plain, foothills, montane zone. Nuratau, Aktau, Nuratau Relic Mountains, Malguzar, North Turkestan, Mirzachul. Dye, oil, weed.

887. *Reseda luteola L.
Biennial. Fine-earth slopes, fields, roadsides, fallow lands, wastelands. Plain, foothills, montane zone. Aktau, Malguzar, North Turkestan. Dye, oil, fodder, weed.

Family 74. Capparaceae

Genus 251. Capparis L.

888. Capparis spinosa L.
Perennial. Sandy, clayey, gravelly soils, banks of rivers, lakes and canals, fallow lands, screes, wastelands. Plain, foothills, montane zone. Nuratau, Aktau, Nuratau Relic Mountains, Malguzar, North Turkestan, Mirzachul, Kyzylkum. Medicinal, food, dye, fodder, ornamental, meliferous, weed.

Family 75. Cleomaceae

Genus 252. Cleome L.

889. Cleome fimbriata Vicary
Annual. Gypsiferous grey soils, limestones, pebbles, dry riverbeds, fallow lands. Foothills. Nuratau, Aktau.

Family 76. Brassicaceae [Cruciferae]

Genus 253. Alliaria Scop.

890. Alliaria petiolata (M. Bieb.) Cavara & Grande
Biennial. Fine-earth and stony slopes, ravines, river valleys, shady wet places, gardens. Montane and alpine zones, Nuratau, Malguzar, North Turkestan. Medicinal, food.

Genus 254. Sisymbrium L.

891. *Sisymbrium altissimum L.
Annual or biennial. Saline lands, wastelands, fallow lands, wastelands, stony slopes, river valleys. Plain, foothills, montane zone. Nuratau, Aktau, Nuratau Relic Mountains, Malguzar, North Turkestan, Mirzachul, Kyzylkum. Food, oil, weed.

892. Sisymbrium brassiciforme C. A. Mey.
Biennial. Stony and gravelly slopes, rocks, screes, dry riverbeds, moraines. Foothills, montane and alpine zones. Nuratau, Malguzar, North Turkestan.

893. Sisymbrium loeselii L.
Annual. Fine-earth slopes, pebbles, meadows, river valleys, settlements, gardens, fields, banks of canals. Plain, foothills, montane zone. Nuratau, Aktau, Nuratau Relic Mountains, Malguzar, North Turkestan, Mirzachul, Kyzylkum. Food, fodder, oil, weed, meliferous.

Genus 255. Neotorularia Hedge & J. Leonard

894. Neotorularia torulosa (Desf.) Hedge & J. Leonard [*Torularia torulosa* (Desf.) O. E. Schulz.]
Annual. Roadsides, fields, fallow lands, wastelands, saline lands, sandy deserts. Plain. Mirzachul, Kyzylkum. Fodder, weed.

Genus 256. Schrenkiella D. A. German & Al-Shehbaz

895. Schrenkiella parvula (Schrenk) D.A. German & Al-Shehbaz [*Arabidopsis parvula* (Schrenk) O. E. Schulz.]
Annual. Clayey soils, saline lands. Plain. Mirzachul, North Turkestan.

Genus 257. Olimarabidopsis Al-Shehbaz, O'Kane & Price

896. Olimarabidopsis pumila (Stephan) Al-Shehbaz, O'Kane & R. A. Price [*Arabidopsis pumila* (Steph.) N. Busch]
Annual. Clayey soils, hills, limestones, saline lands, banks of canals, gardens, stony slopes, ravines. Plain, foothills, montane zone. Nuratau, Aktau, Nuratau Relic Mountains, Malguzar, North Turkestan, Mirzachul, Kyzylkum.

897. Olimarabidopsis umbrosa (Botsch. & Vved.) Al-Shehbaz, O'Kane & R. A. Price [*Trichochiton umbrosum* (Kom.) Botsch. & Vved.]
Annual. Shady wet places, stream banks, screes, rocks. Montane and alpine zones. North Turkestan.

Genus 258. Arabidopsis Heynh.

898. Arabidopsis thaliana (L.) Heynh.
Annual or biennial. Pebbles, banks of rivers, shady places, rocks. Plain, foothills, montane zone. Malguzar, North Turkestan. Food, weed.

Genus 259. Descurainia Webb & Berth.

899. Descurainia sophia (L.) Webb. & Prantl
Annual. Wastelands, settlements, banks of canals, ravines, fields, fallow lands, gardens, saline lands, roadsides. Plain, foothills, montane zone. Nuratau, Aktau, Nuratau Relic Mountains, Malguzar, North Turkestan, Mirzachul, Kyzylkum. Weed.

Genus 260. Smelowskia C. A. Mey. ex Ledeb.

900. Smelowskia sisymbrioides (Regel & Herder) Lipsky ex Paulsen [*Sophiopsis sisymbrioides* (Regel & Herder) O. E. Schulz]
Biennial. Stony and gravelly slopes, rocks, ravines. Montane zone, Malguzar, North Turkestan.

Genus 261. Erysimum L.

901. Erysimum cyaneum Popov
Biennial. Stony slopes. Montane zone. Malguzar. Poisonous.

902. Erysimum marschallianum Andrz. ex DC.
Biennial. Gardens, fallow lands, fine-earth slopes. Foothills, montane and alpine zones. North Turkestan. Poisonous.

903. Erysimum violascens Popov [*Erysimum nuratense* Popovex Botsch. & Vved.]
Biennial. Stony slopes. Montane zone. Nuratau. Medicinal, poisonous.

Genus 262. Barbarea Beck.

904. Barbarea brachycarpa subsp. **minor** (K. Koch) Parolly & Eren [*Barbarea minor* C. Koch]
Perennial. Wet stony slopes, meadows, banks of rivers. Alpine zone. North Turkestan.

905. Barbarea vulgaris R. Br.
Biennial. Wet places, banks of rivers and canals, pebbles, meadows, fields, gardens. Plain, foothills, montane zone. Nuratau, Aktau, Nuratau Relic Mountains, Malguzar, North Turkestan, Mirzachul, Kyzylkum. Medicinal, food, fodder, dye, meliferous, weed.

Genus 263. Rorippa Scop.

906. Rorippa palustris (L.) Besser
Annual or biennial. Swamp meadows, banks of rivers, lakes and canals. Plain. Mirzachul. Food, weed.

Genus 264. Nasturtium R. Br.

907. Nasturtium officinale W. T. Aiton [*Nasturtium fontanum* (Lam.) Aschers.]
Perennial. Swamps, banks of rivers and canals. Plain, foothills. Nuratau, Aktau, Nuratau Relic Mountains, Malguzar, North Turkestan, Mirzachul. Medicinal, food, fodder, meliferous.

Genus 265. Turristis L.

908. Turristis glabra L.
Annual. Fine-earth, gravelly and stony slopes, fallow lands, fields, roadsides, settlements. Foothills, montane zone. Nuratau, Aktau, Malguzar, North Turkestan. Fodder, weed.

Genus 266. Scapiarabis M. A. Koch, R. Karl, D. A. German & Al-Shehbaz

909. Scapiarabis saxicola (Edgew) M. A. Koch, R. Karl, D. A. German & Al. Shehbaz [*Arabis kokanica* Regel & Schmalh.]
Perennial. Stony slopes, rocks. Montane and alpine zones. Nuratau, Malguzar, North Turkestan.

Genus 267. Arabis L.

910. Arabis auriculata Lam
Annual or biennial. Stony slopes, rocks. Foothills, montane zone. Nuratau, Malguzar. Fodder.

911. Arabis montbretiana Boiss.
Annual, biennial. Stony slopes, rocks. Foothills, montane zone. Nuratau, Aktau, Malguzar. Fodder.

Genus 268. Isatis L.

912. Isatis brevipes (Bunge) Jafri [*Pachypterygium brevipes* Bunge]
Annual. Stony slopes. Foothills, montane and alpine zones. Nuratau, Aktau, Nuratau Relic Mountains, Malguzar, North Turkestan.

913. Isatis emarginata Kar. & Kir. [*Isatis violascens* Bunge]
Annual. Sandy deserts. Plain. Kyzylkum.

914. Isatis minima Bunge
Annual. Sandy deserts. Plain. Kyzylkum.

915. Isatis multicaulis (Kar. & Kir.) Jafri [*Pachypterygium densiflorum* Bunge]
Annual. Stony slopes. Foothills, montane zone. Nuratau, Aktau, Nuratau Relic Mountains, Malguzar, North Turkestan.

916. *Isatis tinctoria L.
Biennial. River valleys, pebbles, fallow lands, fields, surroundings of settlements (escaped from culture). Plain, foothills. Nuratau, Malguzar, North Turkestan. Fodder, dye, meliferous.

Genus 269. Tauscheria Fisch.

917. Tauscheria lasiocarpa Fisch. & DC.
Annual. Sandy and clay deserts, fine-earth slopes, banks of rivers, lakes and canals. Plain, foothills, montane zone. Nuratau, Aktau, Nuratau Relic Mountains, Malguzar, North Turkestan, Mirzachul, Kyzylkum. Weed.

Genus 270. Goldbachia DC.

918. Goldbachia laevigata (M. Bieb.) DC.
Annual. Sandy deserts, saline lands, fine-earth, gravelly and stony slopes. Plain, foothills. Nuratau, Aktau, Nuratau Relic Mountains, Malguzar, North Turkestan, Mirzachul, Kyzylkum. Fodder, oil.

919. Goldbachia sabulosa (Kar. & Kir.) D. A. German & Al-Shehbaz [*Spirorrhynchus sabulosus* Kar. & Kir.]
Annual. Sandy desrts. Plain. Kyzylkum. Fodder.

920. Goldbachia torulosa DC.
Annual. Wastelands, fallow lands, fine-earth, gravelly and stony slopes. Foothills. Nuratau, Aktau, Nuratau Relic Mountains, Malguzar, North Turkestan, Mirzachul, Kyzylkum. Fodder, oil, weed.

921. Goldbachia verrucosa Kom.
Annual. Fine-earth, gravelly slopes. Foothills, montane zone. Nuratau. Fodder, oil, weed.

Genus 271. Hesperis L.

922. Hesperis sibirica L.   Figure 285
Biennial or perennial. Shady wet places, banks of rivers, subalpine and alpine meadows. Montane and alpine zones. Malguzar, North Turkestan. Fodder, essential oil, ornamental, meliferous.

Figure 285 Hesperis sibirica (photography by Natalya Beshko)

Genus 272. Parrya R. Br.

923. Parrya fruticulosa Regel & Schmalh. [*Neuroloma fruticulosum* (Regel & Schmalh.) Botsch.]
Semishrub. Stony slopes, rocks, screes. Montane and alpine zones. Nuratau, Malguzar.

924. Parrya hispida (Regel) D. A. German & Al. Shehbaz [*Pseudoclausia hispida* (Regel) Popov ex A.V. Vassil.]
Biennial. Stony slopes, rocks, screes, pebbles. Foothills, montane and alpine zones. Nuratau, Nuratau Relic Mountains, Malguzar, North Turkestan.

925. Parrya khorasanica (Rech. f. & Aellen) D. A. German & Al. Shehbaz [*Pseudoclausia turkestanica* (Lipsky) A.V. Vassil.]   Figure 286
Biennial. Fine-earth and stony slopes, shady places, screes, pebbles, banks of rivers, subalpine meadows. Foothills, montane and alpine zones. Nuratau, Aktau, Malguzar, North Turkestan.

Figure 286 Parrya khorasanica (photography by Natalya Beshko)

926. Parrya nuratensis Botsch. & Vved. [*Neuroloma nuratense* (Botsch. & Vved.) Botsch.]   Figure 287
Semishrub. Fine-earth and stony slopes, rocks. Montane zone. Nuratau. Endemic to the Nuratau Mountains.

Figure 287 **Parrya nuratensis** (photography by Natalya Beshko)

927. Parrya olgae (Regel & Schmalh.) D. A. German & Al. Shehbaz [*Pseudoclausia olgae* (Regel & Schmalh.) Botsch.]   Figure 288

Biennial. Rocks, Fine-earth and stony slopes. Foothills, montane zone. Nuratau, Malguzar, North Turkestan. UzbRDB 2.

Figure 288 **Parrya olgae** (photography by Natalya Beshko)

928. Parrya sarawschanica (Regel & Schmalh.) D. A. German & Al-Shehbaz [*Pseudoclausia sarawschanica* (Regel & Schmalh.) Botsch.]
Biennial. Stony slopes. Montane zone. Nuratau. UzbRDB 2.

Genus 273. Leiospora (C. A. Mey.) Dvorak.

929. Leiospora subscapigera Botsch & Pachom.
Perennial. Moraines, screes, gravelly slopes. Alpine zone. North Turkestan.

Genus 274. Strigosella Boiss.

930. *Strigosella africana (L.) Botsch.
Annual. Sandy and clay deserts, saline lands, ravines, fine-earth and stony slopes, screes, wastelands, fields, fallow lands, gardens. Plain, foothills, montane and alpine zones. Nuratau, Aktau, Nuratau Relic Mountains, Malguzar, North Turkestan, Mirzachul, Kyzylkum. Poisonous, weed.

931. Strigosella brevipes (Bunge) Botsch.
Annual. Sandy and clay deserts, fallow lands. Plain. Kyzylkum.

932. Strigosella grandiflora (Bunge) Botsch.
Annual. Sandy soils, fine-earth and stony slopes, fields. Plain, foothills. Nuratau, Kyzylkum.

933. Strigosella intermedia (C. A. Mey.) Botsch.
Annual. Banks of canals, gravelly slopes, fallow lands, saline lands, sandy and clay deserts. Foothills. Nuratau, Mirzachul, Kyzykum.

934. Strigosella scorpioides (Bunge) Botsch.
Annual. Sandy and clay deserts, dry riverbeds, saline lands, pebbles, fine-earth and stony slopes. Plain, foothills, montane zone. Nuratau Relic Mountains, Malguzar, Mirzachul, Kyzylkum.

935. Strigosella stenopetala (Fisch. & C. A. Mey.) Botsch.
Annual. Sandy and clay deserts, saline lands, fields, banks of canals. Plain, foothills. Nuratau Relic Mountains, Mirzachul, Kyzylkum. Weed.

936. Strigosella trichocarpa (Boiss. & Buhse) Botsch.
Annual. Pebbles, Sandy and clay deserts, fine-earth, gravelly and stony slopes, ravines, fields. Plain, foothills, montane zone. Nuratau, Aktau, Nuratau Relic Mountains, Malguzar, North Turkestan, Mirzachul, Kyzylkum. Weed.

937. Strigosella turkestanica (Litv.) Botsch.   Figure 289
Annual. Sandy and clay deserts, fine-earth and stony slopes, rocks, pebbles, fields, fallow lands, dry riverbeds. Plain, foothills, montane zone. Nuratau, Aktau, Nuratau Relic Mountains, Malguzar, Mirzachul, Kyzylkum. Fodder, weed.

Figure 289 **Strigosella turkestanica** (photography by Natalya Beshko)

Genus 275. Cryptospora Kar. & Kir.

938. Cryptospora falcata Kar. & Kir. [*Cryptospora omissa* Botsch.]
Annual. Sandy and clay deserts, roadsides, fine-earth and stony slopes, fields, fallow lands. Plain, foothills, montane zone. Nuratau, Nuratau Relic Mountains, Malguzar, Mirzachul, Kyzylkum.

939. Cryptospora inconspicua (Kom.) O. E. Schulz [*Trichochiton inconspicuum* Kom.]
Annual. Shady wet places, rocks, screes, pebbles, wet fine-earth and stony slopes, subalpine meadows, stream banks. Montane and alpine zones. North Turkestan. Fodder.

Genus 276. Matthiola R. Br.

940. Matthiola chorassanica Bunge ex Boiss. [*Matthiola integrifolia* Kom.]
Perennial. Stony slopes. Montane and alpine zones. Nuratau, Aktau, North Turkestan.

941. Matthiola obovata Bunge
Perennial. Stony and gravelly slopes, rocks. Montane zone. Nuratau, Malguzar.

Genus 277. Tetracme Bunge

942. Tetracme quadricornis (Steph. ex Willd.) Bunge
Annual. Sandy deserts, saline lands, banks of lakes and canals, fallow lands. Plain. Mirzachul, Kyzylkum. Weed.

943. Tetracme recurvata Bunge
Annual. Sandy and clay deserts, saline lands, dry riverbeds, banks of lakes and canals, fallow lands. Plain, foothills. Mirzachul, Kyzylkum.

Genus 278. Leptaleum DC.

944. Leptaleum filifolium (Willd.) DC.
Annual. Saline lands, Sandy and clay deserts, banks of rivers, lakes and canals, stony slopes, rocks, settlements, roadsides, fallow lands. Plain, foothills, montane zone. Nuratau, Aktau, Nuratau Relic Mountains, Malguzar, North Turkestan, Mirzachul, Kyzylkum.

Genus 279. Streptoloma Bunge

945. Streptoloma desertorum Bunge
Annual. Sandy deserts, saline lands. Plain, foothills, montane zone. Kyzylkum.

Genus 280. Diptychocarpus Trautv.

946. Diptychocarpus strictus (Fisch. ex M. Bieb.) Trautv.
Annual. Gravelly, Fine-earth slopes, rocks, dry riverbeds, banks of rivers and canals, sandy deserts, fallow lands, saline lands, fields, gardens. Plain, foothills, montane zone. Nuratau, Aktau, Nuratau Relic Mountains, Mirzachul, Kyzylkum.

Genus 281. Chorispora R. Br.

947. Chorispora bungeana Fisch. & C. A. Mey.
Perennial. Pebbles, screes, stony and gravelly slopes, rocks, moraines, alpine meadows. Alpine zone. North Turkestan.

948. Chorispora macropoda Trautv.
Perennial. Fine-earth and stony slopes, banks of rivers, pebbles, wet places, screes, rocks, moraines, subalpine and alpine meadows. Montane and alpine zones. North Turkestan.

949. Chorispora sabulosa Cambess. [*Chorispora elegans* Cambess.]
Perennial. Alpine meadows, gravelly and stony slopes. Alpine zone. North Turkestan.

950. Chorispora tenella (Pall.) DC.
Annual. Sandy and clay deserts, saline lands, fine-earth, gravelly and stony slopes, banks of rivers, moraines, fields, fallow lands, dry riverbeds, pebbles, roadsides. Plain, foothills, montane and alpine zones. Nuratau, Aktau, Nuratau Relic Mountains, Malguzar, North Turkestan, Mirzachul, Kyzylkum. Fodder, weed.

Genus 282. *Euclidium R. Br.

951. *Euclidium syriacum (L.) R. Br.
Annual. Saline lands, sandy and clay deserts, ravines, banks of rivers, lakes and canals, fine-earth and stony slopes, shady places, roadsides, fallow lands, fields, wastelands. Plain, foothills, montane zone. Nuratau, Aktau, Nuratau Relic Mountains, Malguzar, North Turkestan, Mirzachul, Kyzylkum. Fodder, weed.

Genus 283. Litwinowia Woron.

952. Litwinowia tenuissima (Pall.) Woronow ex Pavlov
Annual. Saline lands, sandy and clay deserts, pebbles, fine-earth, gravelly and stony slopes, rocks, screes, meadows, settlements, roadsides, banks of canals, gardens, wastelands. Plain, foothills, montane and alpine zones. Nuratau, Aktau, Nuratau Relic Mountains, Malguzar, North Turkestan, Mirzachul, Kyzylkum. Weed.

Genus 284. Octoceras Bunge

953. Octoceras lehmannianum Bunge
Annual. Sandy desert, saline lands, wastelands. Plain. Kyzylkum.

Genus 285. Cithareloma Bunge

954. Cithareloma lehmannii Bunge
Annual. Sandy deserts. Plain. Kyzylkum. Fodder.

Genus 286. Lachnoloma Bunge

955. Lachnoloma lehmanii Bunge
Annual. Sandy and clay deserts, saline lands. Plain. Mirzachul, Kyzylkum.

Genus 287. Alyssum L.

956. Alyssum alyssoides (L.) L. [*Alyssum campestre* L.]   Figure 290
Annual. Fine-earth and stony slopes, piedmont plains, sandy and clay deserts, roadsides, fallow lands, fields. Plain, foothills, montane zone. Nuratau, Aktau, Nuratau Relic Mountains, Malguzar, North Turkestan, Mirzachul, Kyzylkum. Fodder, weed.

Figure 290 Alyssum alyssoides (photography by Natalya Beshko)

957. Alyssum dasycarpum Stephan ex Willd.   Figure 291
Annual. Sandy, clayey and skeleton soils. Plain, foothills. Nuratau, North Turkestan, Kyzylkum. Fodder.

Figure 291 Alyssum dasycarpum (photography by Natalya Beshko)

958. Alyssum desertorum Stapf.
Annual. Sandy and clay deserts, saline lands, banks of rivers, fine-earth and stony slopes, fallow lands. Plain, foothills, montane zone. Nuratau, Aktau, Nuratau Relic Mountains, Malguzar, North Turkestan, Mirzachul, Kyzylkum. Fodder.

959. Alyssum linifolium Stephan ex Willd. [*Meniocus linifolius* (Stephan ex Willd.) DC.]
Annual. Pebbles, gravelly and stony slopes, sandy and clay deserts, dry riverbeds. Plain, foothills, montane zone. Nuratau, Aktau, Nuratau Relic Mountains, Malguzar, North Turkestan, Mirzachul, Kyzylkum. Fodder.

960. Alyssum szovitsianum Fisch. & C. A. Mey. [*Alyssum marginatum* Steud.]
Annual. Fine-earth and stony slopes, rocks. Plain, foothills, montane zone. Nuratau, Malguzar, North Turkestan. Fodder.

Genus 288. Clypeola L.

961. Clypeola jonthlaspi L.
Annual. Pebbles, stony slopes, rocks. Plain, foothills, montane zone. Nuratau, Aktau, Nuratau Relic Mountains, Malguzar, North Turkestan.

Genus 289. Asperuginoides Rauschert

962. Asperuginoides axillaris [Boiss. & Hohen.] Rauschert [*Buchingera axillaris* Boiss.]
Annual. Stony and gravelly slopes, shady wet places. Montane zone. Nuratau, Malguzar, North Turkestan.

Genus 290. Draba L.

963. Draba lanceolata Royle
Perennial. Fine-earth and stony slopes, screes, rocks, subalpine meadows, moraines, wet places near melting snow. Montane and alpine zones. North Turkestan.

964. Draba melanopus Kom
Biennial or perennial. Fine-earth and stony slopes, screes, rocks, alpine meadows, wet places near melting snow. Montane and alpine zones. North Turkestan.

965. Draba nemorosa L.
Annual. Fine-earth, gravelly and stony slopes, pebbles, banks of rivers, subalpine meadows, fallow lands, wet places. Montane zone. Nuratau, Malguzar.

966. Draba nuda (Bel.) Al. Shehbaz & M. Koch [*Drabopsis nuda* (Bel.) Stapf]
Annual. Stony and fine-earth soils, pebbles. Plain, foothills, montane zone. Nuratau, Aktau, Malguzar, North Turkestan. Fodder.

967. Draba stenocarpa Hook. f. & Thomson
Annual. Fine-earth, gravelly and stony slopes, subalpine and alpine meadows, stream banks. Montane and alpine zones. Nuratau, Malguzar, North Turkestan.

Genus 291. Erophila DC.

968. Erophila minima C. A. Mey.
Annual. Fine-earth and stony slopes, fallow lands, fields. Foothills. Nuratau, Aktau, Nuratau Relic Mountains, Malguzar, North Turkestan, Mirzachul. Fodder.

969. Erophila verna (L.) DC.
Annual. Fine-earth, gravelly and stony slopes, sandy and clay deserts, fallow lands, wastelands. Plain, foothills, montane zone. Nuratau, Aktau, Nuratau Relic Mountains, Malguzar, North Turkestan, Mirzachul, Kyzylkum. Fodder.

Genus 292. Brassica L.

970. *Brassica rapa L. [*Brassica campestris* L.]
Annual. Fields, fallow lands, wastelands, gardens, banks of rivers and canals. Plain, foothills, montane zone. Nuratau, Aktau, Malguzar, North Turkestan, Mirzachul. Oil, fodder, weed, meliferous.

971. Brassica elongata Ehrh.
Perennial. Fallow lands, fields, roadsides, wastelands. Plain, foothills, montane zone. Nuratau. Oil, weed.

Genus 293. *Sinapis L.

972. *Sinapis alba L.
Annual. Roadsides, fields, gardens, settlements. Plain, foothills. Nuratau, Aktau, Nuratau Relic Mountains, Malguzar, North Turkestan. Food, medicinal, fodder, oil, weed, meliferous.

973. *Sinapis arvensis L.
Annual. Roadsides, wastelands, fields, gardens, settlements. Plain, foothills. Nuratau, Aktau, Nuratau Relic Mountains, Malguzar, North Turkestan, Mirzachul, Kyzylkum. Food, oil, weed, meliferous.

Genus 294. *Eruca Hill

974. *Eruca vesicaria (L.) Cav. [*Eruca sativa* Mill.]
Annual. Fields, fallow lands, roadsides, sandy deserts, saline lands, stony slopes. Plain, foothills, montane zone. Nuratau, Aktau, Nuratau Relic Mountains, Malguzar, North Turkestan, Mirzachul, Kyzylkum. Oil, food, medicinal, weed.

Genus 295. Crambe L.

975. Crambe cordifolia subsp. **kotschyana** (Boiss.) Jafri [*Crambe kotschyana* Boiss.]
Perennial. Fine-earth, gravelly and stony slopes, fallow lands. Plain, foothills, montane zone. Nuratau, Aktau, Nuratau Relic Mountains, Malguzar, North Turkestan. Fodder, food, meliferous.

Genus 296. Conringia Heist. ex Fabr.

976. Conringia clavata Boiss.   Figure 292
Annual. Stony slopes. Foothills, montane zone. Nuratau, Aktau, Nuratau Relic Mountains, Malguzar, North Turkestan. Fodder.

Figure 292 Conringia clavata (photography by Natalya Beshko)

977. Conringia orientalis (L.) Dumort.
Annual. Fields, fallow lands, roadsides, stony sopes, rocks. Plain, foothills, montane zone. North Turkestan.

978. Conringia persica Boiss.
Annual. Stony slopes. Montane zone. Nuratau, Malguzar, North Turkestan. Fodder.

979. Conringia planisiliqua Fisch. & C. A. Mey.
Annual. Fine-earth slopes, river valleys, dry riverbeds, roadsides. Foothills, montane zone. North Turkestan.

Genus 297. Chalcanthus Boiss.

980. Chalcanthus renifolius [Boiss. & Hohen.] Boiss.   Figure 293
Perennial. Fine-earth and stony slopes, rocks. Montane zone. Nuratau, Malguzar, North Turkestan.

Figure 293 Chalcanthus renifolius (photography by Natalya Beshko)

Genus 298. Lepidium L.

981. Lepidium botschantsevianum Al-Shehbaz [*Stroganowia angustifolia* Botsch. & Vved.]   Figure 294
Perennial. Stony slopes, ravines, rocks. Foothills, montane zone. Nuratau.

Figure 294 Lepidium botschantsevianum (photography by Natalya Beshko)

982. Lepidium draba subsp. **chalepense** L. (P. Fourn.) [*Cardaria repens* (Schrenk) Jarm.]
Perennial. Wet places, fallow lands, fields, wastelands, roadsides, river valleys, banks of lakes and canals. Plain, foothills, montane zone. Nuratau, Aktau, Nuratau Relic Mountains, Malguzar, North Turkestan, Mirzachul, Kyzylkum. Meliferous, weed.

983. Lepidium ferganense Korsh.
Perennial. Stony slopes, rocks, banks of rivers. Foothills, montane zone. Nuratau, Nuratau Relic Mountains.

984. Lepidium lacerum C. A. Mey.
Perennial. Fine-earth and stony slopes, rocks, banks of rivers. Foothills, montane zone. Nuratau.

985. Lepidium latifolium L.
Perennial. Saline lands, swamps, meadows, banks of rivers and canals, pebbles, stony slopes, fields, settlements. Plain, foothills, montane zone. Nuratau, Aktau, Nuratau Relic Mountains, Malguzar, North Turkestan, Mirzachul. Medicinal, meliferous, insecticide, weed.

986. Lepidium lipskyi (N. Busch) Al-Shehbaz & Mummenhoff [*Stubendorffia lipskyi* N. Busch]
Perennial. Fine-earth and stony slopes. Montane zone. Nuratau.

987. Lepidium olgae (R. Vinogradova) Al-Shehbaz & Mummenhoff [*Stubendorffia olgae* R. Vinogradova]   Figure 295
Perennial. Stony slopes. Montane zone. Nuratau. Medicinal. Threatened species, UzbRDB 2. Endemic to the Nuratau ridge.

Figure 295 Lepidium olgae (photography by N. Yu. Beshko)

988. Lepidium orientale (Schrenk) Al-Shehbaz & Mummenhoff [*Stubendorffia orientalis* Schrenk]   Figure 296

Perennial. Fine-earth and stony slopes, rocks, river valleys. Montane zone. Nuratau. Meliferous.

Figure 296 Lepidium orientale (photography by Natalya Beshko)

989. Lepidium perfoliatum L.

Annual or biennial. River valleys, saline lands, fields, roadsides, wastelands. Plain, foothills, montane zone. Nuratau, Nuratau Relic Mountains, Mirzachul, Kyzylkum. Medicinal, poisonous, weed.

Genus 299. Megacarpaea DC.

990. Megacarpaea gigantea Regel

Perennial. Fine-earth and stony slopes, rocks. Montane zone. Nuratau, Malguzar, North Turkestan. Fodder, meliferous.

991. Megacarpaea orbiculata B. Fedtsch.

Perennial. Fine-earth and stony slopes, rocks, ravines, banks of rivers, subalpine meadows. Montane zone. Nuratau, Malguzar, North Turkestan. Fodder, meliferous.

Genus 300. Hornungia Rchb.

992. Hornungia procumbens (L.) Hayek [*Hymenolobus procumbens* (L.) Nutt.]

Annual. Saline lands, banks of rivers and canals, gardens, fallow lands. Plain, foothills, montane zone. Nuratau, Malguzar, North Turkestan, Mirzachul.

Genus 301. Iberidella Boiss.

993. Iberidella trinervia (DC.) Boiss.   Figure 297
Perennial. Gravelly and stony slopes, rocks. Montane and alpine zones. Nuratau, Malguzar, North Turkestan.

Figure 297 Iberidella trinervia (photography by Natalya Beshko)

Genus 302. Didymophysa Boiss.

994. Didymophysa fedtschenkoana Regel
Perennial. Screes, pebbles, moraines, stony slopes. Alpine zone. North Turkestan.

Genus 303. Thlaspi L.

995. *Thlaspi arvense L.
Annual. Fine-earth and stony slopes, banks of rivers, wet places, fields, fallow lands, roadsides, dry riverbeds, river valleys. Plain, foothills, montane zone. Nuratau, Nuratau Relic Mountains, Malguzar, North Turkestan, Mirzachul, Kyzylkum. Medicinal, oil, weed.

996. Thlaspi perfoliatum L. [*Microthlaspi perfoliatum* (L.) F. K. Mey.]
Annual. Fine-earth and stony slopes, pebbles, rocks, saline lands, settlements, fields. Plain, foothills,

montane zone. Nuratau, Aktau, Nuratau Relic Mountains, Malguzar, North Turkestan, Mirzachul, Kyzylkum. Fodder.

Genus 304. Neurotropis (DC.) F. K. Mey.

997. Neurotropis platycarpa (Fisch. & C. A. Mey.) F. K. Mey. [*Thlaspi kotschyanum* Boiss. & Hohen.]
Annual. Gravelly and stony slopes, screes. Montane zone. Malguzar, North Turkestan.

Genus 305. Aethionema W. T. Aiton

998. Aethionema carneum (Banks & Soland.) B. Fedtsch. [*Campyloptera carnea* (Banks & Soland.) Botsch. & Vved.]   Figure 298
Annual. Fine-earth and stony slopes, screes, pebbles, river valleys, dry riverbeds. Foothills, montane zone. Nuratau, Nuratau Relic Mountains, Malguzar, North Turkestan.

Figure 298 Aethionema carneum (photography by Natalya Beshko)

Genus 306. Camelina Crantz.

999. Camelina microcarpa Andrz. ex DC. [*Camelina sylvestris* Wallr.]
Annual. Fine-earth and stony slopes, banks of rivers, lakes and canals, fields, fallow lands, roadsides. Plain, foothills, montane zone. Nuratau, Aktau, Nuratau Relic Mountains, Malguzar, Mirzachul, Kyzylkum. Oil, weed.

1000. Camelina rumelica Velen
Annual. Fine-earth and stony slopes, ravines, banks of rivers, wastelands, fallow lands. Plain, foothills, montane zone. Nuratau, Nuratau Relic Mountains, Malguzar, Mirzachul. Oil, weed.

Genus 307. Neslia Desv.

1001. Neslia paniculata subsp. **thracica** (Velen.) Bornm. [*Neslia apiculata* Fisch. & C. A. Mey.]
Annual. Fine-earth and stony slopes, ravines, pebbles, saline lands, fields, fallow lands. Foothills, montane zone. Nuratau, Aktau, Nuratau Relic Mountains, Malguzar, North Turkestan, Mirzachul. Dye, weed.

Genus 308. *Capsella Medik.

1002. *Capsella bursa-pastoris (L.) Medik.
Annual. Wastelands, fields, gardens, roadsides, fine-earth and stony slopes, pebbles, banks of rivers. Plain, foothills, montane and alpine zones. Nuratau, Aktau, Nuratau Relic Mountains, Malguzar, North Turkestan, Mirzachul, Kyzylkum. Medicinal.

Family 77. Santalaceae

Genus 309. Arceuthobium M. Bieb.

1003. Arceuthobium oxycedri (DC.) M. Bieb.
Shrub. Parasite of juniper. Montane zone. Malguzar, North Turkestan. Medicinal.

Family 78. Frankeniaceae

Genus 310. Frankenia L.

1004. Frankenia bucharica Basil.
Semishrub. Saline lands, banks of rivers, lakes and canals, fallow lands. Plain. Mirzachul.

1005. Frankenia bucharica subsp. **vvedenskyi** (Botsch.) Chrtek. [*Frankenia vvedenskyi* Botsch.]
Semishrub. Saline lands, banks of salt lakes. Plain. Mirzachul, Kyzylkum.

1006. Frankenia hirsuta L.
Semishrub. Saline lands, banks of rivers, lakes and canals. Plain. Kyzylkum.

1007. Frankenia pulverulenta L.
Annual. Saline lands, fallow lands, sandy deserts. Plain. Mirzachul, Kyzylkum.

Family 79. Tamaricaceae

Genus 311. Reaumuria L.

1008. Reaumuria alternifolia (Labill.) Britten [*Reaumuria reflexa* Lipsky; *Reaumuria sogdiana* Kom.; *Reaumuria turkestanica* Gorschk.]
Semishrub. Saline and gypsiferous soils, sandy desert. Plain, foothills, montane zone. Nuratau, Nuratau Relic Mountains, Malguzar, North Turkestan, Mirzachul, Kyzylkum.

1009. Reaumuria fruticosa Boiss.
Shrub. Saline lands, sandy deserts. Plain. Kyzylkum.

Genus 312. Tamarix L.

1010. Tamarix androssowii var. **transcaucassica** (Bunge) Qaiser [*Tamarix litvinowii* Gorschk.]
Shrub. Banks of rivers, lakes and canals, saline lands, dry riverbeds. Plain. Mirzachul, Kyzylkum. Meliferous, ornamental, afforestation, tanniferous.

1011. Tamarix arceuthoides Bunge   Figure 299
Shrub. River valleys, pebbles. Foothills, montane zone. Nuratau, Malguzar, North Turkestan. Meliferous, ornamental, afforestation.

Figure 299 **Tamarix arceuthoides** (photography by Natalya Beshko)

1012. Tamarix elongata Ledeb.

Shrub. Saline lands, depressions, sandy deserts, dry riverbeds. Plain. Mirzachul, Kyzylkum. Meliferous, ornamental, afforestation, dye.

1013. Tamarix hispida Willd.   Figure 300

Shrub. Saline lands, depressions, banks of rivers, lakes and canals. Plain. Mirzachul, Kyzylkum. Meliferous, ornamental, afforestation, dye.

Figure 300 Tamarix hispida (photography by Natalya Beshko)

1014. Tamarix laxa Willd.   Figure 301

Shrub. Saline lands, banks of rivers, lakes and canals. Plain. Mirzachul, Kyzylkum. Meliferous, ornamental, afforestation.

Figure 301 Tamarix laxa (photography by Natalya Beshko)

1015. Tamarix ramosissima Ledeb.

Shrub. Banks of rivers, lakes and canals, saline lands, sandy and clay deserts, fallow lands. Plain. Mirzachul, Kyzylkum. Meliferous, ornamental, afforestation, dye.

1016. Tamarix smyrnensis Bunge [*Tamarix hohenackeri* Bunge]

Shrub. Banks of rivers. Plain. Mirzachul, Kyzylkum. Meliferous, ornamental, afforestation.

Genus 313. Myricaria Desv.

1017. Myricaria germanica (L.) Desv. [*Myricaria bracteata* Royle]

Shrub. Banks of rivers, ravines, pebbles. Foothills, montane zone. North Turkestan. Dye, tanniferous, ornamental.

Family 80. Plumbaginaceae

Genus 314. Acantholimon Boiss.

1018. Acantholimon erythraeum Bunge   Figure 302

Semishrub. Stony slopes. Montane and alpine zones. Nuratau, Aktau, Malguzar, North Turkestan.

Figure 302 Acantholimon erythraeum (photography by Natalya Beshko)

1019. Acantholimon nuratavicum Zakirov ex Lincz.　　Figure 303

Semishrub. Stony slopes. Montane zone. Nuratau. UzbRDB 2. Endemic to the Nuratau ridge.

Figure 303 Acantholimon nuratavicum (photography by Natalya Beshko)

1020. Acantholimon subavenaceum Lincz. Figure 304
Semishrub. Stony slopes. Montane zone. Nuratau.

Figure 304 Acantholimon subavenaceum (photography by Natalya Beshko)

1021. Acantholimon tataricum Boiss. ex DC. Figure 305
Semishrub. Stony slopes, rocks. Montane and alpine zones. Nuratau, Aktau, Malguzar, North Turkestan. Endemic to the Nuratau ridge.

Figure 305 Acantholimon tataricum (photography by Natalya Beshko)

1022. Acantolimon zakirovii Beshko
Semishrub. Stony slopes. Montane zone. Nuratau. Endemic to the Nuratau ridge.

Genus 315. Chaetolimon (Bunge) Lincz.

1023. Chaetolimon setiferum (Bunge) Lincz.   Figure 306
Perennial. Fine-earth and stony slopes, piedmont plains. Foothills, montane zone. Nuratau, Aktau, Nuratau Relic Mountains, Malguzar, North Turkestan.

Figure 306 Chaetolimon setiferum (photography by Natalya Beshko)

Genus 316. Limonium Mill.

1024. Limonium otolepis (Schrenk) Kuntze   Figure 307

Perennial. Saline lands, banks of rivers, lakes and canals. Plain. Nuratau Relic Mountains, Mirzachul, Kyzylkum. Tanniferous, dye, meliferous, ornamental.

Figure 307 Limonium otolepis (photography by Natalya Beshko)

1025. Limonium reniforme (Girard) Lincz.
Perennial. Saline lands, banks of rivers, lakes and canals. Plain. Mirzachul. Tanniferous, weed.
1026. Limonium suffruticosum (L.) Kuntze   Figure 308
Semishrub. Saline lands. Plain. Mirzachul. Tanniferous, dye, meliferous, fodder.

Figure 308 **Limonium suffruticosum** (photography by Natalya Beshko)

Genus 317. Eremolimon Lincz.

1027. Eremolimon sogdianum (Popov) Lincz.
Perennial. Saline lands. Plain. Mirzachul. Dye.

Genus 318. Psylliostachys (Jaub. & Spach) Nevski

1028. Psylliostachys leptostachya (Boiss.) Roshkova
Annual. Saline lands, fallow lands, fields. Plain, foothills. North Turkestan, Mirzachul. Dye.
1029. Psylliostachys × myosuroides (Regel) Roshkova
Annual. Saline lands. Plain. Mirzachul. Meliferous, ornamental.
1030. Psylliostachys suworowii (Regel) Roshkova
Annual. Saline lands, river valleys, banks of canals, fallow lands, fields. Plain. Mirzachul. Meliferous, ornamental.

Family 81. Polygonaceae

Genus 319. Oxyria Hill

1031. Oxyria digyna (L.) Hill   Figure 309
Perennial. Banks of rivers, pebbles, wet stony slopes and screes. Alpine zone. North Turkestan. Dye, food.

Figure 309 Oxyria digyna (photography by Natalya Beshko)

Genus 320. Rumex L.

1032. Rumex acetosa L.
Perennial. Banks of rivers, wet places, gardens, meadows. Plain, foothills, montane zone. Nuratau, Mirzachul. Medicinal, food, fodder, dye.

1033. Rumex chalepensis Mill. [*Rumex drobovii* Korovin]
Perennial. Meadows, wet places, roadsides, banks of rivers and canals, fine-earth and stony slopes, gardens, fields, wastelands. Plain, foothills, montane zone. Nuratau, Malguzar, North Turkestan. Tanniferous, weed.

1034. *Rumex conglomeratus Murray
Perennial. Banks of rivers and canals, wet places, swamps, roadsides, wastelands. Plain, foothills, montane zone. Nuratau, Aktau, Nuratau Relic Mountains, Malguzar, North Turkestan. Medicinal, tanniferous, weed.

1035. *Rumex crispus L.
Perennial. Meadows, fine-earth and stony slopes, banks of rivers, lakes and canals. Plain, foothills, montane zone. Nuratau, Aktau, Nuratau Relic Mountains, Malguzar, North Turkestan. Medicinal, tanniferous, weed, food.

1036. Rumex dentatus subsp. **halacsyi** (Rech.) Rech. f. [*Rumex halacsyi* Rech.]
Annual or biennial. River valleys, wet places, saline lands, banks of canals, fallow lands. Plain, foothills. Malguzar, Mirzachul. Weed.

1037. Rumex marschallianus Rchb.
Annual. Banks of rivers, lakes and canals, saline lands. Plain. Mirzachul.

1038. Rumex pamiricus Rech. f.
Perennial. Fine-earth and stony slopes, screes, wet places, banks of rivers and canals, ravines, fallow lands, wastelands. Plain, foothills, montane and alpine zones. Nuratau, Aktau, Malguzar, North Turkestan. Tanniferous, weed.

1039. Rumex paulsenianus Rech. f.
Perennial. Banks of rivers, ravines, fine-earth and stony slopes, fallow lands. Montane and alpine zones. North Turkestan. Tanniferous, weed.

1040. Rumex rectinervis Rech. f.
Perennial. Banks of rivers and canals, swamps. Plain, foothills, montane zone. Nuratau, Aktau, Malguzar, North Turkestan. Medicinal, tanniferous.

1041. *Rumex syriacus Meissn
Perennial. Banks of rivers and canals, wet places, pebbles, fallow lands, fields, gardens. Plain, foothills. Nuratau Relic Mountains, Mirzachul. Tanniferous, dye, weed.

1042. Rumex tianschanicus Losinsk.
Perennial. Fine-earth and stony slopes, meadows, banks of rivers. Montane and alpine zones. Nuratau, Malguzar, North Turkestan. Medicinal, tanniferous, fodder, weed.

Genus 321. Rheum L.

1043. Rheum cordatum Losinsk.

Perennial. Fine-earth, gravelly and stony slopes. Montane zone. Nuratau. Food, medicinal, tanniferous, dye.

1044. Rheum macrocarpum Losinsk.

Perennial. Fine-earth, gravelly and stony slopes. Montane zone. Nuratau, Malguzar, North Turkestan. Food, tanniferous, dye.

1045. Rheum maximowiczii Losinsk.   Figure 310

Perennial. Fine-earth, gravelly and stony slopes. Foothills, montane zone. Nuratau, Aktau, Nuratau Relic Mountains, Malguzar, North Turkestan. Food, medicinal, tanniferous, dye.

Figure 310 Rheum maximowiczii (photography by Natalya Beshko)

1046. Rheum turkestanicum Janisch.　Figure 311
Perennial. Sandy deserts. Plain. Kyzylkum. Food, medicinal, tanniferous, dye, fodder.

Figure 311 **Rheum turkestanicum** (photography by Natalya Beshko)

Genus 322. Atraphaxis L.

1047. Atraphaxis compacta Ledeb.
Shrub. Stony slopes. Foothills. Nuratau Relic Mountains. Tanniferous, ornamental, meliferous.
1048. Atraphaxis karataviensis Lipsch. & Pavlov.
Shrub. Stony slopes, dry riverbeds. Montane zone. Nuratau. Tanniferous, ornamental, meliferous.
1049. Atraphaxis pyrifolia Bunge　Figure 312
Shrub. Fine-earth and stony slopes, ravines. Foothills, montane zone. Nuratau, Aktau, Nuratau Relic Mountains, Malguzar, North Turkestan. Tanniferous, dye, ornamental, meliferous.

Figure 312 **Atraphaxis pyrifolia** (photography by Natalya Beshko)

1050. Atraphaxis seravschanica Pavlov

Shrub. Fine-earth and stony slopes, rocks, ravines. Montane zone. Nuratau, Aktau, Nuratau Relic Mountains, Malguzar, North Turkestan. Tanniferous, ornamental, meliferous.

1051. Atraphaxis spinosa L.

Shrub. Fine-earth, gravelly and stony slopes. Foothills, montane zone. Nuratau. Tanniferous, dye, ornamental, fodder, meliferous.

1052. Atraphaxis virgata (Regel) Krasn.   Figure 313

Shrub. Stony slopes, pebbles, dry riverbeds. Foothills, montane zone. Nuratau, Malguzar, North Turkestan. Tanniferous, ornamental, meliferous.

Figure 313 Atraphaxis virgata (photography by Natalya Beshko)

Genus 323. Calligonum L.

1053. Calligonum aphyllum (Pall.) Gürke

Shrub. Sandy deserts. Plain. Kyzylkum. Fodder, tanniferous, dye, afforestation, meliferous.

1054. Calligonum caput-medusae Schrenk   Figure 314

Shrub. Sandy deserts. Plain. Kyzylkum. Fodder, tanniferous, dye, afforestation, meliferous.

Figure 314 **Calligonum caput-medusae** (photography by Natalya Beshko)

1055. Calligonum leucocladum (Schrenk) Bunge   Figure 315
Shrub. Sandy deserts. Plain. Kyzylkum. Fodder, tanniferous, dye, afforestation, meliferous.

Figure 315 Calligonum leucocladum (photography by Natalya Beshko)

1056. Calligonum macrocarpum I. G. Borshch.   Figure 316
Shrub. Sandy deserts. Plain. Kyzylkum. Fodder, afforestation, meliferous.

Figure 316 Calligonum macrocarpum (photography by Natalya Beshko)

1057. Calligonum microcarpum I. G. Borshch.   Figure 317
Shrub. Sandy deserts. Plain. Kyzylkum. Fodder, afforestation, meliferous.

Figure 317 Calligonum microcarpum (photography by Natalya Beshko)

Genus 324. Polygonum L.

1058. Polygonum acetosum M. Bieb.
Annual. Sandy deserts. Plain. Kyzylkum.

1059. Polygonum argyrocoleon Steud. ex Kunze
Annual. Sandy soils, saline lands, wet places, roadsides, fields. Plain, foothills. Nuratau, Aktau, Mirzachul, Kyzylkum. Medicinal, tanniferous, weed.

1060. Polygonum aviculare L.
Annual. Wastelands, fields, fallow lands, roadsides, banks of canals. Plain, foothills, montane and alpine zones. Nuratau, Aktau, Nuratau Relic Mountains, Malguzar, North Turkestan, Mirzachul, Kyzylkum. Medicinal, dye, food, fodder, weed.

1061. Polygonum biaristatum Aitch. & Hemsl
Semishrub. Stony slopes, screes. Montane and alpine zones. Nuratau, Malguzar, North Turkestan. Fodder.

1062. Polygonum coriarium Grig. [*Aconogonon coriarum* (Grig.) Sojăk]   Figure 318
Perennial. Fine-earth and stony slopes, subalpine meadows. Montane and alpine zones. North Turkestan. Medicinal, tanniferous, dye, food, meliferous.

Figure 318 Polygonum coriarium (photography by Natalya Beshko)

1063. Polygonum fibrilliferum Kom.
Perennial. Gravelly and stony slopes. Montane and alpine zones. Malguzar, North Turkestan. Fodder,

weed.

1064. Polygonum hissaricum Popov [*Aconogonon hissaricum* (Popov) Sojăk]
Perennial. Fine-earth and stony slopes. Alpine zone. North Turkestan. Medicinal, tanniferous, food, meliferous.

1065. Polygonum inflexum Kom.
Annual. Banks of rivers and canals, fine-earth slopes, saline lands, fallow lands, roadsides, fields. Plain, foothills. Mirzachul.

1066. Polygonum molliiforme Boiss.
Annual. Stony slopes, pebbles, banks of rivers. Montane and alpine zones. Malguzar, North Turkestan.

1067. Polygonum paronychioides C. A. Mey.
Semishrub. Stony slopes, screes, rocks, pebbles. Foothills, montane and alpine zones. Nuratau, Aktau, Nuratau Relic Mountains, Malguzar, North Turkestan. Tanniferous.

1068. Polygonum patulum M. Bieb.
Annual. Stony slopes, screes, saline lands, roadsides, fields, fallow lands. Plain, foothills, montane zone. Nuratau, Malguzar, North Turkestan. Fodder, weed.

1069. Polygonum polychnemoides Jaub. & Spach.
Annual. Stony slopes, screes, pebbles, roadsides, fields. Plain, foothills, montane and alpine zones. Nuratau, Malguzar, North Turkestan.

1070. Polygonum pulvinatum Kom.
Semishrub. Stony slopes. Montane zone. Nuratau.

1071. Polygonum rottboellioides Jaub. & Spach.
Annual. Fine-earth and stony slopes, pebbles. Plain, foothills, montane zone. Nuratau, Nuratau Relic Mountains.

1072. Polygonum subaphyllum Sumnev.
Annual. Fine-earth and stony slopes, pebbles, fallow lands, banks of canals, fields. Foothills, montane zone. Nuratau.

1073. Polygonum vvedenskyi Sumnev.   Figure 319
Semishrub. Stony slopes, screes. Montane and alpine zones. Nuratau, Malguzar, North Turkestan.

Figure 319 Polygonum vvedenskyi (photography by Natalya Beshko)

Genus 325. Persicaria Hill

1074. Persicaria amphibia (L.) Delarbre [*Polygonum amphibium* L]
Perennial. Banks of rivers, lakes, ponds and canals. Plain, foothills, montane and alpine zones. Nuratau, Nuratau Relic Mountains, Malguzar, North Turkestan, Mirzachul, Kyzylkum. Medicinal, tanniferous.

1075. Persicaria bistorta (L.) Samp. [*Bistortia elliptica* (Willd. ex Spreng.) Kom.; *Polygonum nitens* (Fish. & C. A. Mey.) V. Petr. ex Kom.]   Figure 320
Perennial. Fine-earth and stony slopes, screes, banks of rivers, swamps, subalpine and alpine meadows. Montane and alpine zones. Malguzar, North Turkestan. Medicinal, tanniferous, dye, meliferous.

Figure 320 Persicaria bistorta (photography by Natalya Beshko)

1076. Persicaria hydropiper (L.) Delarbre
Annual. Banks of rivers, ponds and canals, swamps. Foothills, montane zone. Nuratau, Aktau, Nuratau Relic Mountains, Malguzar, North Turkestan, Mirzachul, Kyzylkum. Medicinal, dye.

1077. Persicaria lapathifolia (L.) Delarbre
Annual. Banks of rivers, ponds and canals, wet places, fields. Plain, foothills, montane zone. Nuratau, North Turkestan. Medicinal, tanniferous, food, fodder, meliferous.

1078. Persicaria maculata (Raf.) Gray.
Annual. Banks of rivers and canals, wet places, fields, wasteands. Plain, foothills, montane and alpine zones. Nuratau, Aktau, Nuratau Relic Mountains, Malguzar, North Turkestan, Mirzachul, Kyzylkum. Medicinal, essential oil, dye, weed, meliferous.

1079. Persicaria minor (Huds.) Opiz.
Annual. Banks of rivers and canals. Plain, foothills, montane zone. Nuratau, Malguzar, North Turkestan. Medicinal, weed.

Genus 326. Fallopia Adans.

1080. Fallopia convolvulus (L.) A. Löve
Annual. Woodlands, banks of rivers, fields, gardens, wastelands. Plain, foothills, montane and alpine zones. Nuratau, Aktau, Nuratau Relic Mountains, Malguzar, North Turkestan, Mirzachul, Kyzylkum. Medicinal, tanniferous, food, meliferous, weed.

Family 82. Caryophyllaceae

Genus 327. Stellaria L.

1081. Stellaria alsinoides Boiss. & Buhse [*Tytthostemma alsinoides* (Boiss. & Buhse) Nevski]
Annual. Stony slopes, pebbles. Foothills, montane zone. Nuratau, Malguzar.

1082. Stellaria brachypetala Bunge
Perennial. Wet places, banks of rivers, pebbles, meadows, fields, fine-earth slopes. Montane and alpine zones. Malguzar, North Turkestan. Poisonous.

1083. *Stellaria media (L.) Vill.
Annual. Roadsides, banks of rivers and canals, shady wet places, settlements, gardens, wastelands. Plain, foothills, montane zone. Nuratau. Weed, medicinal, dye, saponin-bearing.

1084. Stellaria neglecta Weihe
Annual. River valleys, gardens, fields, wastelands, roadsides, swamps, shady wet places. Plain, foothills, montane zone. Nuratau, Aktau, Nuratau Relic Mountains, Malguzar, North Turkestan, Mirzachul. Weed.

1085. Stellaria pallida (Dumort.) Crep.
Annual. Shady wet places, gardens, fields, fallow lands. Nuratau.

1086. Stellaria turkestanica Schischk.
Perennial. Screes, fine-earth and stony slopes. Alpine zone. Malguzar, North Turkestan.

Genus 328. Mesostemma Vved.

1087. Mesostemma karatavica (Schischk.) Vved. [*Stellaria karatavica* Schischk.]   Figure 321
Perennial. Stony slopes, rocks. Montane zone. Nuratau, North Turkestan.

Figure 321 Mesostemma karatavica (photography by Natalya Beshko)

Genus 329. Cerastium L.

1088. Cerastium arvense L.
Perennial. Fine-earth slopes, meadows. Montane zone. North Turkestan. Medicinal, saponin-bearing, fodder.

1089. Cerastium cerastoides (L.) Britton.
Perennial. Fine-earth, gravelly and stony slopes, rocks, alpine meadows, places near melting snow. Alpine zone. North Turkestan. Fodder, saponin-bearing.

1090. Cerastium dichotomum L.
Annual. Fine-earth, gravelly and stony slopes. Foothills, montane zone. Nuratau, Aktau, Nuratau Relic Mountains, Malguzar, North Turkestan.

1091. Cerastium dichotomum subsp. **inflatum** Cullen [*Cerastium inflatum* Link. ex Desf.]
Annual. Fine-earth, gravelly and stony slopes. Foothills, montane zone. Nuratau, Aktau, Nuratau Relic Mountains, Malguzar, North Turkestan.

1092. Cerastium falcatum (Gren.) Bunge ex Fenzl [*Cerastium bungeanum* Vved.]
Perennial. Fine-earth, gravelly and stony slopes, rocks, saline lands. Plain, foothills, montane zone. Nuratau, Aktau, Nuratau Relic Mountains, Malguzar, North Turkestan.

1093. Cerastium fontanum subsp. **vulgare** (Hartm.) Greuter & Burdet [*Cerastium holosteoides* Fries]
Annual, biennial, perennial. Montane and alpine zones. North Turkestan. Saponin-bearing.

1094. Cerastium glomeratum Thuill
Annual. River valleys, pebbles, roadsides, gardens, fields, fallow lands. Plain, foothills, montane zone. Nuratau, Aktau, Nuratau Relic Mountains, Malguzar, North Turkestan.

1095. Cerastium lithospermifolium Fisch.
Perennial. Screes. Alpine zone. North Turkestan.

1096. Cerastium perfoliatum L.
Annual. Fine-earth and stony slopes, gardens, fields, fallow lands. Foothills, montane zone. Nuratau, Aktau, Nuratau Relic Mountains, Malguzar, North Turkestan. Saponin-bearing.

1097. Cerastium pusillum Ser.
Perennial. Stony slopes, banks of rivers, swamps, pebbles. Montane and alpine zones. Malguzar, North Turkestan.

1098. Cerastium semidecandrum L. [*Cerastium pentandrum* L.; *Cerastium dentatum* Möeschl.]
Annual. Clayey soils. Plain, foothills. Nuratau, Aktau, Nuratau Relic Mountains, Malguzar, North Turkestan, Mirzachul.

Genus 330. Holosteum L.

1099. Holosteum umbellatum L.
Annual. Piedmont plains, fine-earth, gravelly and stony slopes, pebbles, fallow lands, fields. Plain, foothills, montane zone. Nuratau, Aktau, Nuratau Relic Mountains, Malguzar, North Turkestan, Mirzachul, Kyzylkum. Weed, saponin-bearing.

1100. Holosteum umbellatum subsp. **glutinosum** (M. Bieb.) Nyman [*Holosteum glutinosum* (M. Bieb.) Fisch. & C. A. Mey.; *Holosteum polygamum* C. Koch]
Annual. Sandy and clay deserts, piedmont plains, fine-earth, gravelly and stony slopes, pebbles, fallow lands. Plain, foothills, montane zone. Nuratau, Aktau, Nuratau Relic Mountains, Malguzar, North Turkestan, Mirzachul, Kyzylkum. Weed, saponin-bearing.

Genus 331. Sagina L.

1101. Sagina saginoides (L.) H. Karst.
Perennial. Stony slopes, pebbles, screes, wet places, banks of rivers, swamps. Alpine zone. North Turkestan.

Genus 332. Bufonia L.

1102. Bufonia oliveriana Ser.
Semishrub. Fine-earth and stony slopes. Foothills, montane zone. Nuratau, Aktau.

Genus 333. Lepyrodiclis Fenzl.

1103. Lepyrodiclis holosteoides (C. A. Mey.) Fenzl ex Fisch. & C. A. Mey.
Annual. River valleys, fine-earth and stony slopes, shady wet places, fallow lands, fields, gardens. Plain, foothills, montane and alpine zones, Nuratau, Aktau, Nuratau Relic Mountains, Malguzar, North Turkestan. Weed, saponin-bearing.

1104. Lepyrodiclis stellarioides Fisch. & C. A. Mey.
Annual. River valleys, fine-earth and stony slopes, woodlands, shady wet places, gardens. Foothills, montane and alpine zones. Nuratau, Aktau, Nuratau Relic Mountains, Malguzar, North Turkestan. Saponin-bearing.

Genus 334. Minuartia L.

1105. Minuartia hamata (Hausskn.) Mattf. [*Queria hispanica* L.]
Annual. Pebbles, fine-earth, gravelly and stony slopes. Foothills. Nuratau, Aktau, Nuratau Relic Mountains, Malguzar, North Turkestan.

1106. Minuartia kryloviana Schischk.
Perennial. Stony slopes, rocks. Alpine zone. North Turkestan.

1107. Minuartia meyeri (Boiss.) Bornm.
Annual. Fine-earth, gravelly and stony slopes, pebbles. Foothills, montane zone. Nuratau, Aktau, Nuratau Relic Mountains, Malguzar, North Turkestan.

Genus 335. Eremogone Fenzl

1108. Eremogone griffithii (Boiss.) Ikonn. [*Arenaria griffithii* Boiss.]   Figure 322
Perennial. Stony slopes, rocks. Montane and alpine zones. Nuratau, Malguzar, North Turkestan. Saponin-bearing.

Figure 322 Eremogone griffithii (photography by Natalya Beshko)

Genus 336. Arenaria L.

1109. Arenaria rotundifolia M. Bieb.
Perennial. Stony slopes, banks of rivers, alpine meadows, swamps. Alpine zone. North Turkestan.

1110. Arenaria serpyllifolia L.
Annual. Fine-earth, gravelly and stony slopes, fields, fallow lands. Foothills, montane zone. Nuratau, Aktau, Nuratau Relic Mountains, Malguzar, North Turkestan. Fodder.

1111. Arenaria serpyllifolia subsp. **leptoclados** (Rchb.) Nyman [*Arenaria leptoclados* (Rchb.) Guss.]
Annual. Clay deserts, fine-earth, gravelly and stony slopes. Plain, foothills, montane zone. Nuratau, Aktau, Nuratau Relic Mountains, Malguzar, North Turkestan, Mirzachul, Kyzylkum. Fodder.

Genus 337. Spergularia L.

1112. Spergularia diandra (Guss.) Heldr.
Annual. Saline lands. Plain. Mirzachul, Kyzylkum.

1113. *Spergularia marina (L.) Besser [*Spergularia salina* J. Presl & C. Presl]
Annual, biennial or perennial. Saline lands, banks of rivers, lakes and canals. Plain, foothills. Nuratau, Mirzachul, Kyzylkum.

1114. Spergularia media (L.) C. Presl.
Annual or biennial. Saline lands, wastelands, banks of canas. Plain. Mirzachul, Kyzylkum. Weed.

1115. *Spergularia rubra J. Presl & C. Presl.
Annual or biennial. Saline lands, wastelands, roadsides, fields, banks of rivers. Plain, foothills. Nuratau, Mirzachul, Kyzylkum. Weed.

Genus 338. Herniaria L.

1116. Herniaria glabra L.
Annual. Fine-earth, gravelly and stony slopes, banks of rivers, roadsides. Plain, foothills, montane zone. Nuratau, Aktau, Nuratau Relic Mountains, Malguzar, North Turkestan, Mirzachul. Weed, poisonous.

1117. Herniaria hirsuta L.
Annual. River valleys, dry riverbeds, pebbles, roadsides. Plain, foothills, montane zone. Nuratau, Nuratau Relic Mountains, Malguzar, North Turkestan, Mirzachul. Weed, poisonous.

Genus 339. *Agrostemma L.

1118. *Agrostemma githago L.
Annual. Fallow lands, fields, roadsides. Plain, foothills, montane zone. Nuratau, Aktau, Nuratau Relic Mountains, Malguzar, North Turkestan, Mirzachul. Medicinal, weed, poisonous.

Genus 340. Silene L.

1119. Silene brahuica Boiss.
Perennial. Fine-earth, gravelly and stony slopes, ravines, rocks, pebbles. Foothills, montane zone, Nuratau, Aktau, Nuratau Relic Mountains, Malguzar, North Turkestan. Medicinal, saponin-bearing.

1120. Silene claviformis Litv.
Perennial. Fine-earth, gravelly and stony slopes. Foothills, montane zone. Nuratau, Malguzar, North Turkestan.

1121. Silene conica L.   Figure 323
Annual. River valleys, pebbles, fine-earth and stony slopes, fallow lands, fields. Foothills, montane zone. Nuratau, Aktau, Nuratau Relic Mountains, Malguzar, North Turkestan. Saponin-bearing.

Figure 323 Silene conica (photography by Natalya Beshko)

1122. Silene coniflora Nees ex Otth.   Figure 324

Annual. Fine-earth, gravelly and stony slopes, river valleys, banks of canals, roadsides. Foothills, montane zone. Nuratau, Aktau, Nuratau Relic Mountains, Malguzar, North Turkestan.

Figure 324 Silene coniflora (photography by Natalya Beshko)

1123. Silene conoidea L.   Figure 325

Annual. Fine-earth, gravelly and stony slopes, river valleys, banks of canals, wastelands, roadsides, gardens, fields. Plain, foothills, montane zone. Nuratau, Aktau, Nuratau Relic Mountains, Malguzar, North Turkestan, Kyzylkum. Weed.

Figure 325 Silene conoidea (photography by Natalya Beshko)

1124. Silene graminifolia Otth [*Silene schischkinii* Sobolevsk.]
Perennial. Fine-earth, gravelly and stony slopes. Alpine zone. North Turkestan.

1125. Silene guntensis Schischk.
Perennial. Stony slopes, rocks, screes, pebbles. Montane and alpine zones. Nuratau, Aktau, Malguzar, North Turkestan.

1126. Silene incurvifolia Kar. & Kir.
Perennial. Stony slopes, rocks, pebbles. Foothills, montane and alpine zones. North Turkestan.

1127. Silene kuschakewiczii Regel & Schmalh.
Perennial. Stony slopes, rocks, screes. Montane and alpine zones. Nuratau, Malguzar, North Turkestan.

1128. Silene latifolia Poir. [*Melandrium album* (Mill.) Garcke]
Annual or biennial. Fine-earth slopes, river valleys, ravines. Montane zone. North Turkestan. Weed, saponin-bearing.

1129. Silene longicalycina Kom.   Figure 326
Perennial. Stony slopes, screes. Montane and alpine zones. Nuratau, Malguzar, North Turkestan.

Figure 326 Silene longicalycina (photography by Natalya Beshko)

1130. Silene nana Kar. & Kir.   Figure 327

Annual. Sandy deserts. Plain. Kyzylkum.

Figure 327 **Silene nana** (photography by Natalya Beshko)

1131. Silene nevskii Schischk.

Perennial. Stony slopes, rocks. Montane and alpine zones. Nuratau, Malguzar, North Turkestan.

1132. Silene obtusidentata B. Fedtsch. & Popov

Perennial. Stony slopes. Montane zone. Nuratau.

1133. Silene paranadena Bondarenko & Vved.   Figure 328

Perennial. Fine-earth, gravelly and stony slopes. Montane zone. Nuratau, Malguzar. UzbRDB 2.

Figure 328 **Silene paranadena** (photography by N. Yu. Beshko)

1134. Silene plurifolia Schischk.
Perennial. Fine-earth, gravelly and stony slopes. Montane zone. Nuratau, Malguzar.

1135. Silene praelonga Ovcz.
Perennial. Stony slopes, rocks. Montane zone. Nuratau.

1136. Silene praemixta Popov
Perennial. Sandy and clay deserts, piedmont plains, foothills, pebbles, river valleys, fine-earth, gravelly and stony slopes. Plain, foothills, montane zone. Nuratau, Nuratau Relic Mountains, Malguzar, North Turkestan. Saponin-bearing.

1137. Silene pugionifolia Popov
Perennial. Fine-earth, gravelly and stony slopes. Foothills, montane zone. Nuratau.

1138. Silene ruinarum Popov [*Melandrium ruinarium* Popov]
Perennial. Fine-earth, gravelly and stony slopes, river valleys, wet places. Foothills, montane zone. Nuratau.

1139. Silene stenantha Ovcz.
Perennial. Fine-earth, gravelly and stony slopes. Foothills, montane zone. North Turkestan.

1140. Silene tachtensis Franch.
Perennial. Stony slopes, rocks, screes. Montane and alpine zones. North Turkestan.

1141. Silene turkestanica Regel [*Melandrium turkestanicum* Vved.]
Perennial. Stony slopes, screes, woodlands. Montane zone. North Turkestan.

1142. Silene uralensis subsp. **apetala** (L.) Bocquet [*Gastrolychnis apetala* (L.) Tolm. & Kozhanch]
Perennial. Stony slopes, screes, swamps, pebbles. Alpine zone. North Turkestan. Saponin-bearing.

1143. Silene vulgaris (Moench) Garcke   Figure 329
Perennial. Fine-earth, gravelly and stony slopes, screes, river valleys, ravines, subalpine meadows. Montane and alpine zones. Malguzar, North Turkestan. Medicinal, saponin-bearing, ornamental.

Figure 329 Silene vulgaris (photography by Natalya Beshko)

Genus 341. Gastrolychnis (Fenzl.) Rchb.

1144. Gastrolychnis longicarpophora (Kom.) Czerep. [*Silene longicarpophora* (Kom.) Bocquet]
Perennial. Stony slopes, rocks, screes, alpine meadows, places near melting snow. Alpine zone. North Turkestan.

Genus 342. Gypsophila L.

1145. Gypsophila cephalotes (Schrenk) F. N. Williams
Perennial. Stony and gravelly slopes, pebbles, river valleys. Montane and alpine zones. Malguzar, North Turkestan. Saponin-bearing.

1146. Gypsophila herniarioides Boiss.
Perennial. Moraines, screes, stony slopes, pebbles. Alpine zone. North Turkestan.

1147. *Gypsophila paniculata L.
Perennial. Roadsides, dry riverbeds, wastelands. Plain, foothills, montane zone. Malguzar. Medicinal, poisonous, ornamental.

Genus 343. Bolbosaponaria Bondar.

1148. Bolbosaponaria sewerzowii Bondar.   Figure 330
Perennial. Stony slopes, rocks. Foothills, montane zone. North Turkestan. Saponin-bearing.

Figure 330 Bolbosaponaria sewerzowii (photography by Natalya Beshko)

Genus 344. Kuhitangia Ovcz.

1149. Kuhitangia knorringiana Bondarenko
Perennial. Stony slopes, rocks. Montane zone. Nuratau, Malguzar, North Turkestan. Saponin-bearing.

Genus 345. Petrorhagia (Ser. ex DC.) Link.

1150. Petrorhagia alpina (Hablitz) P. W. Ball & Heywood
Annual or biennial. Stony slopes, screes, river valleys, ravines, rocks, subalpine and alpine meadows. Montane and alpine zones. North Turkestan. Saponin-bearing.

Genus 346. Acanthophyllum C. A. Mey.

1151. Acanthophyllum aculeatum Schischk.
Perennial. Sandy deserts, piedmont plains, fine-earth and stony slopes, screes. Plain, foothills, montane zone. Nuratau, Aktau, Nuratau Relic Mountains, Malguzar. Medicinal, meliferous, saponin-bearing.

1152. Acanthophyllum elatius Bunge
Perennial. Sandy deserts. Plain. Kyzylkum. Saponin-bearing.

1153. Acanthophyllum pungens (Bunge) Boiss.   Figure 331
Perennial. Sandy deserts, piedmont plains, fine-earth, gravelly and stony slopes, screes. Plain, foothills, montane zone. Nuratau, Aktau, Nuratau Relic Mountains, Malguzar, North Turkestan. Medicinal, meliferous, saponin-bearing.

Figure 331 Acanthophyllum pungens (photography by Natalya Beshko)

Genus 347. Allochrusa Bunge ex Boiss.

1154. Allochrusa gypsophiloides (Regel) Schischk. [*Acanthophyllum gypsophiloides* Regel]
Perennial. Fine-earth, gravelly and stony slopes, dry riverbeds, fallow lands. Foothills, montane zone. Nuratau, Malguzar, North Turkestan. Medicinal, food, saponin-bearing. UzbRDB 3.

Genus 348. *Vaccaria Medik.

1155. *Vaccaria hispanica (Mill.) Rauschert
Annual. Fields, fallow lands, banks of canals, meadows, wastelands. Plain, foothills, montane zone.

Nuratau, Aktau, Nuratau Relic Mountains, Malguzar, North Turkestan, Mirzachul. Weed, ornamental, medicinal.

Genus 349. Dianthus L.

1156. Dianthus baldzhuanicus Lincz.
Perennial. Fine-earth, gravelly and stony slopes. Montane zone. Nuratau, Malguzar, North Turkestan.

1157. Dianthus brevipetalus Vved.
Perennial. Fine-earth, gravelly and stony slopes. Montane zone. North Turkestan.

1158. Dianthus crinitus subsp. **tetralepis** (Nevski) Rech.f. [*Dianthus tetralepis* Nevski]
Perennial. Fine-earth, gravelly and stony slopes, river valleys, piedmont plains. Plain, foothills, montane zone. Nuratau, Aktau, Nuratau Relic Mountains, Malguzar, North Turkestan. Ornamental, saponin-bearing.

1159. Dianthus darvazicus Lincz.   Figure 332
Perennial. Fine-earth, gravelly and stony slopes. Montane zone. Nuratau, Malguzar. Saponin-bearing.

Figure 332 **Dianthus darvazicus** (photography by Natalya Beshko)

1160. Dianthus helenae Vved.   Figure 333

Perennial. Fine-earth, gravelly and stony slopes. Montane zone. Nuratau, Aktau. Endemic to the Nuratau Mountains.

Figure 333 Dianthus helenae (photography by Natalya Beshko)

1161. Dianthus subscabridus Lincz.
Perennial. Stony slopes, rocks, dry riverbeds. Montane and alpine zones. Malguzar, North Turkestan.

Genus 350. Pleioneura Rech. f.

1162. Pleioneura griffithiana (Boiss.) Rech. f. [*Saponaria griffithiana* Boiss.]
Perennial. Stony slopes, rocks, screes, pebbles. Montane and alpine zones. Malguzar, North Turkestan.

Genus 351. Velezia L.

1163. Velezia rigida L.
Annual. Fine-earth and stony slopes, screes, pebbles, dry riverbeds. Foothills, montane zone. Nuratau, Aktau, Malguzar, North Turkestan.

Family 83. Amaranthaceae

Genus 352. *Amaranthus L.

1164. *Amaranthus albus L.
Annual. Fields, fallow lands, wastelands, roadsides, banks of canals. Plain, foothills, montane zone. Nuratau, Aktau, Nuratau Relic Mountains, Malguzar, North Turkestan, Mirzachul, Kyzylkum. Weed, fodder.
1165. *Amaranthus blitoides S. Watson
Annual. Fields, fallow lands, wastelands, roadsides. Plain, foothills, montane zone. North Turkestan. Weed, fodder.
1166. *Amaranthus deflexus L.
Annual. Gardens, fields, fallow lands, wastelands. Plain. Mirzachul, Kyzylkum. Weed.
1167. *Amaranthus retroflexus L.
Annual. Settlements, gardens, fields, fallow lands, roadsides, wastelands. Plain, foothills, montane zone. Nuratau, Aktau, Nuratau Relic Mountains, Malguzar, North Turkestan, Mirzachul, Kyzylkum. Weed, fodder, medicinal.

Genus 353. Polycnemum L.

1168. Polycnemum arvense L.
Annual. Stony slopes, pebbles, fallow lands, wastelands. Plain, foothills, montane zone. Nuratau,

Malguzar. Weed, fodder.

1169. Polycnemum perenne Litv.   Figure 334
Perennial. Stony slopes, rocks, screes. Foothills, montane zone. North Turkestan.

Figure 334 Polycnemum perenne (photography by Natalya Beshko)

Genus 354. Chenopodium L.

1170. Chenopodium album L.
Annual. Wastelands, fields, gardens, fallow lands, roadsides. Plain, foothills, montane and alpine

zones. Nuratau, Aktau, Nuratau Relic Mountains, Malguzar, North Turkestan, Mirzachul, Kyzylkum. Weed, fodder, medicinal, food, dye, essential oil, meliferous.

1171. Chenopodium botrys L.

Annual. Dry riverbeds, pebbles, stony slopes, wastelands, fields. Plain, foothills, montane zone. Nuratau, Aktau, Nuratau Relic Mountains, Malguzar, North Turkestan, Mirzachul, Kyzylkum. Medicinal, dye, essential oil.

1172. Chenopodium chenopodioides (L.) Aellen

Annual. Saline lands. Plain, foothills. Kyzylkum. Weed.

1173. Chenopodium foliosum (Moench) Aschers   Figure 335

Annual. Stony slopes, pebbles, screes, rocks, wastelands, roadsides. Plain, foothills, montane and alpine zones. Nuratau, Aktau, Nuratau Relic Mountains, Malguzar, North Turkestan, Mirzachul, Kyzylkum. Food, medicinal, dye.

Figure 335 Chenopodium foliosum (photography by Natalya Beshko)

1174. Chenopodium glaucum L.
Annual. Saline lands, banks of rivers, lakes and canals, fields, wastelands. Plain, foothills, montane and alpine zones. Nuratau, Aktau, Nuratau Relic Mountains, Malguzar, North Turkestan, Mirzachul, Kyzylkum. Weed.

1175. *Chenopodium murale L.
Annual. Wastelands. Foothills, montane zone. Nuratau, Malguzar, North Turkestan. Food, medicinal.

1176. Chenopodium rubrum L.
Annual. Saline lands, wastelands, banks of rivers, lakes and canals, river valleys, fields. Plain, foothills, montane zone. Nuratau, Aktau, Nuratau Relic Mountains, Malguzar, North Turkestan, Kyzylkum. Food, medicinal, weed.

1177. *Chenopodium vulvaria L.
Annual. Gravelly slopes, pebbles, dry riverbeds, wastelands. Plain, foothills, montane and alpine zones. Nuratau, Aktau, Nuratau Relic Mountains, Malguzar, North Turkestan, Kyzylkum. Weed, medicinal, dye.

Genus 355. Spinacia L.

1178. Spinacia turkestanica Iljin
Annual. Fields, fallow lands, wastelands, banks of canals. Plain, foothills, montane zone. Nuratau, Nuratau Relic Mountains, Malguzar, North Turkestan, Mirzachul, Kyzylkum. Food, fodder, weed.

Genus 356. Atriplex L.

1179. Atriplex aucherii Moq.
Annual. Saline lands, roadsides, wastelands, surroundings of wells and sheepyards. Plain, foothills. Nuratau Relic Mountains, Kyzylkum. Fodder.

1180. Atriplex dimorphostegia Kar. & Kir.
Annual. Sandy deserts, saline lands. Plain. Kyzylkum. Fodder.

1181. Atriplex flabellum Bunge
Annual. Gravelly and fine-earth slopes, roadsides, wastelands, surroundings of wells and sheepyards. Plain, foothills, montane zone. Nuratau, Aktau, Nuratau Relic Mountains, Malguzar, North Turkestan, Kyzylkum, Mirzachul. Fodder.

1182. Atriplex micrantha Ledeb.
Annual. Saline lands, banks of canals, wastelands, fallow lands. Plain, foothills. Malguzar, Mirzachul. Fodder, weed.

1183. Atriplex ornata Iljin
Annual. Saline lands, fallow lands, banks of canals, sandy deserts. Plain. Mirzachul, Kyzyzlkum. Fodder, weed.

1184. Atriplex prostrata subsp. **calotheca** (Rafn) M. A. Gust. [*Atriplex hastata* L.]
Annual. Saline lands, banks of canals, roadsides, fields, fallow lands. Plain. Mirzachul. Fodder, weed.

1185. Atriplex tatarica L.
Annual. Saline lands, banks of canals, roadsides, surroundings of wells and sheepyards, fields, wastelands. Plain, foothills, montane and alpine zones. Nuratau, Aktau, Nuratau Relic Mountains, Malguzar, North Turkestan, Mirzachul, Kyzylkum. Fodder, food, medicinal.

Genus 357. Krascheninnikovia Gueldenst.

1186. Krascheninnikovia ceratoides (L.) Gueldenst. [*Ceratoides latens* Reveal & Holmgren; *Eurotia ceratoides* (L.) C. A. Mey.]   Figure 336
Semishrub. Stony, gravelly and fine-earth slopes, sandy soils, pebbles, dry riverbeds. Plain, foothills, montane and alpine zones. Malguzar, North Turkestan, Mirzachul, Kyzylkum. Fodder.

Figure 336 Krascheninnikovia ceratoides (photography by Natalya Beshko)

1187. Krascheninnikovia ewersmanniana (Stschegl. ex Losinsk.) Grubov [*Ceratoides ewersmanniana* (Stschegl. ex Losinsk.) Botsch. & Ikonn.; *Eurotia ewersmanniana* Stschegl. ex Losinsk.]
Semishrub. Stony slopes, screes, sandy soils, saline lands, dry riverbeds. Plain, foothills, montane and alpine zones. Malguzar, North Turkestan, Mirzachul, Kyzylkum. Fodder, oil, dye.

Genus 358. Ceratocarpus L.

1188. Ceratocarpus arenarius L. [*Ceratocarpus utriculosus* Bluket ex Krylov]
Annual. Sandy and clay deserts, saline lands, pebbles, dry riverbeds, fine-earth slopes, roadsides, wastelands, fallow lands. Plain, foothills, montane zone. Nuratau, Aktau, Nuratau Relic Mountains, Malguzar, Mirzachul, Kyzylkum. Weed, fodder.

Genus 359. Camphorosma L.

1189. Camphorosma monspeliaca L.
Semishrub. Sandy deserts, saline lands, banks of lakes. Plain. Kyzylkum. Fodder, essential oil, medicinal.

Genus 360. Bassia All.

1190. Bassia crassifolia (Pall.) Soldano [*Suaeda crassifolia* Pall.]
Annual. Saline lands, banks of lakes and canals. Plain. Mirzachul. Fodder.

1191. Bassia eriophora (Schrad.) Asch. [*Kirilovia eriantha* Bunge; *Londesia eriantha* Fisch. et C. A. Mey.]
Annual. Sandy and clay deserts, surroundings of wells and sheepyards. Plain. Kyzylkum.

1192. Bassia hyssopifolia (Pall.) Kuntze
Annual. Banks of canals, saline lands, wastelands, fields, surroundings of wells and sheepyards. Plain. Mirzachul, Kyzylkum. Weed.

Genus 361. Kochia Roth.

1193. Kochia iranica Bornm.
Annual. Sandy soils, saline lands. Plain. Kyzylkum. Fodder.

1194. Kochia prostrata (L.) Schrad.
Semishrub. Sandy soils, saline lands, stony and gravelly slopes. Plain, foothills, montane and alpine zones, Nuratau, Aktau, Nuratau Relic Mountains, Malguzar, North Turkestan, Kyzylkum. Fodder.

1195. Kochia scoparia (L.) Schrad.
Annual. Fields, gardens, roadsides, wastelands. Plain, foothills, montane zone. Nuratau, Aktau, Nuratau Relic Mountains, Malguzar, Mirzachul, Kyzylkum. Fodder, medicinal, ornamental, weed.

Genus 362. Corispermum L.

1196. Corispermum lehmanninum Bunge
Annual. Sandy deserts. Plain. Kyzylkum. Fodder.

Genus 363. Agriophyllum M. Bieb.

1197. Agriophyllum lateriflorum (Lam.) Moq.   Figure 337
Annual. Sandy deserts. Plain. Kyzylkum. Fodder.

Figure 337 Agriophyllum lateriflorum (photography by Natalya Beshko)

Genus 364. Kalidium Moq.

1198. Kalidium caspicum (L.) Ung.-Sternb.
Semishrub. Saline lands, banks of lakes. Plain. Kyzylkum. Industrial.

Genus365. Halopeplis Bunge

1199. Halopeplis pygmaea (Pall.) Bunge ex Ung.-Sternb.
Annual. Saline lands, banks of lakes and canals. Plain. Mirzachul, Kyzylkum.

Genus 366. Halostachys C. A. Mey.

1200. Halostachys belangeriana (Moq.) Botsch.   Figure 338
Shrub. Saline lands, banks of lakes and canals, fallow lands. Plain. Nuratau Relic Mountains, Mirzachul, Kyzylkum. Industrial.

Figure 338 Halostachys belangeriana (photography by Natalya Beshko)

Genus 367. Halocnemum M. Bieb.

1201. Halocnemum strobilaceum (Pall.) M. Bieb.
Semishrub. Saline lands, banks of lakes and canals. Plain. Mirzachul, Kyzylkum. Alkaloid-bearing, insecticide, industrial.

Genus 368. Salicornia L.

1202. Salicornia europaea L.
Annual. Saline lands, banks of lakes and canals, depressions, fallow lands. Plain. Nuratau Relic

Mountains, Mirzachul, Kyzylkum. Fodder, insecticide, food, medicinal, industrial.

Genus 369. Suaeda Forsk.

1203. Suaeda acuminata (C. A. Mey.) Moq.
Annual. Saline lands, wastelands, fallow lands, fields. Plain. Mirzachul, Kyzylkum. Fodder.

1204. Suaeda altissima (L.) Pall.
Annual. Wastelands, fallow lands, fields, saline lands, banks of canals, roadsides. Plain. Mirzachul, Kyzylkum. Fodder.

1205. Suaeda arcuata Bunge
Annual. Saline lands, wastelands, fallow lands, fields. Plain. Mirzachul, Kyzylkum. Fodder, alkaloid-bearing.

1206. Suaeda dendroides (C. A. Mey.) Moq.
Shrub. Saline lands, banks of lakes and canals. Plain. Mirzachul, Kyzylkum. Fodder, industrial.

1207. Suaeda heterophylla Bunge ex Boiss.
Annual. Saline lands, fallow lands, fields, roadsides, banks of canals, wastelands. Plain. Mirzachul, Kyzylkum. Fodder, weed.

1208. Suaeda linifolia Pall.
Annual. Saline lands, banks of lakes and canals, fallow lands, fields. Plain. Mirzachul, Kyzylkum. Fodder, weed.

1209. Suaeda paradoxa (Bunge) Bunge
Annual. Saline lands, wastelands, fallow lands, fields. Plain. Mirzachul, Kyzylkum. Fodder, weed.

1210. Suaeda physophora Pall.
Semishrub. Saline lands, banks of lakes and canals. Plain. Mirzachul. Industrial, poisonous.

Genus 370. Salsola L.

1211. Salsola arbuscula Pall.
Shrub. Sandy deserts, stony and gravelly slopes. Plain, foothills. Nuratau Relic Mountains, Malguzar, Mirzachul, Kyzylkum. Fodder, dye, tanniferous, afforestation.

1212. Salsola arbusculiformis Drobow
Shrub. Sandy deserts, clayey and gravelly soils, pebbles. Plain, foothills. Nuratau Relic Mountains, Malguzar, Kyzylkum. Fodder.

1213. Salsola collina Pall.
Annual. Sandy and clay deserts, piedmont plain, fallow lands, saline lands, screes, roadsides, dry riverbeds, surroundings of settlements, wells and sheepyards. Plain, foothills. Nuratau Relic Mountains, Kyzylkum. Fodder, medicinal.

1214. Salsola dendroides Pall.
Semishrub. Saline lands, gravelly slopes, river valleys, fields. Plain, foothills. Nuratau Relic

Mountains, Mirzachul, Kyzylkum. Fodder, afforestation.

1215. Salsola foliosa (L.) Schrad. ex Schult.

Annual. Sandy and clay deserts, saline lands. Plain. Mirzachul, Kyzylkum. Fodder.

1216. Salsola iberica Sennen & Pau.

Annual. Sandy, clayey and skeleton soils, saline lands, wastelands, fields. Plain, foothills. Nuratau, Aktau, Nuratau Relic Mountains, Mirzachul, Kyzylkum. Fodder.

1217. Salsola leptoclada Gand.

Annual. Sandy, clayey and gravelly soils. Plain. Nuratau Relic Mountains, Mirzachul, Kyzylkum. Fodder.

1218. Salsola micranthera Botschantz.

Annual. Saline lands, sandy and clay deserts. Plain, foothills. Nuratau Relic Mountains, Kyzylkum. Fodder.

1219. Salsola orientalis S. G. Gmel.

Semishrub. Clay desert, saline lands, piedmont plains, gravelly slopes. Plain, foothills. Nuratau Relic Mountains, Mirzachul, Kyzylkum. Fodder.

1220. Salsola paletzkiana Litv.

Tree or shrub. Sandy deserts. Plain. Kyzylkum. Dye, afforestation, ornamental.

1221. Salsola paulsenii Litv.   Figure 339

Annual. Sandy and clay deserts, saline lands. Plain. Mirzachul, Kyzylkum. Fodder.

Figure 339 Salsola paulsenii (photography by Natalya Beshko)

1222. Salsola richteri (Moq.) Kar. ex Litv.   Figure 340

Tree or shrub. Sandy deserts. Plain. Kyzylkum. Fodder, dye, afforestation, medicinal, alkaloid-bearing.

Figure 340 **Salsola richteri** (photography by Natalya Beshko)

1223. Salsola sclerantha C. A. Mey.

Annual. Sandy deserts, saline lands. Plain. Mirzachul, Kyzylkum. Fodder.

1224. Salsola titovii Botsch.

Semishrub. Stony slopes. Montane zone. Aktau. UzbRDB 1.

1225. Salsola turkestanica Litv.

Annual. Saline lands. Plain. Mirzachul, Kyzylkum. Fodder.

1226. Salsola vvedenskyi Iljin & Popov

Annual. Saline lands, sandy deserts. Plain. Kyzylkum. Fodder.

Genus 371. Climacoptera Botsch.

1227. Climacoptera lanata (Pall.) Botsch.

Annual. Saline lands, sandy and clay deserts, banks of lakes and canals, fields, fallow lands. Plain, foothills. Nuratau Relic Mountains, Mirzachul, Kyzylkum. Fodder.

1228. Climacoptera longistylosa (Iljin) Botsch.

Annual. Saline lands. Plain, foothills. Nuratau Relic Mountains, Mirzachul, Kyzylkum. Fodder.

1229. Climacoptera minkvitziae (Korovin) Botsch.
Annual. Saline lands. Plain. Kyzylkum. Fodder.

1230. Climacoptera obtusifolia (Schrenk) Botsch.
Annual. Saline lands. Plain. Mirzachul, Kyzylkum. Fodder.

1231. Climacoptera olgae (Iljin) Botsch.
Annual. Saline lands. Plain. Mirzachul, Kyzylkum. Fodder.

1232. Climacoptera transoxana (Iljin) Botsch.
Annual. Saline lands, roadsides, banks of lakes and canals, fallow lands, fields. Plain. Mirzachul, Kyzylkum. Fodder.

Genus 372. Halothamnus Jaub. et Spach

1233. Halothamnus glaucus (M. Bieb.) Botsch. [*Aellenia glauca* (M. Bieb.) Aellen]
Semishrub. Sandy deserts, saline lands. Plain. Kyzylkum. Fodder, medicinal.

1234. Halothamnus iliensis (Lipsky) Botsch. [*Aellenia iliensis* (Lipsky) Aellen]
Annual. Saline lands, Sandy and clay deserts. Plain. Kyzylkum. Fodder.

1235. Halothamnus subaphyllus (C.A.Mey.) Botsch. [*Aellenia subaphylla* (C.A. Mey.) Aellen]
Semishrub. Sandy soils, saline lands. Plain. Kyzylkum. Fodder, medicinal, dye.

Genus 373. Girgensohnia Bunge

1236. Girgensohnia oppositiflora (Pall.) Fenzl
Annual. Saline, sandy, clayey and gravelly soils, fields, roadsides, fallow lands, dry riverbeds. Plain, foothills. Nuratau, Aktau, Nuratau Relic Mountains, Mirzachul, Kyzylkum. Fodder.

1237. Girgensohnia diptera Bunge
Annual. Saline lands. Plain. Kyzylkum. Fodder, dye.

Genus 374. Gamanthus Bunge

1238. Gamanthus gamocarpus (Moq.) Bunge
Annual. Saline, sandy, clayey and gravelly soils, fields, roadsides, fallow lands, dry riverbeds. Plain, foothills. Nuratau, Aktau, Nuratau Relic Mountains, Malguzar, Mirzachul, Kyzylkum. Fodder.

Genus 375. Anabasis (Schrenk) Benth.

1239. Anabasis eriopoda (Schrenk) Paulsen   Figure 341
Semishrub. Gravelly and stony slopes, saline lands. Plain, foothills. Nuratau Relic Mountains, Kyzylkum. Fodder, meliferous.

Figure 341 **Anabasis eriopoda** (photography by Natalya Beshko)

1240. Anabasis jaxartica (Bunge) Benth. ex Iljin
Semishrub. Saline lands. Plain. Mirzachul.

1241. Anabasis salsa (Ledeb.) Benth. ex Volkens   Figure 342
Semishrub. Saline lands. Plain. Kyzylkum. Fodder, medicinal.

Figure 342 **Anabasis salsa** (photography by Natalya Beshko)

Genus 376. Haloxylon Bunge

1242. Haloxylon ammodendron (C. A. Mey.) Bunge ex Fenzl [*Haloxylon aphyllum* (Minkw.) Iljin]   Figure 343

Tree. Saline lands, sandy and clay deserts. Plain, foothills. Kyzylkum. Fodder, afforestation.

Figure 343 **Haloxylon ammodendron** (photography by Natalya Beshko)

1243. Haloxylon persicum Bunge   Figure 344
Tree. Sandy deserts. Plain. Kyzylkum. Fodder, afforestation, tanniferous.

Figure 344 Haloxylon persicum (photography by Natalya Beshko)

Genus 377. Nanophyton Botsch.

1244. Nanophyton erinaceum (Pall.) Bunge   Figure 345
Semishrub. Gravelly and stony slopes, piedmont plains. Plain, foothills, montane zone. Nuratau, Aktau, Nuratau Relic Mountains.

Figure 345 Nanophyton erinaceum (photography by Natalya Beshko)

1245. Nanophyton saxatile Botsch.   Figure 346
Semishrub. Gravelly and stony slopes, piedmont plains. Plain, foothills. Nuratau, Malguzar.

Figure 346 Nanophyton saxatile (photography by Natalya Beshko)

Genus 378. Halocharis Moq.

1246. Halocharis hispida (Schrenk) Bunge
Annual. Sandy, clayey and gravelly soils, saline lands, fields, fallow lands, banks of canals, roadsides. Plain, foothills. Nuratau, Aktau, Nuratau Relic Mountains, Mirzachul, Kyzylkum. Fodder, weed.

Genus 379. Halimocnemis C. A. Mey.

1247. Halimocnemis mollissima Bunge
Annual. Saline lands. Plain. Mirzachul, Kyzylkum. Fodder.

1248. Halimocnemis villosa Kar. & Kir.
Annual. Saline lands, sandy and clay deserts. Plain. Mirzachul, Kyzylkum. Fodder.

Genus 380. Halogeton C. A. Mey.

1249. Halogeton glomeratus (M. Bieb.) Ledeb.   Figure 347
Annual. Saline lands, sandy soils, roadsides. Plain. Mirzachul, Kyzylkum.

Figure 347 Halogeton glomeratus (photography by Natalya Beshko)

Family 84. *Portulacaceae

Genus 381. *Portulaca L.

1250. *Portulaca oleracea L.
Annual. Fallow lands, wastelands, banks of rivers and canals, fields. Plain, foothills, montane zone. Nuratau, Aktau, Nuratau Relic Mountains, Malguzar, North Turkestan, Mirzachul. Food, medicinal, weed, alkaloid-bearing.

Family 85. Balsaminaceae

Genus 382. Impatiens L.

1251. Impatiens parviflora DC.   Figure 348

Annual. Shady wet places, banks of rivers, ravines, stony slopes, screes, gardens. Foothills, montane zone. Nuratau, Malguzar, North Turkestan. Medicinal, poisonous, weed.

Figure 348 Impatiens parviflora (photography by Natalya Beshko)

MAGNOLIOPSIDA
[EUDICOTS]   347

Family 86. Primulaceae

Genus 383. Primula L.

1252. Primula algida Adams
Perennial. Alpine meadows, swamps, moraines, stream banks, places near melting snow. Alpine zone. North Turkestan. Ornamental.

1253. Primula fedtschenkoi Regel   Figure 349
Perennial. Fine-earth and stony slopes. Foothills, montane zone. Nuratau, Nuratau Relic Mountains, Malguzar, North Turkestan. Ornamental.

Figure 349 **Primula fedtschenkoi** (photography by Natalya Beshko)

1254. Primula iljinskii Al. Fed. ex Schischk. & Bobrov
Perennial. Wet meadows, swamps, banks of rivers. Montane and alpine zones. North Turkestan. Ornamental.

1255. Primula matthioli (L.) V. A. Richt. [*Cortusa turkestanica* Losinsk.]   Figure 350
Perennial. Stream banks, shady wet places, subalpine and alpine meadows, swamps. Montane and alpine zones. Malguzar, North Turkestan. Ornamental.

Figure 350 Primula matthioli (photography by Natalya Beshko)

1256. Primula olgae Regel

Perennial. Banks of rivers, swamps, subalpine and alpine meadows, moraines. Montane and alpine zones, Nuratau, Malguzar, North Turkestan. Ornamental.

1257. Primula pamirica Al. Fed. ex Schischk. & Bobrov

Perennial. Alpine meadows, banks of rivers. Alpine zone. North Turkestan. Ornamental.

1258. Primula turkeviczii V. V. Byalt [*Primula lactiflora* Turkev.]

Perennial. Rocks, places near melting snow. Montane and alpine zones. North Turkestan. Ornamental.

Genus 384. Androsace L.

1259. Androsace dasyphylla Bunge ex Ledeb.

Perennial. Stony slopes, screes, rocks. Alpine zone. North Turkestan.

1260. Androsace maxima L.

Annual. Pebbles, rocks, fine-earth and stony slopes, meadows, fallow lands, fields, roadsides. Foothills, montane zone. North Turkestan.

1261. Androsace sericea Ovcz.

Perennial. Fine-earth and stony slopes, alpine meadows, moraines, rocks, screes. Alpine zone. North Turkestan.

Genus 385. Samolus L.

1262. Samolus valerandi L.

Perennial. Wet places, banks of rivers and canals, fallow lands. Plain, foothills. Mirzachul.

Genus 386. Lysimachia L.

1263. Lysimachia dubia Soland. ex Aiton

Perennial. Wet places, banks of rivers and canals, fields. Plain. Mirzachul.

1264. Lysimachia maritima (L.) Galasso, Banfi & Soldano [*Glaux maritima* L.]

Perennial. Wet places, saline lands, banks of rivers, swamps, meadows. Plain, foothills, montane and alpine zones. Nuratau, Nuratau Relic Mountains, North Turkestan. Dye.

Genus 387. *Anagallis L.

1265. *Anagallis arvensis L. [*Anagallis foemina* Mill.]   Figure 351

Annual. Banks of rivers and canals, roadsides, gardens, fields, fallow lands, wastelands, pebbles, dry riverbeds, stony slopes. Plain, foothills, montane zone. Nuratau, Aktau Nuratau Relic Mountains, Malguzar, North Turkestan, Mirzachul. Medicinal, poisonous, weed.

Figure 351 Anagallis arvensis (above) and **Anagallis arvensis** subsp. **foemina** (below) (photography by Natalya Beshko)

Family 87. Ericaceae

Genus 388. Pyrola L.

1266. Pyrola rotundifolia L.
Perennial. Fine-earth, gravelly and stony slopes, screes, banks of rivers. Plain, foothills, montane and alpine zones. North Turkestan. Medicinal, ornamental.

Family 88. Rubiaceae

Genus 389. Crucianella L.

1267. Crucianella chlorostachys Fisch. & C. A. Mey.
Annual. Pebbles, stony slopes, fallow lands. Plain, foothills, montane zone. Nuratau, Nuratau Relic Mountains, North Turkestan.

1268. Crucianella exasperata Fisch. & C. A. Mey.
Annual. Fine-earth, gravelly and stony slopes, rocks, screes, pebbles. Montane zone. Nuratau, Aktau, Nuratau Relic Mountains, Malguzar, North Turkestan.

1269. Crucianella filifolia Regel & Schmalh.
Annual. Fine-earth, gravelly and stony slopes, banks of canals, pebbles, fallow lands. Plain, foothills, montane zone. Nuratau, Aktau, Nuratau Relic Mountains, North Turkestan.

Genus 390. Asperula L.

1270. Asperula glabrata Tschern.
Semishrub. Stony slopes, rocks. Montane zone. Nuratau, Malguzar, North Turkestan.

1271. Asperula oppositifolia Regel & Schmalh.
Semishrub. Fine-earth, gravelly and stony slopes, rocks, screes. Montane and alpine zones. North Turkestan.

1272. Asperula setosa Jaub. & Spach.
Annual. Stony slopes, screes, rocks, alpine meadows, wet places near springs. Montane and alpine zones. Nuratau, Malguzar, North Turkestan.

1273. Asperula trichodes J. Gay ex DC.
Annual. Fine-earth, gravelly and stony slopes, pebbles, dry riverbeds, banks of rivers. Foothills, montane zone. North Turkestan.

Genus 391. Galium L.

1274. Galium aparine L.
Annual. Fine-earth, gravelly and stony slopes, screes, rocks, woodlands, river valleys, gardens, fields, fallow lands, wastelands. Plain, foothills, montane and alpine zones. Nuratau, Aktau, Nuratau Relic Mountains, Malguzar, North Turkestan, Mirzachul. Medicinal, dye, fodder, weed.

1275. Galium ceratopodum Boiss.
Annual. Gravelly and stony slopes. Foothills, montane zone. Nuratau, Malguzar.

1276. Galium humifusum M. Bieb. [*Asperula humifusa* (M. Bieb.) Besser]
Perennial. Fine-earth, gravelly and stony slopes, pebbles, saline lands, meadows, swamps, banks of canals, fields, fallow lands, wastelands. Plain, foothills, montane zone. Nuratau, Aktau, Nuratau Relic Mountains, Malguzar, North Turkestan. Medicinal, dye, fodder, weed.

1277. Galium karakulense Pobed.
Perennial. Wet places, swamps, banks of rivers. Montane zone. Nuratau.

1278. Galium pamiroalaicum Pobed.   Figure 352
Perennial. Fine-earth, gravelly and stony slopes, subalpine and alpine meadows, banks of rivers. Foothills, montane and alpine zones. Nuratau, Aktau, Malguzar, North Turkestan. Medicinal, fodder, meliferous.

Figure 352 Galium pamiroalaicum (photography by Natalya Beshko)

1279. Galium pseudorivale Tzvelev [*Asperula aparine* M. Bieb.]
Perennial. Woodlands, river valleys, gardens, fine-earth and stony slopes. Foothills, montane and alpine zones. Nuratau, Malguzar, North Turkestan.

1280. Galium setaceum Lam. [*Galium decaisnei* Boiss.]
Annual. Fine-earth, gravelly and stony slopes, screes, rocks, banks of rivers, fallow lands. Foothills, montane and alpine zones. Nuratau, Malguzar, North Turkestan.

1281. Galium songaricum Schrenk
Annual. Juniper forests, wet shady places. Montane and alpine zones. North Turkestan.

1282. Galium spurium L.
Annual. Fine-earth, gravelly and stony slopes, screes, rocks, river valleys, woodlands, gardens, fields, fallow lands. Plain, foothills, montane and alpine zones, Nuratau, Aktau, Nuratau Relic Mountains, Malguzar, North Turkestan.

1283. Galium spurium subsp. **ibicinum** (Boiss. & Hausskn.) Ehrend. [*Galium ibicinum* Boiss. & Hausskn.]
Annual. Fine-earth slopes, river valleys, wet shady places. Montane and alpine zones. North Turkestan.

1284. Galium tenuissimum M. Bieb.
Annual. Fine-earth, gravelly and stony slopes, pebbles, river valleys, piedmont plains, clay deserts, fallow lands. Plain, foothills, montane zone. Nuratau, Aktau, Nuratau Relic Mountains, Malguzar, North Turkestan, Mirzachul. Fodder.

1285. Galium tianschanicum Popov
Perennial. Gravelly and stony slopes, screes, rocks, moraines, pebbles, dry riverbeds. Montane and alpine zones. Nuratau, Malguzar, NorthTurkestan.

1286. Galium tricornutum Dandy
Annual. Gravelly and stony sloper, screes, rocks, banks of rivers and canals, fields, fallow lands. Plain, foothills, montane zone. Nuratau, Aktau, Nuratau Relic Mountains, Malguzar, North Turkestan.

1287. Galium turkestanicum Pobed.
Perennial. Fine-earth slopes, meadows, river valleys, woodlands. Montane and alpine zones. Nuratau, Malguzar, North Turkestan. Medicinal.

1288. Galium verticellatum Danthoine ex Lam
Perennial. Fine-earth, gravelly and stony slopes, screes, rocks, fields, fallow lands. Foothills, montane zone. Nuratau, Aktau, Nuratau Relic Mountains, Malguzar, North Turkestan.

1289. Galium verum L. [*Galium ruthenicum* Willd.]
Perennial. Fine-earth, gravelly and stony slopes, dry riverbeds, banks of rivers, subalpine and alpine meadows, banks of rivers. Foothills, montane and alpine zones. Nuratau, Malguzar, North Turkestan. Medicinal, poisonous, dye, ornamental, meliferous.

Genus 392. Cruciata Mill.

1290. Cruciata pedemontana (Bellardi) Ehrend.
Annual. Fine-earth slopes, banks of rivers, fallow lands. Foothills, montane zone. Nuratau.

Genus 393. Rubia L.

1291. Rubia regelii Pojark.   Figure 353
Semishrub. Gravelly and stony slopes, screes. Montane zone. Nuratau.

Figure 353 Rubia regelii (photography by Natalya Beshko)

Genus 394. Callipeltis Stev.

1292. Callipeltis cucularis (L.) DC.
Annual. Gravelly and stony sloper, screes, rocks, pebbles. Plain, foothills, montane zone. Nuratau, Aktau, Nuratau Relic Mountains, Malguzar, North Turkestan.

Family 89. Gentianaceae

Genus 395. Centaurium Hill

1293. Centaurium erythraea subsp. **turcicum** (Velen.) Melderis [*Centaurium turcicum* (Velen.) Bornm.]   Figure 354

Biennial. Banks of rivers and canals, swamps, meadows. Montane zone. Nuratau. Medicinal, ornamental.

Figure 354 **Centaurium erythraea** subsp. **turcicum** (photography by Natalya Beshko)

1294. Centaurium pulchellum (Sw.) Druce

Annual. Meadows, saline lands, wet places, banks of rivers and canals, sandy deserts, fallow lands. Plain, foothills, montane zone. Nuratau, Aktau, Nuratau Relic Mountains, Malguzar, North Turkestan, Mirzachul, Kyzylkum. Medicinal, ornamental.

1295. Centaurium spicatum (L.) Fritsch

Annual. Meadows, saline lands, wet places, banks of rivers and canals, fallow lands. Plain. Mirzachul. Medicinal, ornamental.

Genus 396. Gentiana L.

1296. Gentiana leucomelaena Maxim.
Annual or biennial. Meadows, swamps, banks of rivers. Montane and alpine zones. Malguzar, North Turkestan.

1297. Gentiana olivieri Griseb.   Figure 355
Perennial. Fine-earth slopes, piedmont plains, clay deserts, banks of rivers, fallow lands. Plain, foothills, montane and alpine zones. Nuratau, Aktau, Nuratau Relic Mountains, Malguzar, North Turkestan, Mirzachul. Medicinal, ornamental.

Figure 355 Gentiana olivieri (photography by Natalya Beshko)

1298. Gentiana prostrata Haenke

Annual or biennial. Subalpine and alpine meadows, swamps, wet places, juniper forests, banks of rivers. Montane and alpine zones. North Turkestan.

1299. Gentiana riparia Kar. & Kir.

Annual or biennial. Swamps, meadows, banks of rivers. Montane and alpine zones. North Turkestan.

Genus 397. Gentianella Moench

1300. Gentianella sibirica (Kuzn.) Holub [*Gentiana sibirica* (Kuzn.) Grossh.]

Annual or biennial. Alpine meadows, moraines, pebbles, shady wet places, banks of rivers. Montane and alpine zones. North Turkestan.

1301. Gentianella turkestanorum (Gand.) Holub [*Gentiana turkestanorum* Gand.]

Annual or biennial. Meadows, swamps, banks of rivers, woodlands, wet places near springs. Montane and alpine zones. Malguzar, North Turkestan. Medicinal, ornamental.

Genus 398. Gentianopsis Ma

1302. Gentianopsis barbata (Froel.) Ma [*Gentiana barbata* Froel.].

Annual or biennial. Meadows, fine-earth slopes, moraines, swamps, banks of rivers. Montane and alpine zones. North Turkestan. Medicinal, ornamental.

Genus 399. Lomatogonium A. Braun

1303. Lomatogonium carinthiacum (Wulfen) A. Braun

Annual. Wet stony slopes, alpine meadows, swamps, moraines, banks of streams. Alpine zone. North Turkestan. Medicinal, meliferous.

Genus 400. Swertia L.

1304. Swertia lactea Bunge

Perennial. Banks of rivers, wet stony slopes and rocks, woodlands, subalpine meadows, swamps, moraines. Montane and alpine zones. Malguzar, North Turkestan. Ornamental.

Family 90. Apocynaceae

Genus 401. Cynanchum L.

1305. Cynanchum acutum subsp. **sibiricum** (Willd.) Rech. f. [*Cynanchum sibiricum* Willd.]   Figure 356
Perennial. Riparian forests, banks of rivers, lakes and canals, wet stony slopes, pebbles, saline lands, sandy deserts. Plain, foothills, montane zone. Nuratau, Aktau, Nuratau Relic Mountains, Malguzar, North Turkestan, Mirzachul, Kyzylkum. Medicinal, weed.

Figure 356 Cynanchum acutum subsp. **sibiricum** (photography by Natalya Beshko)

Genus 402. Trachomitum Woodson

1306. Trachomitum scabrum (Russanov) Pobed.   Figure 357
Perennial. Riparian forests, banks of rivers, lakes and canals, saline lands. Plain, foothills, montane zone. Nuratau, Nuratau Relic Mountains, Malguzar, North Turkestan, Mirzachul, Kyzylkum. Medicinal, poisonous, industrial, ornamental, meliferous.

Figure 357 **Trachomitum scabrum** (photography by Natalya Beshko)

Family 91. Boraginaceae

Genus 403. Heliotropium L.

1307. Heliotropium arguzioides Kar. & Kir.

Perennial. Sandy deserts. Kyzylkum. Poisonous, fodder, meliferous.

1308. Heliotropium bogdanii Czukav.

Annual. Stony banks of rivers, pebbles, conglomerates. Foothills. North Turkestan.

1309. Heliotropium dasycarpum Ledeb.

Perennial. Sandy, clayey and gravelly soils. Plain, foothills. Nuratau, Malguzar, North Turkestan, Mirzachul, Kyzylkum. Poisonous, fodder, weed.

1310. Heliotropium ellipticum Ledeb.

Annual. Pebbles, dry riverbeds, gravelly soils, fields, fallow lands. Plain, foothills. Nuratau. Poisonous, weed.

1311. Heliotropium lasiocarpum Fisch. & C. A. Mey.

Annual. Sandy, clayey and gravelly soils, wastelands, roadsides, fallow lands. Foothills. Nuratau, Aktau, Nuratau Relic Mountains, Malguzar, North Turkestan, Mirzachul, Kyzylkum. Poisonous, weed.

1312. Heliotropium olgae Bunge
Annual. Sandy, clayey and gravelly soils, pebbles. Plain, foothills. Nuratau, Nuratau Relic Mountains, Malguzar, North Turkestan, Mirzachul, Kyzylkum. Poisonous.

Genus 404. Lithospermum L.

1313. Lithospermum officinale L.
Perennial. Fine-earth slopes, river valleys, banks of canals. Plain, foothills, montane zone. Nuratau, Malguzar, North Turkestan, Mirzachul. Medicinal, dye, weed, fodder.

Genus 405. Buglossoides Moench

1314. Buglossoides arvensis (L.) J. M. Johnst.
Annual. Fine-earth, gravelly and stony slopes, river valleys, fallow lands, fields. Nuratau, Aktau, Nuratau Relic Mountains, Malguzar, North Turkestan, Mirzachul. Dye, weed.

Genus 406. Arnebia Forsk.

1315. Arnebia coerulea Schipcz.   Figure 358
Annual. Fine-earth, gravelly and stony slopes, sandy deserts, wastelands. Plain, foothills, montane zone. Nuratau, Aktau, Nuratau Relic Mountains, Malguzar, North Turkestan, Mirzachul, Kyzylkum. Dye, fodder, ornamental.

Figure 358 **Arnebia coerulea** (photography by Natalya Beshko)

1316. Arnebia decumbens (Vent.) Coss. & Kralik.   Figure 359
Annual. Sandy and clay deserts, gravelly and stony slopes, fields, fallow lands, banks of canals. Plain, foothills. Nuratau, Aktau, Nuratau Relic Mountains, Malguzar, North Turkestan, Mirzachul, Kyzylkum. Dye, fodder, weed.

Figure 359 Arnebia decumbens (photography by Natalya Beshko)

1317. Arnebia euchroma (Royle) I. M. Johnst. [*Macrotomia euchroma* (Royle) Paulsen]
Perennial. Stony slopes, rocks. Alpine zone. North Turkestan. Dye.

1318. Arnebia obovata Bunge
Perennial. Stony slopes. Montane zone. Malguzar, North Turkestan. Dye, fodder.

1319. Arnebia transcaspica Popov   Figure 360
Annual. Sandy, clayey and gravelly soils, stony slopes. Plain, foothills. Nuratau, Nuratau Relic Mountains, Kyzylkum. Dye, fodder.

Figure 360 **Arnebia transcaspica** (photography by Natalya Beshko)

Genus 407. Onosma L.

1320. Onosma dichroanta Boiss.
Perennial. Fine-earth, gravelly and stony slopes. Foothills, montane zone. Nuratau, Nuratau Relic Mountains, Malguzar, North Turkestan. Medicinal, dye, meliferous.

1321. Onosma maracandica Zakirov
Perennial. Rocks, screes, stony slopes. Montane zone. North Turkestan.

Genus 408. Echium L.

1322. Echium biebersteinii Lacaita

Biennial. Fields, wastelands, fallow lands, roadsides, fine-earth and stony slopes, river valleys. Foothills, montane zone. Nuratau, Nuratau Relic Mountains, Malguzar, North Turkestan, Mirzachul, Kyzylkum. Medicinal, dye, weed, meliferous.

1323. *Echium vulgare L.

Biennial. Wastelands, fallow lands, roadsides, fine-earth and stony slopes. Foothills. Malguzar, North Turkestan. Medicinal, dye, weed, meliferous.

Genus 409. *Anchusa L.

1324. *Anchusa arvensis subsp. **orientalis** (L.) Nordh. [*Lycopsis orientalis* L.]

Annual. Fine-earth, gravelly and stony slopes, wastelands, fallow lands, fields, roadsides. Plains, foothills, montane zone. Nuratau, Aktau, Nuratau Relic Mountains, Malguzar, North Turkestan, Mirzachul, Kyzylkum. Weed.

1325. *Anchusa azurea Mill. [*Anchusa italica* Retz.]   Figure 361

Perennial. Ravines, river valleys, fine-earth, gravelly and stony slopes, pebbles, wastelands, fallow lands, fields, roadsides, banks of canals. Foothills, montane zone. Nuratau, Aktau, Nuratau Relic Mountains, Malguzar, North Turkestan, Mirzachul, Kyzylkum. Dye, medicinal, weed, meliferous, ornamental.

Figure 361 Anchuza azurea (photography by Natalya Beshko)

Genus 410. Nonea Medik.

1326. Nonea caspica (Willd.) G. Don.   Figure 362
Annual. Sandy and clay deserts, gravelly and stony slopes, roadsides, fallow lands. Plain, foothills, montane and alpine zones. Nuratau, Aktau, Nuratau Relic Mountains, Malguzar, North Turkestan, Mirzachul, Kyzylkum. Weed.

Figure 362 Nonea caspica (photography by Natalya Beshko)

1327. Nonea melanocarpa Boiss.
Annual. Fields, fallow lands, fields. North Turkestan, Mirzachul. Weed.

1328. Nonea pulla (L.) DC.
Perennial. Fallow lands, wastelands, roadsides, river valleys. North Turkestan (this species can be found in the Zaamin National Park, in basin of the river Uriklisay, because it was reported from the neighbouring area of Tajikistan, from the basin of the river Kusavli). Medicinal, meliferous.

Genus 411. Myosotis L.

1329. Myosotis alpestris F. W. Schmidt
Perennial. Subalpine and alpine meadows, gravelly and stony slopes, moraines. Montane and alpine zones, North Turkestan. Fodder, ornamental.

1330. Myosotis laxa subsp. **caespitosa** (Schultz) Hyl. ex Nordh. [*Myosotis caespitosa* Schultz]
Perennial. River valleys, swamps, wet places near springs. Nuratau, Nuratau Relic Mountains, Malguzar, North Turkestan.

1331. Myosotis refracta Boiss.
Annual. Shady wet places, fine-earth, gravelly and stony slopes, pebbles. Foothills, montane zone. Nuratau, Aktau, Nuratau Relic Mountains, Malguzar, North Turkestan. Weed.

1332. Myosotis stricta Link. ex Roem. & Schult. [*Myosotis micrantha* Pall. & Lehm.]
Annual. River valleys, fine-earth, gravelly and stony slopes, pebbles, rocks, fallow lands. Plain, foothills, montane zone. Nuratau, Aktau, Nuratau Relic Mountains, Malguzar, North Turkestan, Mirzachul. Weed.

Genus 412. Lappula Moench

1333. Lappula brachycentra (Ledeb.) Gürke
Annual or biennial. Gravelly and stony slopes. Foothills, montane and alpine zones. Nuratau, Nuratau Relic Mountains, Malguzar, North Turkestan.

1334. Lappula microcarpa (Ledeb.) Gürke
Annual or biennial. Sandy and clay deserts, piedmont plains, fine-earth, gravelly and stony slopes, fallow lands. Plain, foothills, montane zone. Nuratau, Aktau, Nuratau Relic Mountains, Malguzar, North Turkestan, Mirzachul, Kyzylkum. Weed.

1335. Lappula myosotis Moench
Annual or biennial. Wastelands, roadsides, fallow lands, surroundings of wells and sheepyards, fine-earth, gravelly and stony slopes. Plain, foothills. Nuratau, Nuratau Relic Mountains, Malguzar, North Turkestan, Mirzachul, Kyzylkum. Weed.

1336. Lappula nuratavica Nabiev & Zakirov   Figure 363
Annual. Stony slopes, rocks. Montane zone. Nuratau. Endemic to the Nuratau ridge. UzbRDB 2.

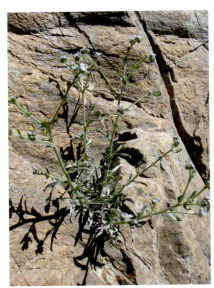

Figure 363 Lappula nuratavica (photography by Natalya Beshko)

1337. Lappula occultata Popov
Annual. Gravelly and stony slopes, screes, rocks. Foothills, montane zone. Malguzar, North Turkestan.

1338. Lappula marginata (M. Bieb.) Gürke [*Lappula patula* (Lehm.) Asch. ex Gurke; *Lappula semiglabra* (Ledeb.) Gürke]
Annual. Gardens, fields, wastelands, roadsides, fallow lands, pebbles, fine-earth and gravelly slopes, river valleys, sandy and clay deserts. Plain, foothills. Nuratau, Aktau, Nuratau Relic Mountains, Malguzar, North Turkestan, Mirzachul, Kyzylkum. Weed.

1339. Lappula sarawschanica (Lipsky) Nabiev
Perennial. Stony slopes, screes, rocks. Montane and alpine zones. Malguzar, North Turkestan.

1340. Lappula semialata Popov
Biennial. Gravelly and stony slopes. Foothills, montane and alpine zones. Malguzar, North Turkestan.

1341. Lappula sessiliflora Gürke
Annual. Gravelly and stony slopes. Foothills, montane zone. Nuratau, Nuratau Relic Mountains.

1342. Lappula sinaica (A. DC.) Asch. & Schweinf.
Annual. Fine-earth, gravelly and stony slopes, rocks, fallow lands. Foothills, montane zone. Nuratau, Aktau, Nuratau Relic Mountains, Malguzar, North Turkestan. Weed.

1343. Lappula spinocarpos (Forssk.) Asch. ex Kuntze
Annual. Sandy and clay deserts, piedmont plains, gravelly and stony slopes. Plain, foothills. North Turkestan, Mirzachul, Kyzylkum.

1344. Lappula squarrosa (Retz.) Dumort. [*Lappula consanguinea* (Fisch. & C. A. Mey.) Gürke]
Annual or biennial. Fields, fallow lands, wastelands. Plain, foothills, montane and alpine zones. Nuratau, Aktau, Nuratau Relic Mountains, Malguzar, North Turkestan, Mirzachul, Kyzylkum. Weed.

Genus 413. Asperugo L.

1345. Asperugo procumbens L.
Annual. Fields, river valleys, wastelands, roadsides, settlements, surroundings of wells and sheepyards, gardens, banks of rivers, lakes and canals, fallow lands, shady wet places. Plain, foothills, montane and alpine zones. Nuratau, Aktau, Nuratau Relic Mountains, Malguzar, North Turkestan, Mirzachul. Medicinal, weed.

Genus 414. Heterocaryum DC.

1346. Heterocaryum laevigatum (Kar. & Kir.) A. DC.
Annual. Gravelly and stony slopes, river valleys. Foothills. Nuratau. Weed.

1347. Heterocaryum macrocarpum Zakirov
Annual. Fine-earth and stony slopes, wastelands, roadsides, fallow lands. Foothills, montane zone. Malguzar, North Turkestan. Weed.

1348. Heterocaryum subsessile Vatke
Annual. Gravelly slopes, wastelands. Foothills, montane zone. Nuratau, Aktau, Malguzar, North Turkestan, Mirzachul. Weed.

1349. Heterocaryum szovitsianum (Fisch. & C. A. Mey.) A. DC. [*Heterocaryum rigidum* A. DC.]
Annual. Fine-earth, gravelly and stony slopes, wastelands, fallow lands, roadsides. Plain, foothills, montane zone. Nuratau, Aktau, Nuratau Relic Mountains, Malguzar, North Turkestan, Mirzachul. Weed.

Genus 415. Rochelia Rchb.

1350. Rochelia bungei Trautv.
Annual. Fine-earth, gravelly and stony slopes, dry riverbeds, sandy deserts, fields, fallow lands. Plain, foothills, montane zone. Nuratau, Nuratau Relic Mountains, Kyzylkum. Weed.

1351. Rochelia cardiosepala Bunge
Annual. Fine-earth, gravelly and stony slopes, fallow lands, fields, dry riverbeds, sandy deserts. Plain, foothills, montane zone. Nuratau, Aktau, Nuratau Relic Mountains, Malguzar, North Turkestan, Mirzachul, Kyzylkum. Weed.

1352. Rochelia disperma subsp. **retorta** (Pall.) Kotejowa [*Rochelia retorta* (Pall.) Lipsky]
Annual. Clay deserts, river valleys, piedmont plains, fine-earth, gravelly and stony slopes, fields, fallow lands, wastelands. Plain, foothills, montane zone. Nuratau, Aktau, Nuratau Relic Mountains, Malguzar, North Turkestan, Mirzachul. Weed.

1353. Rochelia leiocarpa Ledeb.
Annual. Fine-earth, gravelly and stony slopes, pebbles. Plain, foothills, montane zone. Nuratau,

Aktau, Nuratau Relic Mountains, Malguzar, North Turkestan. Weed.

1354. Rochelia peduncularis Boiss.
Annual. Fine-earth, gravelly and stony slopes. Plain, foothills, montane zone. Nuratau, Aktau, Malguzar, North Turkestan. Weed.

Genus 416. Rindera Pall.

1355. Rindera tetraspis Pall.   Figure 364
Perennial. Rocks, fine-earth, gravelly and stony slopes, piedmont plains. Plain, foothills, montane zone. Nuratau, Aktau, Nuratau Relic Mountains, Malguzar, North Turkestan.

Figure 364 Rindera tetraspis (photography by Natalya Beshko)

Genus 417. Hackelia Opiz.

1356. Hackelia uncinata (Benth.) C. E. C. Fisch. [*Paracaryum glochidiatum* (A. DC.) Benth. & Hook. f.]
Annual. Clayey and gravelly soils. Mirzachul.

Genus 418. Mattiastrum Brand

1357. Mattiastrum himalayense (Klotzsch) Brand [*Paracaryum himalayense* (Klotzsch) C. B. Clarke]
Perennial. Stony slopes, screes. Montane and alpine zones. North Turkestan.

Genus 419. Lindelofia Lehm.

1358. Lindelofia olgae Brand
Perennial. Fine-earth, gravelly and stony slopes. Alpine zone. North Turkestan. Dye.

1359. Lindelofia macrostyla (Bunge) Popov   Figure 365
Perennial. Fine-earth, gravelly and stony slopes. Foothills, montane zone. Nuratau, Aktau, Nuratau Relic Mountains, Malguzar, North Turkestan. Dye, weed.

Figure 365 **Lindelofia macrostyla** (photography by Natalya Beshko)

1360. Lindelofia stylosa (Kar. & Kir.) Brand
Perennial. Fine-earth, gravelly and stony slopes. Montane and alpine zones. North Turkestan. Ornamental.

Genus 420. Solenanthus Ledeb.

1361. Solenanthus circinatus Ledeb.
Perennial. Fine-earth, gravelly and stony slopes, shady wet places. Montane zone. Nuratau, Aktau, Nuratau Relic Mountains, Malguzar, North Turkestan. Ornamental, meliferous.

Genus 421. Kuschakewiczia Regel & M. Smirn.

1362. Kuschakewiczia turkestanica Regel & Smirn. [*Solenanthus turkestanicus* (Regel & Smirn.) Kuzn.]
Perennial. Clayey and fine-earth soils. Foothills. Nuratau, Aktau, Nuratau Relic Mountains, Malguzar, North Turkestan, Mirzachul. Meliferous.

Genus 422. *Cynoglossum L.

1363. *Cynoglossum creticum Mill.
Annual or biennial. Banks of rivers and canals, fallow lands, fields, wastelands. Plain, foothills, montane zone. Nuratau, North Turkestan. Medicinal, meliferous, weed.

Genus 423. Trichodesma R. Br.

1364. Trichodesma incanum (Bunge) A. DC.
Perennial. Fine-earth, gravelly and stony slopes, pebbles, river valleys, fields, fallow lands, settlements. Foothills, montane and alpine zones. Nuratau, Aktau, Nuratau Relic Mountains, Malguzar, North Turkestan, Mirzachul. Weed, poisonous, medicinal.

Family 92. Convolvulaceae

Genus 424. Cuscuta L.

1365. Cuscuta approximata Bab. [*Cuscuta cupulata* Engelm.]
Annual. Fields, fallow lands, gardens. Plain, foothills, montane zone. Nuratau, Aktau, Nuratau Relic Mountains, Malguzar, North Turkestan, Mirzachul. Poisonous. Parasite. Host plants are *Medicago sativa*, *Setaria viridis*, *Cynodon dactylon*, *Convolvulus arvensis*, *Lactuca scariola*. Quarantine invasive weed.

1366. Cuscuta brevistyla A. Braun ex A. Rich.
Annual. Fine-earth, gravelly and stony slopes. Foothills, montane and alpine zones. Nuratau, Malguzar, North Turkestanl. Poisonous. Parasite. Host plants are different wild herbs and semishrubs. Quarantine invasive weed.

1367. *Cuscuta campestris Yunck.
Annual. Gardens, fields, vineyards. Plain, foothills. Nuratau, Aktau, Malguzar, North Turkestan, Mirzachul. Parasite. Host plants are different crops, vegetables and ornamentals, as well as weeds

and wild species of *Polygonum, Zygophyllum, etc.* Quarantine invasive weed.

1368. Cuscuta europaea L.

Annual. Woodlands, wet places, banks of rivers, lakes and canals, fine-earth, gravelly and stony slopes. Plain, foothills, montane and alpine zones. Nuratau, Aktau, Nuratau Relic Mountains, Malguzar, North Turkestan, Mirzachul. Medicinal, poisonous. Parasite. Host plants are different wild herbs, semishrubs and shrubs (*Glycyrrhiza glabra, Ferula* sp., *Artemisia dracunculus, etc.*). Quarantine invasive weed.

1369. Cuscuta lehmanniana Bunge

Annual. Riparian forests, banks of rivers, lakes and canals, gardens, fields, vineyards. Plain, foothills, montane zone. Nuratau, Aktau, Nuratau Relic Mountains, Malguzar, North Turkestan, Mirzachul, Kyzylkum. Medicinal. Parasite. Host plants are different cultivated and wild herbs, semishrubs, shrubs and trees (more than 200 species). Quarantine invasive weed.

1370. Cuscuta monogyna Vahl.

Annual. Woodlands, gardens, vineyards, river valleys, banks of lakes and canals. Plain, foothills, montane zone. Nuratau, Nuratau Relic Mountains, Malguzar, North Turkestan, Mirzachul, Kyzylkum. Parasite. Host plants are different cultivated and wild herbs, shrubs and trees (species of genera *Rhamnus, Salix, Ziziphus, Paliurus, Tamarix, Berberis, Rosa, Fraxinus, Pistacia, Vitis, etc.*). Quarantine invasive weed.

1371. Cuscuta pedicellata Ledeb.

Annual. Sandy and clay deserts, piedmont plains, fine-earth, gravelly and stony slopes. Plain, foothills. Nuratau, Aktau, Nuratau Relic Mountains, Malguzar, North Turkestan, Mirzachul, Kyzylkum. Medicinal. Parasite. Host plants are different wild herbs and shrubs (*Poa bulbosa, Peganum harmala, Alhagi* sp., *Trifolium* sp., *Trigonella* sp., *Galium* sp., *Artemisia* sp., *etc.*). Quarantine invasive weed.

1372. Cuscuta pellucida Butkov

Annual. Piedmont plains, fine-earth, gravelly and stony slopes. Plain, foothills, montane zone. Nuratau, Nuratau Relic Mountains, North Turkestan. Parasite. Host plants are different wild herbs and semishrubs (*Alhagi* sp., *Artemisia* sp., *Origanum* sp., *etc.*). Quarantine invasive weed.

Genus 425. Convolvulus L.

1373. *Convolvulus arvensis L.

Perennial. A weed widespread in different habitats. Plain, foothills, montane zone. Nuratau, Aktau, Nuratau Relic Mountains, Malguzar, North Turkestan, Mirzachul, Kyzylkum. Medicinal, alkaloid-bearing, weed, fodder, meliferous.

1374. Convolvulus fruticosus Pall.   Figure 366

Semishrub. Sandy deserts. Plain. Mirzachul, Kyzylkum. Meliferous.

Figure 366 Convolvulus fruticosus (photography by Natalya Beshko)

1375. Convolvulus hamadae (Vved.) V. Petrov
Semishrub. Sandy and clay deserts, gravelly and stony slopes of relic mountains. Plain, foothills. Nuratau Relic Mountains, Mirzachul, Kyzylkum. Fodder, meliferous.

1376. Convolvulus korolkowii Regel & Schmalh.
Perennial. Sandy deserts. Plain. Kyzylkum. Meliferous.

1377. Convolvulus lineatus L.   Figure 367
Perennial. Fine-earth, gravelly and stony slopes, fallow lands, roadsides, banks of rivers and canals, wet places. Foothills, montane and alpine zones. Nuratau, Aktau, Malguzar, North Turkestan. Fodder, weed, meliferous.

Figure 367 Convolvulus lineatus (photography by Natalya Beshko)

1378. Convolvulus pilosellifolius Desr.
Perennial. Fallow lands, clayey and skeleton soils, banks of canals. Plain. Mirzachul. Fodder, meliferous.

1379. Convolvulus pseudocantabricus Schrenk ex Fisch. & C. A. Mey.
Perennial. Piedmont plains, fine-earth, gravelly and stony slopes. Foothills, montane and alpine zones. Nuratau, Aktau, Nuratau Relic Mountains, Malguzar, North Turkestan. Medicinal, poisonous, meliferous.

1380. Convolvulus subhirsutus Regel & Schmalh.
Perennial. Piedmont plains, fine-earth, gravelly and stony slopes. Foothills, montane zone. Nuratau, Aktau, Malguzar, North Turkestan. Medicinal, fodder, weed, meliferous.

Genus 426. Calystegia R. Br.

1381. Calystegia sepium (L.) R. Br.
Perennial. Riparian forests, gardens, banks of rivers and canals, stony wet places, settlements. Plain, foothills, montane zone. Nuratau, Nuratau Relic Mountains, Malguzar, North Turkestan, Mirzachul, Kyzylkum. Medicinal, poisonous, ornamental, meliferous.

Family 93. Solanaceae

Genus 427. Solanum L.

1382. *Solanum americanum Mill. (*Solanum nigrum* L.)
Annual. Gardens, fields, fallow lands, settlements, river valleys, wastelands. Plain. Nuratau, Aktau, Nuratau Relic Mountains, Malguzar, North Turkestan, Mirzachul, Kyzylkum. Medicinal, alkaloid-bearing, oil, poisonous, food, weed, meliferous.

1383. Solanum asiae-mediae Pojark.
Semishrub. Riparian forests, ravines, stream banks. Plain, foothills, montane zone. Nuratau, Malguzar, North Turkestan, Mirzachul.

1384. Solanum kitagawae Schonb.-Tem.
Semishrub. Riparian forests, stony slopes, ravines, banks of rivers, lakes and canals. Plain, foothills, montane zone. Nuratau, North Turkestan, Mirzachul.

1385. Solanum olgae Pojark.
Annual. Fields, fallow lands, wastelands, gardens, banks of rivers, lakes and canals, stony slopes. Plain, foothills, montane zone. Nuratau, Aktau, Nuratau Relic Mountains, Malguzar, North Turkestan, Mirzachul, Kyzylkum.

Genus 428. Lycium L.

1386. Lycium dasystemum Pojark.
Shrub. Sandy and clay deserts, saline lands, banks of rivers, lakes and canals. Plain, foothills. Nuratau Relic Mountains, Mirzachul, Kyzylkum. Medicinal, oil, afforestation, meliferous.

1387. Lycium ruthenicum Murray   Figure 368
Shrub. Sandy and clay deserts, saline lands, banks of rivers, lakes and canals. Nuratau Relic Mountains, Mirzachul, Kyzylkum. Medicinal, oil, afforestation, weed, meliferous.

Figure 368 Lycium ruthenicum (photography by Natalya Beshko)

Genus 429. Hyoscyamus L.

1388. *Hyoscyamus niger L.   Figure 369

Biennial. Fine-earth, gravelly and stony slopes, wastelands, fallow lands, fields, gardens, settlements, roadsides. Nuratau, Aktau, Nuratau Relic Mountains, Malguzar, North Turkestan, Mirzachul, Kyzylkum. Medicinal, poisonous, insecticide, weed, alkaloid-bearing.

Figure 369 **Hyoscyamus niger** (photography by Natalya Beshko)

1389. Hyoscyamus pusillus L.   Figure 370

Annual. Fine-earth, gravelly and stony slopes, screes, rocks, sandy and clay deserts, saline lands, wastelands, fallow lands. Plain, foothills, montane and alpine zones. Nuratau, Aktau, Nuratau Relic Mountains, Malguzar, North Turkestan, Mirzachul, Kyzylkum. Poisonous, alkaloid-bearing, weed.

Figure 370 Hyoscyamus pusillus (photography by Natalya Beshko)

Genus 430. *Datura L.

1390. *Datura stramonium L.
Annual. Settlements, wastelands, fields, fallow lands, roadsides, banks of canals. Nuratau, Nuratau Relic Mountains, Malguzar, North Turkestan, Mirzachul, Kyzylkum. Medicinal, poisonous, insecticide, weed, alkaloid-bearing.

Genus 431. *Nicandra Adans.

1391. *Nicandra physaloides (L.) Gaertn
Annual. Settlements, wastelands, fields, fallow lands, roadsides, banks of canals. Mirzachul, North

Turkestan. Medicinal, oil, fodder, weed.

Family 94. Oleaceae

Genus 432. Fraxinus L.

1392. Fraxinus sogdiana Bunge
Tree. River valleys, fine-earth slopes. Foothills, montane zone. Nuratau, Malguzar, North Turkestan. IUCN NT. Industrial, afforestation, ornamental, medicinal.

1393. *Fraxinus angustifolia subsp. **syriaca** (Boiss.) Yalt. [*Fraxinus syriaca* Boiss.]
Tree. River valleys. Montane zone. Nuratau. Industrial, afforestation, ornamental, medicinal.

Family 95. Plantaginaceae

Genus 433. *Kickxia Dumort.

1394. *Kickxia elatine (L.) Dumort.
Annual. Fine-earth slopes, pebbles, riparian forests, fields, fallow lands. Plain, foothills, montane zone. North Turkestan, Mirzachul. Medicinal.

Genus 434. Veronica L.

1395. *Veronica anagallis-aquatica L.
Perennial. Wet places near spings, banks of rivers and canals. Plain, foothills, montane zone. Nuratau, Aktau, Malguzar, North Turkestan. Medicinal, meliferous.

1396. Veronica anagalloides Guss
Perennial. Banks of rivers, lakes and canals, swamps, meadows, fallow lands, pebbles, saline lands. Plain, foothills, montane zone. Nuratau, Aktau, Malguzar, North Turkestan, Mirzachul, Kyzylkum.

1397. Veronica arguteserrata Regel & Schmalh.
Perennial. Fine-earth and stony slopes, screes, rocks, piedmont plains, fields, fallow lands. Plain, foothills, montane and alpine zones. Nuratau, Aktau, Nuratau Relic Mountains, Malguzar, North Turkestan, Mirzachul.

1398. Veronica arvensis L.
Annual. Fine-earth slopes, wet meadows, river valleys, roadsides, fields, fallow lands. Foothills, montane zone. North Turkestan. Medicinal.

1399. Veronica beccabunga L.
Perennial. Banks of rivers and canals, swamps, wet places near springs. Plain, foothills, montane and alpine zones. Nuratau, Malguzar, North Turkestan. Medicinal.

1400. Veronica biloba Schreb. ex L.
Annual. Sandy and clay deserts, fine-earth, gravelly and stony slopes, pebbles, rocks, screes, moraines. Plain, foothills, montane and alpine zones. Nuratau, Malguzar, North Turkestan, Mirzachul, Kyzylkum.

1401. Veronica campylopoda Boiss.
Annual. Sandy and clay deserts, piedmont plains, fine-earth, gravelly and stony slopes, rocks, pebbles, fields, fallow lands, saline lands, river valleys. Plain, foothills, montane zone. Nuratau, Aktau, Malguzar, North Turkestan, Mirzachul, Kyzylkum.

1402. Veronica capillipes Nevski
Annual. Fine-earth, gravelly and stony slopes, screes, rocks, river valleys. Foothills, montane zone. Nuratau, Aktau, Nuratau Relic Mountains, Malguzar, North Turkestan.

1403. Veronica cardiocarpa (Kar. & Kir.) Walp.
Annual. Sandy and clay deserts, fine-earth, gravelly and stony slopes, rocks, screes. Plain, foothills, montane zone. Nuratau, Malguzar, North Turkestan, Mirzachul, Kyzylkum.

1404. Veronica hederifolia L.
Annual. Settlements, fields, fallow lands, gardens, wastelands, roadsides, river valleys, banks of canals, stony slopes. Plain, foothills, montane and alpine zones. Nuratau, Malguzar, North Turkestan, Mirzachul, Kyzylkum. Weed.

1405. Veronica intercedens Bornm.
Annual. Fine-earth and stony slopes, screes. Foothills, montane and alpine zones. Nuratau, Aktau, Nuratau Relic Mountains, Malguzar, North Turkestan.

1406. Veronica oxycarpa Boiss.
Perennial. Banks of rivers, lakes and canals, wet places near springs, gardens, meadows. Plain, foothills, montane and alpine zones. Nuratau, Aktau, Nuratau Relic Mountains, Malguzar, North Turkestan, Mirzachul.

1407. *Veronica persica Poir.
Annual. Settlements, fields, fallow lands, wastelands, gardens, banks of canals, river valleys. Plain, foothills, montane and alpine zones. Nuratau, Aktau, Nuratau Relic Mountains, Malguzar, North Turkestan, Mirzachul, Kyzylkum. Weed.

1408. *Veronica polita Fr.
Annual. Settlements, banks of rivers and canals, fields, fallow lands, fine-earth slopes, pebbles, river valleys. Plain, foothills, montane zone. Nuratau, Aktau, Nuratau Relic Mountains, Malguzar, North Turkestan, Mirzachul, Kyzylkum.

1409. Veronica verna L.
Annual. Roadsides, pebbles, meadows, gravelly and fine-earth slopes. Foothills, montane and alpine zones. Nuratau, North Turkestan.

Genus 435. Lagotis Gaertn.

1410. Lagotis korolkowii (Regel & Schmalh.) Maxim.   Figure 371
Perennial. Fine-earth and stony slopes, alpine meadows. Montane and alpine zones. North Turkestan.

Figure 371 Lagotis korolkowii (photography by Natalya Beshko)

Genus 436. Plantago L.

1411. Plantago gentianoides Sm.
Perennial. Swamps, stream banks. Alpine zone. North Turkestan.

1412. Plantago lagocephala Bunge
Annual. Sandy and clay deserts, saline lands. Plain. Kyzylkum.

1413. Plantago lanceolata L.
Perennial. Fine-earth slopes, river valleys, banks of lakes and canals, roadsides, wastelands, gardens, wet places. Plain, foothills, montane and alpine zones. Nuratau, Aktau, Nuratau Relic Mountains, Malguzar, North Turkestan, Mirzachul, Kyzylkum. Medicinal, fodder.

1414. Plantago major L.
Biennial or perennial. Wet places, swamps, meadows, banks of rivers, lakes and canals, riparian forests, roadsides, settlements, fields, gardens. Plain, foothills, montane zone. Nuratau, Aktau, Nuratau Relic Mountains, Malguzar, North Turkestan, Mirzachul, Kyzylkum. Medicinal, fodder.

Genus 437. Hippuris L.

1415. Hippuris vulgaris L.
Perennial. Standing and slowly flowing water. Nuratau Relic Mountains, North Turkestan, Mirzachul, Kyzylkum. Medicinal, fodder.

Family 96. Scrophulariaceae

Genus 438. Verbascum L.

1416. Verbascum blattaria L.
Annual or biennial. Fine-earth slopes, piedmont plains, sandy and clay deserts, banks of rivers, lakes and canals, saline lands, riparian forests, fallow lands, fields, gardens, meadows. Plain, foothills. Nuratau, Malguzar, North Turkestan, Mirzachul, Kyzylkum. Medicinal, meliferous, weed, insecticide.

1417. Verbascum erianthum Benth. [*Verbascum bactrianum* Bunge]
Biennial. Fields, fallow lands, banks of rivers and canals, saline lands, stony slopes, pebbles. Plain, foothills, montane and alpine zones. Nuratau, Aktau, Malguzar, North Turkestan, Mirzachul. Medicinal, meliferous, insecticide.

1418. Verbascum songaricum Schrenk   Figure 372
Biennial. Sandy deserts, roadsides, banks of canals, dry riverbeds, river valleys, ravines, fallow lands, fields, fine-earth, gravelly and stony slopes. Plain, foothills, montane and alpine zones. Nuratau, Nuratau Relic Mountains, Malguzar, North Turkestan, Mirzachul, Kyzylkum. Medicinal, meliferous, insecticide.

Figure 372 Verbascum songaricum (photography by Natalya Beshko)

1419. Verbascum turkestanicum Franch.

Annual or biennial. Fine-earth and stony slopes. Montane zone. North Turkestan.

Genus 439. Linaria Mill.

1420. Linaria popovii Kuprian.   Figure 373

Perennial. Fine-earth, gravelly and stony slopes. Foothills, montane zone. Nuratau, Aktau, Malguzar, North Turkestan. Medicinal.

Figure 373 **Linaria popovii** (photography by Natalya Beshko)

1421. Linaria sessilis Kuprian.

Perennial. Stony slopes, screes, pebbles. Montane and alpine zones. Malguzar, North Turkestan.

Genus 440. Scrophularia L.

1422. Scrophularia gontscharovii Gorschk.

Perennial. Stony slopes. Montane and alpine zones. North Turkestan.

1423. Scrophularia griffithii Benth. ex DC.

Perennial. Fine-earth, gravelly and stony slopes. Montane and alpine zones. North Turkestan.

1424. Scrophularia heucheriiflora Schrenk

Perennial. Stony and gravelly slopes, meadows, river valleys. Foothills, montane zone. North Turkestan.

1425. Scrophularia incisa Weinm.

Perennial. Stony slopes, pebbles, screes. Plain, foothills, montane and alpine zones. North Turkestan. Medicinal.

1426. Scrophularia integrifolia Pavlov

Perennial. Gravelly and stony slopes, rocks. Montane zone. North Turkestan.

1427. Scrophularia kiriloviana Schischk.
Perennial. Fine-earth, gravelly and stony slopes, screes, ravines, river valleys. Montane zone. North Turkestan.

1428. Scrophularia leucoclada Bunge
Perennial. Sandy soils, stony and gravelly slopes, dry riverbeds, saline lands. Plain, foothills, montane zone. Nuratau, Nuratau Relic Mountains, Mirzachul, Kyzylkum.

1429. Scrophularia scoparia Pennell
Perennial. Stony and gravelly slopes, pebbles. Foothills, montane and alpine zones. Nuratau, Malguzar, North Turkestan.

1430. Scrophularia striata Boiss.
Perennial. Fine-earth, gravelly and stony slopes, dry riverbeds, pebbles. Foothills, montane zone. North Turkestan.

1431. Scrophularia umbrosa Dumort.
Perennial. Fine-earth, gravelly and stony slopes, stream banks, wet places, meadows. Montane zone. Nuratau, Aktau, Malguzar, North Turkestan. Medicinal, poisonous, meliferous.

1432. Scrophularia vvedenskyi Bondarenko & Filatova
Biennial. Rocks, fine-earth, gravelly and stony slopes, pebbles. Foothills, montane zone. Nuratau, Aktau, Malguzar, North Turkestan.

1433. Scrophularia xanthoglossa Boiss.
Perennial. Rocks, stony slopes. Montane and alpine zones. Nuratau, Malguzar, North Turkestan.

Family 97. Lentibulariaceae

Genus 441. Utricularia L.

1434. Utricularia minor L.
Perennial. Standing and slowly flowing water. Montane zone. North Turkestan.

Family 98. Verbenaceae

Genus 442. Verbena L.

1435. Verbena officinalis L.
Perennial. River valleys, ravines, gardens, meadows, banks of rivers, lakes and canals, fallow lands, fine-earth slopes, roadsides. Plain, foothills, montane zone. Nuratau, Aktau, Nuratau Relic Mountains, Malguzar, North Turkestan, Mirzachul, Kyzylkum. Medicinal, poisonous, weed.

Family 99. Lamiaceae [Labiatae]

Genus 443. Teucrium L.

1436. Teucrium scordium subsp. **scordioides** (Schreb.) Arcang. [*Teucrium scordioides* Schreb.]
Perennial. Wet places, banks of rivers, lakes and canals. Plain. Mirzachul. Medicinal, essential oil, poisonous, insecticide.

Genus 444. Scutellaria L.

1437. Scutellaria adenostegia Briq.
Perennial. Fine-earth and stony slopes, rocks, screes, dry riverbeds. Montane zone. Nuratau, North Turkestan.

1438. Scutellaria comosa Juz.   Figure 374
Semishrub. Stony slopes, rocks, screes. Montane zone. Nuratau, Aktau, Malguzar, North Turkestan.

Figure 374 Scutellaria comosa (photography by Natalya Beshko)

1439. Scutellaria cordifrons Juz.   Figure 375
Semishrub. Stony slopes, rocks, screes, ravines. Montane and alpine zones. Malguzar, North Turkestan.

Figure 375 **Scutellaria cordifrons** (photography by Natalya Beshko)

1440. Scutellaria galericulata L.
Perennial. Banks of rivers, riparian forests, fields, wet slopes. Plain, foothills, montane zone. Nuratau, Malguzar, North Turkestan. Medicinal, essential oil, dye, meliferous.

1441. Scutellaria glabrata Vved.

Semishrub. Stony slopes, screes. Montane and alpine zones. Nuratau, Malguzar, North Turkestan.

1442. Scutellaria immaculata Nevski ex Juz.   Figure 376

Semishrub. Stony slopes, rocks. Montane zone. Nuratau, Malguzar, North Turkestan.

Figure 376 Scutellaria immaculata (photography by Natalya Beshko)

1443. Scutellaria intermedia Popov

Semishrub. Stony and gravelly slopes, rocks. Montane zone. Aktau, Malguzar, North Turkestan.

1444. Scutellaria oxystegia Juz.

Semishrub. Fine-earth and stony slopes, rocks, screes, banks of rivers. Montane zone. Nuratau, Aktau, Malguzar. Essential oil, meliferous, alkaloid-bearing.

1445. Scutellaria phyllostachya Juz.
Semishrub. Stony slopes, banks of rivers. Montane and alpine zones. North Turkestan.

1446. Scutellaria ramosissima Popov   Figure 377
Semishrub. Stony slopes, screes, dry riverbeds. Montane zone. Nuratau, Malguzar, North Turkestan.

Figure 377 Scutellaria ramosissima (photography by Natalya Beshko)

1447. Scutellaria schachristanica Juz.
Perennial. Gravelly and stony slopes, screes. Alpine zone. Malguzar, North Turkestan.

1448. Scutellaria squarrosa Nevski
Semishrub. Rocks, stony slopes, screes. Montane zone. Malguzar, North Turkestan.

Genus 445. Drepanocaryum Pojark.

1449. Drepanocaryum sewerzowii (Regel) Pojark.
Annual. Fine-earth, gravelly and stony slopes, rocks, screes, shady places, gardens. Foothills, montane zone. Nuratau, Aktau, Nuratau Relic Mountains, Malguzar, North Turkestan. Weed.

Genus 446. Marrubium L.

1450. Marrubium anisodon K. Koch
Perennial. Fine-earth and stony slopes, rocks, screes, pebbles, fallow lands, roadsides, banks of canals, wastelands, fields, surroundings of wells and sheepyards. Plain, foothills, montane zone. Nuratau, Aktau, Nuratau Relic Mountains, Malguzar, North Turkestan, Mirzachul, Kyzylkum. Medicinal, essential oil, meliferous, weed.

Genus 447. Sideritis L.

1451. Sideritis montana L.
Annual. Fine-earth and stony slopes, screes, rocks, pebbles, fields. Foothills, montane zone. Nuratau, Aktau, Nuratau Relic Mountains, Malguzar, North Turkestan. Medicinal, essential oil, food, meliferous.

Genus 448. Lophanthus Adans.

1452. Lophanthus ouroumitanensis (Franch.) Kochk. & Zuckerw.
Perennial. Fine-earth and stony slopes, river valleys, juniper forests. Montane zone. Malguzar, North Turkestan. Essential oil, meliferous.

1453. Lophanthus schtschurowskianus (Regel) Lipsky   Figure 378
Perennial. Fine-earth, gravelly and stony slopes, dry riverbeds. Montane zone. Nuratau, Malguzar, North Turkestan. Essential oil, meliferous.

Figure 378 Lophanthus schtschurowskianus (photography by Natalya Beshko)

1454. Lophanthus subnivalis Lipsky
Perennial. Rocks, stony slopes, pebbles, moraines. Montane and alpine zone. North Turkestan. Essential oil, meliferous.

Genus 449. Nepeta L.

1455. Nepeta alatavica Lipsky
Perennial. Stony slopes, screes, pebbles. Montane and alpine zones. North Turkestan. Essential oil, meliferous.

1456. Nepeta bracteata Benth.
Annual. Stony and gravelly slopes. Montane and alpine zones. North Turkestan. Essential oil,

meliferous.

1457. Nepeta cataria L.
Perennial. Meadows, banks of rivers and canals, stony slopes. Plain, foothills, montane zone. Nuratau, Aktau, Nuratau Relic Mountains, Malguzar, North Turkestan. Essential oil, meliferous.

1458. Nepeta kokanica Regel
Perennial. Stony slopes, screes, rocks. Alpine zone. North Turkestan. Essential oil, meliferous.

1459. Nepeta mariae Regel
Perennial. Stony slopes, screes, banks of rivers. Montane and alpine zones. Malguzar, North Turkestan. Essential oil, meliferous.

1460. Nepeta maussarifii Lipsky
Perennial. Stony slopes. Montane zone. North Turkestan. Essential oil, meliferous.

1461. Nepeta micrantha Bunge
Annual. Gravelly and stony slopes. Plain, foothills, montane zone. Nuratau, Aktau, Nuratau Relic Mountains, Malguzar, North Turkestan. Essential oil, meliferous.

1462. Nepeta nuda L. (*Nepeta pannonica* L.)
Perennial. Meadows, pebbles, fine-earth slopes. Foothills, montane zone. North Turkestan. Essential oil, medicinal, meliferous.

1463. Nepeta olgae Regel
Perennial. Stony slopes, pebbles, rocks. Foothills, montane zone. Nuratau, Aktau, Malguzar, North Turkestan. Essential oil, medicinal, meliferous.

1464. Nepeta podostachys Benth.
Perennial. Stony slopes, rocks, screes, pebbles. Montane and alpine zones. Malguzar, North Turkestan. Essential oil, meliferous.

1465. Nepeta pungens (Bunge) Benth.
Annual. Stony and gravelly slopes, dry riverbeds. Foothills, montane zone. Nuratau, Malguzar, North Turkestan. Essential oil, meliferous.

1466. Nepeta saturejoides Boiss.
Annual. Stony slopes, rocks, screes, pebbles. Foothills, montane zone. Nuratau, Malguzar, North Turkestan. Essential oil, meliferous.

1467. Nepeta ucranica L.
Perennial. Pebbles, banks of rivers. Montane zone. Malguzar, North Turkestan. Essential oil, meliferous.

Genus 450. Dracocephalum L.

1468. Dracocephalum diversifolium Rupr.
Perennial. Fine-earth and stony slopes, screes, pebbles. Montane and alpine zones. Malguzar, North Turkestan. Medicinal, meliferous.

1469. Dracocephalum imberbe Bunge
Perennial. Rocks, screes, moraines, stony slopes, wet places near melting snow. Alpine zone. North

Turkestan.

1470. Dracocephalum nodulosum Rupr.

Semishrub. Rocks, stony slopes, screes, moraines. Montane and alpine zones, Nuratau, North Turkestan. Medicinal, meliferous.

1471. Dracocephalum nuratavicum Adylov   Figure 379

Semishrub. Stony and gravelly slopes, rocks. Montane zone. Nuratau. Ornamental. Endemic to the Nuratau Mountains.

Figure 379 Dracocephalum nuratavicum (photography by N. Yu. Beshko)

1472. Dracocephalum scrobiculatum Regel

Semishrub. Stony slopes, moraines. Alpine zone. North Turkestan.

Genus 451. Kudrjaschevia Pojark.

1473. Kudrjaschevia jacubi (Lipsky) Pojark.

Annual. Fine-earth, gravelly and stony slopes, screes. Montane zone. Nuratau, Malguzar, North Turkestan.

Genus 452. Lallemantia Fisch. & C. A. Mey.

1474. Lallemantia royleana (Benth.) Benth.   Figure 380

Annual. Sandy and clay deserts, fine-earth, gravelly and stony slopes, fields, fallow lands. Nuratau, Aktau, Nuratau Relic Mountains, Malguzar, North Turkestan, Mirzachul, Kyzylkum. Medicinal, essential oil.

Figure 380 Lallemantia royleana (photography by Natalya Beshko)

Genus 453. Hypogomphia Bunge

1475. Hypogomphia turkestana Bunge [*Hypogomphia turkestanica* Regel]   Figure 381
Annual. Clay deserts, fine-earth, gravelly and stony slopes, river valleys, fields, fallow lands. Foothills, montane zone. Nuratau, Aktau, Nuratau Relic Mountains, Malguzar, North Turkestan, Mirzachul.

Figure 381 *Hypogomphia turkestana* (photography by Natalya Beshko)

Genus 454. Prunella L.

1476. Prunella vulgaris L.
Perennial. Swamps, meadows, banks of rivers, lakes and canals, pebbles, stony and gravelly slopes, wet places. Plain, foothills, montane and alpine zones. Nuratau, Aktau, Nuratau Relic Mountains,

Malguzar, North Turkestan, Mirzachul. Medicinal, essential oil, fodder, ornamental, meliferous.

Genus 455. Phlomoides Moench

1477. Phlomoides ambigua (Popov ex Pazij & Vved.) Adylov, Kamelin & Makhm. Figure 382
Perennial. Fine-earth, gravelly and stony slopes. Montane and alpine zones. Nuratau, Malguzar, North Turkestan. Meliferous, ornamental.

Figure 382 Phlomoides ambigua (photography by Natalya Beshko)

1478. Phlomoides anisochila (Pazij & Vved.) Salmaki   Figure 383
Perennial. Stony slopes. Montane zone. Nuratau. Endemic to the Nuratau ridge. UzbRDB 1.

Figure 383 **Phlomoides anisochila** (photography by N. Yu. Beshko)

1479. Phlomoides canescens (Regel) Adylov, Kamelin & Makhm. Figure 384
Perennial. Fine-earth, gravelly and stony slopes. Montane and alpine zones. Malguzar, North Turkestan. Meliferous, ornamental.

Figure 384 Phlomoides canescens (photography by Natalya Beshko)

1480. Phlomoides eriocalyx (Regel) Adylov, Kamelin & Makhm.   Figure 385
Perennial. Fine-earth, gravelly and stony slopes. Foothills, montane zone. Nuratau, Aktau, Nuratau Relic Mountains, Malguzar. Meliferous, ornamental.

Figure 385 Phlomoides eriocalyx (photography by Natalya Beshko)

1481. Phlomoides kaufmanniana (Regel) Adylov, Kamelin & Makhm.   Figure 386
Perennial. Fine-earth, gravelly and stony slopes. Montane zone. Nuratau, Aktau, Malguzar, North Turkestan. Meliferous, ornamental.

Figure 386 Phlomoides kaufmanniana (photography by Natalya Beshko)

1482. Phlomoides labiosa (Bunge) Adylov, Kamelin & Makhm.
Perennial. Sandy and clay deserts. Plain. Mirzachul, Kyzylkum. Meliferous, ornamental.

1483. Phlomoides napuligera (Franch.) Adylov, Kamelin & Makhm.   Figure 387
Perennial. Piedmont plains, gravelly, fine-earth and stony slopes. Plain, foothills. Nuratau, Aktau, Nuratau Relic Mountains, Malguzar, North Turkestan. Meliferous, ornamental.

Figure 387 Phlomoides napuligera (photography by Natalya Beshko)

1484. Phlomoides oreophila (Kar. & Kir.) Adylov, Kamelin & Makhm. [*Phlomoides dszumrutensis* (Afan.) Adylov, Kamelin & Makhm.]
Perennial. Fine-earth, gravelly and stony slopes. Montane zone. North Turkestan. Meliferous, ornamental.

1485. Phlomoides sogdiana (Pazij & Vved.) Salmaki   Figure 388
Perennial. Fine-earth, gravelly and stony slopes. Foothills, montane zone. Nuratau, Malguzar, North Turkestan. Meliferous, ornamental.

Figure 388 **Phlomoides sogdiana** (photography by Natalya Beshko)

1486. Phlomoides speciosa (Rupr.) Adylov, Kamelin & Makhm.   Figure 389
Perennial. Fine-earth, gravelly and stony slopes. Montane and alpine zones. Nuratau, Malguzar, North Turkestan. Meliferous, ornamental.

Figure 389 **Phlomoides speciosa** (photography by Natalya Beshko)

1487. Phlomoides uniflora (Regel) Adylov, Kamelin & Makhm.   Figure 390
Perennial. Sandy, clayey and gravelly soils. Plain, foothills. Nuratau, Nuratau Relic Mountains, Mirzachul, Kyzylkum. Meliferous, ornamental.

Figure 390 **Phlomoides uniflora** (photography by Natalya Beshko)

Genus 456. Phlomis L.

1488. Phlomis linearifolia Zakirov   Figure 391
Perennial. Fine-earth, gravelly and stony slopes, rocks. Foothills, montane zone. Nuratau, Malguzar, North Turkestan. Essential oil, meliferous.

Figure 391 **Phlomis linearifolia** (photography by Natalya Beshko)

1489. Phlomis nubilans Zakirov   Figure 392

Perennial. Fine-earth, gravelly and stony slopes. Montane zone, Nuratau. Essential oil, meliferous. Threatened species, UzbRDB 2. Endemic to the Nuratau Mountains.

Figure 392 Phlomis nubilans (photography by N. Yu. Beshko)

1490. Phlomis olgae Regel
Perennial. Fine-earth and stony slopes. Montane zone. Malguzar, North Turkestan. Essential oil, meliferous.

1491. Phlomis salicifolia Regel
Perennial. Fine-earth, gravelly and stony slopes. Foothills, montane zone. Nuratau, Malguzar, North Turkestan. Essential oil, meliferous.

1492. Phlomis thapsoides Bunge   Figure 393
Perennial. Piedmont plains, fine-earth, gravelly and stony slopes. Foothills. Nuratau, Aktau, Nuratau Relic Mountains, Malguzar, North Turkestan. Essential oil, meliferous.

Figure 393 Phlomis thapsoides (photography by Natalya Beshko)

Genus 457. Stachyopsis Popov & Vved.

1493. Stachyopsis oblongata (Schrenk) Popov & Vved.   Figure 394
Perennial. Stony slopes. Montane and alpine zones. North Turkestan.

Figure 394 Stachyopsis oblongata (photography by Natalya Beshko)

Genus 458. Lamium L.

1494. Lamium album L. Figure 395

Perennial. Shady wet places, banks of rivers and canals, gardens, stony slopes, rocks. Montane and alpine zones. Nuratau, Malguzar, North Turkestan. Medicinal, essential oil, ornamental, meliferous.

Figure 395 **Lamium album** (photography by Natalya Beshko)

1495. Lamium amplexicaule L. Figure 396

Annual or biennial. Stony slopes, rocks, screes, fallow lands, fields, gardens, roadsides, banks of rivers and canals. Plain, foothills, montane zone. Nuratau, Aktau, Nuratau Relic Mountains, Malguzar, North Turkestan, Mirzachul, Kyzylkum. Medicinal, meliferous.

Figure 396 **Lamium amplexicaule** (photography by Natalya Beshko)

Genus 459. Leonurus L.

1496. Leonurus glaucescens Bunge
Perennial. Fine-earth, gravelly and stony slopes, rocks, ravines, banks of rivers. Montane zone. North Turkestan.

1497. Leonurus turkestanicus V. I. Krecz. & Kuprian.   Figure 397
Perennial. Fine-earth, gravelly and stony slopes, ravines, river valleys, shady wet places. Montane zone. Nuratau, Malguzar, North Turkestan. Medicinal, meliferous.

Figure 397 Leonurus turkestanicus (photography by Natalya Beshko)

Genus 460. Lagochilus Bunge ex Benth.

1498. Lagochilus inebrians Bunge   Figure 398
Semishrub. Clay deserts, piedmont plains, stony slopes, pebbles. Plain, foothills. Nuratau, Aktau, Nuratau Relic Mountains, Malguzar. Medicinal. Threatened species, UzbRDB 2.

Figure 398 Lagochilus inebrians (photography by Natalya Beshko)

1499. Lagochilus olgae Kamelin   Figure 399

Semishrub. Stony and gravelly slopes. Montane zone. Nuratau. UzbRDB 2. Endemic to the Nuratau ridge.

Figure 399 **Lagochilus olgae** (photography by N. Yu. Beshko)

1500. Lagochilus proskorjakovii Ikramov   Figure 400

Semishrub. Stony slopes, rocks, screes. Montane zone. Nuratau. Medicinal. Threatened species, UzbRDB 1. Endemic to the Nuratau ridge.

Figure 400 **Lagochilus proskorjakovii** (photography by N. Yu. Beshko)

1501. Lagochilus seravschanicus Knorring
Semishrub. Stony slopes, screes. Montane zone. Malguzar, North Turkestan. Medicinal.

Genus 461. Stachys L.

1502. Stachys hissarica Regel   Figure 401
Perennial. Fine-earth, gravelly and stony slopes, screes, river valleys. Montane and alpine zones. Malguzar, North Turkestan. Medicinal, essential oil, meliferous.

Figure 401 Stachys hissarica (photography by Natalya Beshko)

1503. Stachys setifera C. A. Mey.   Figure 402
Perennial. Wet places, pebbles, river valleys, fine-earth and stony slopes. Foothills, montane zone. Nuratau, Malguzar, North Turkestan. Medicinal, essential oil, meliferous.

Figure 402 Stachys setifera (photography by Natalya Beshko)

Genus 462. Chamaesphacos Schrenk

1504. Chamaesphacos ilicifolius Schrenk   Figure 403
Annual. Sandy deserts. Plain. Kyzylkum.

Figure 403 Chamaesphacos ilicifolius (photography by Natalya Beshko)

Genus 463. Salvia L.

1505. Salvia aequidens Botsch.
Perennial. Stony slopes, rocks. Montane zone. Aktau, North Turkestan.

1506. Salvia deserta Schangin   Figure 404
Perennial. Banks of rivers and canals, fine-earth slopes, meadows. Montane zone. Nuratau, Malguzar, North Turkestan. Medicinal, essential oil, ornamental, meliferous.

Figure 404 Salvia deserta (photography by Natalya Beshko)

1507. Salvia glabricaulis Pobed.
Perennial. Fine-earth, gravelly and stony slopes. Montane zone. North Turkestan. Essential oil, ornamental, meliferous.

1508. Salvia macrosiphon Boiss.
Perennial. River valleys, dry riverbeds, fine-earth, gravelly and stony slopes. Montane zone. Nuratau, Malguzar, North Turkestan. Essential oil, ornamental, meliferous.

1509. Salvia sarawschanica Regel & Schmalh.
Perennial. Dry riverbeds, rocks, fine-earth, gravelly and stony slopes. Montane zone. Nuratau, Malguzar, North Turkestan. Medicinal, essential oil, ornamental, meliferous.

1510. Salvia sclarea L.   Figure 405
Perennial. River valleys, banks of canals, dry riverbeds, ravines, fine-earth, gravelly and stony slopes,

wastelands, fallow lands, gardens. Plain, foothills, montane zone. Nuratau, Nuratau Relic Mountains, Malguzar, North Turkestan, Mirzachul. Medicinal, essential oil, food, ornamental, meliferous.

Figure 405 **Salvia sclarea** (photography by Natalya Beshko)

1511. Salvia spinosa L.   Figure 406

Perennial. Clayey and gravelly soils, pebbles, dry riverbeds, stony slopes. Plain, foothills, montane zone. Nuratau, Nuratau Relic Mountains, Malguzar, North Turkestan, Mirzachul. Medicinal, essential oil, ornamental, meliferous.

Figure 406 **Salvia spinosa** (photography by Natalya Beshko)

1512. Salvia submutica Botsch. & Vved.   Figure 407
Perennial. Stony slopes, rocks. Montane zone. Nuratau. Essential oil, ornamental, meliferous. Threatened species, UzbRDB 2. Endemic to the Nuratau ridge.

Figure 407 Salvia submutica (photography by N. Yu. Beshko)

1513. Salvia virgata Jacq.
Perennial. River valleys, ravines, gardens, meadows, stream banks, fine-earth slopes. Montane zone. Nuratau, Malguzar, North Turkestan. Medicinal, essential oil, ornamental, meliferous.

Genus 464. Perovskia Kar.

1514. Perovskia angustifolia Kudrjasch.
Semishrub. Dry riverbeds, pebbles, stony slopes. Foothills, montane zone. Nuratau, Nuratau Relic Mountains, Malguzar, North Turkestan. Medicinal, essential oil, dye, meliferous.

1515. Perovskia botschantzevii Kovalevsk. & Koczk.
Semishrub. Stony slopes, pebbles, river valleys. Foothills, montane zone. Nuratau. Medicinal, essential oil, dye, meliferous.

1516. Perovskia kudrjaschevii Gorschk. & Pjataeva
Semishrub. Pebbles, banks of rivers, stony slopes. Foothills, montane zone. Nuratau, North Turkestan. Medicinal, essential oil, dye, meliferous.

1517. Perovskia scrophulariifolia Bunge
Semishrub. Banks of rivers, rocks, screes, stony slopes, pebbles, dry riverbeds. Foothills, montane zone. Nuratau, Aktau, Nuratau Relic Mountains, Malguzar, North Turkestan. Medicinal, essential oil, dye, meliferous.

Genus 465. Ziziphora L.

1518. Ziziphora clinopodioides Lam.   Figure 408
Semishrub. Rocks, stony slopes, screes, banks of rivers, pebbles. Foothills, montane and alpine zones. Nuratau, Aktau, Malguzar, North Turkestan. Medicinal, essential oil, food (spice-aromatic), meliferous.

Figure 408 Ziziphora clinopodioides (photography by Natalya Beshko)

1519. Ziziphora pamiroalaica Juz.
Semishrub. Fine-earth and stony slopes, rocks, screes, pebbles. Plain, foothills, montane and alpine zones. Nuratau, Malguzar, North Turkestan. Medicinal, essential oil, food (spice-aromatic), meliferous.

1520. Ziziphora suffruticosa Pazij & Vved.
Semishrub. Stony slopes, pebbles. Montane zone. Malguzar, North Turkestan. Medicinal, essential oil, food (spice-aromatic), meliferous.

1521. Ziziphora tenuior L.
Annual. Sandy and clay deserts, piedmont plains, river valleys, fine-earth, gravelly and stony slopes. Plain, foothills, montane zone. Nuratau, Aktau, Nuratau Relic Mountains, Malguzar, North Turkestan, Mirzachul, Kyzylkum. Medicinal, essential oil, food (spice-aromatic), meliferous.

Genus 466. Clinopodium L.

1522. Clinopodium alpinum (L.) Kuntze [*Acinos rotundiflorus* Friv. ex Walp.]
Annual. Stony slopes, pebbles. Montane zone. Malguzar, North Turkestan. Medicinal.

1523. Clinopodium debile (Bunge) Kuntze [*Antonia debilis* (Bunge) Vved.]
Annual. Stony slopes, shady wet places, dry riverbeds. Montane and alpine zones. Malguzar, North Turkestan.

Genus 467. Hyssopus L.

1524. Hyssopus seravschanicus (Dubj.) Pazij   Figure 409
Semishrub. Stony slopes, rocks, pebbles, screes. Montane and alpine zones. Malguzar, North Turkestan. Medicinal, essential oil, food (spice-aromatic), meliferous.

Figure 409 Hyssopus seravschanicus (photography by Natalya Beshko)

Genus 468. Origanum L.

1525. Origanum vulgare subsp. **gracile** (K. Koch) Ietsw. [*Origanum tyttanthum* Gontsch.]
Perennial. Fine-earth and stony slopes, river valleys, pebbles. Montane zone. Nuratau, Aktau, Malguzar, North Turkestan. Medicinal, essential oil, food, meliferous.

Genus 469. Thymus L.

1526. Thymus seravschanicus Klokov   Figure 410
Semishrub. Stony and gravelly slopes, screes, pebbles, rocks, alpine meadows. Montane and

alpine zones. Nuratau, Malguzar, North Turkestan. Medicinal, essential oil, food (spice-aromatic), meliferous.

Figure 410 Thymus seravschanicus (photography by Natalya Beshko)

1527. Thymus subnervosus Vved., Nabiev & Tulyag.
Semishrub. Rocks, stony slopes. Montane zone. Nuratau. Essential oil, meliferous.

Genus 470. Lycopus L.

1528. Lycopus europaeus L.
Perennial. Wet places, swamps, banks of rivers and canals. Plain, foothills, montane zone. Nuratau, Malguzar, North Turkestan, Mirzachul, Kyzylkum. Medicinal, essential oil, tanniferous, dye, meliferous.

Genus 471. Mentha L.

1529. Mentha longifolia var. **asiatica** (Boriss.) Rech. f.
Perennial. Banks of rivers and canals, wet places. Plain, foothills, montane zone. Nuratau, Nuratau Relic Mountains, Malguzar, North Turkestan, Mirzachul. Medicinal, essential oil, food, fodder,

ornamental, meliferous.

1530. Mentha pamiroalaica Boriss.
Perennial. Wet places, banks of rivers. Montane zone. Nuratau, Malguzar, North Turkestan. Medicinal, essential oil, food, fodder, ornamental, meliferous.

Family 100. Mazaceae

Genus 472. Dodartia L.

1531. Dodartia orientalis L.   Figure 411
Perennial. Sandy and clay deserts, piedmont plains, fine-earth, gravelly and stony slopes, river valleys, wastelands, fallow lands, fields, roadsides. Plain, foothills, montane zone. Nuratau, Aktau, Nuratau Relic Mountains, Malguzar, North Turkestan, Mirzachul, Kyzylkum. Medicinal, meliferous, ornamental, insecticide.

Figure 411 Dodartia orientalis (photography by Natalya Beshko)

Family 101. Orobanchaceae

Genus 473. Orobanche L.

1532. *Orobanche aegyptiaca Pers. [*Phelipanche aegyptiaca* (Pers.) Pom.]
Perennial. Fields, gardens, fallow lands. Plain, foothills, montane zone Nuratau, Aktau, Malguzar, North Turkestan, Mirzachul. Poisonous, weed. Parasite. Host plants are different vegetables, field crops, ornamental plants, and rarely wild growing species.

1533. Orobanche amoena C. A. Mey. ex Ledeb.
Perennial. Piedmont plains, fine-earth and stony slopes, fallow lands, rocks, screes. Plain, foothills, montane zone. Nuratau, Nuratau Relic Mountains, Malguzar, North Turkestan. Weed. Parasite. Host plants are representatives of genera *Artemisia* and *Alhagi*.

1534. Orobanche cernua Loefl.
Perennial. Sandy deserts, piedmont plains, banks of canals, fields, roadsides, riparian forests, screes, fine-earth and stony slopes. Plain, foothills, montane and alpine zones. Nuratau, Aktau, Nuratau Relic Mountains, Malguzar, North Turkestan, Mirzachul. Weed. Parasite. Host plants are sunflowers, *Karelinia caspica*, representatives of genera *Artemisia, Lactuca, Scariola, Cousinia, Carthamus,* and species of Solanaceae.

1535. Orobanche coelestis (Reut.) Boiss. & Reut. ex Beck [*Phelipanche coelestis* (Reut.) Soják]
Perennial. Fine-earth, gravelly and stony slopes. Foothills, montane zone. Nuratau, Malguzar, North Turkestan. Weed. Parasite. Host plants are representatives of genera *Artemisia, Cousinia, Scorsonera* and *Centaurea*.

1536. Orobanche elatior Sutton
Perennial. Stony slopes, screes, piedmont plains. Plain, foothills, montane zone. Malguzar, North Turkestan. Weed. Parasite. Host plants are representatives of family Asteraceae (*Echinops, Centaurea, Artemisia,* etc.).

1537. Orobanche gigantea (Beck) Gontsch.
Perennial. Rocks, screes, fine-earth and stony slopes. Montane zone. Nuratau. Weed. Parasite. Host plants are representatives of family Apiaceae (*Ferula, Schrenkia,* etc.).

1538. Orobanche hansii A. Kern
Perennial. Piedmont plains, fallow lands, pebbles, stony slopes, sandy and clay deserts. Plain, foothills, montane zone. Nuratau, Aktau, Nuratau Relic Mountains, Malguzar, North Turkestan, Mirzachul, Kyzylkum. Weed. Parasite. Host plants are representatives of genera *Artemisia, Inula, Tussilago* and *Cousinia*.

1539. Orobanche orientalis Beck [*Phelipanche orientalis* (Beck) Soják]
Perennial. Fine-earth, gravelly and stony slopes. Montane zone. Nuratau, Malguzar, North Turkestan. Weed. Parasite. Host plants are wild almonds (*Prunus bucharica, Prunus spinosissima*).

1540. Orobanche pallens F. W. Schultz [*Phelipanche pallens* (Bunge) Soják]
Perennial. Fine-earth, gravelly and stony slopes. Montane zone. Nuratau, Malguzar, North Turkestan. Weed. Parasite. Host plants are species of *Artemisia*.

1541. Orobanche sulphurea Gontsch.
Perennial. Fine-earth, gravelly and stony slopes. Foothills, montane zone. Nuratau. Weed. Parasite. Host plant is *Crambe kotschyana*.

Genus 474. Cistanche Hoffmanns. & Link.

1542. Cistanche mongolica Beck
Perennial. Saline lands, banks of lakes and canals. Plain. Mirzachul, Kyzylkum. Weed. Parasite. Host plants are species of *Tamarix*.

1543. Cistanche salsa (C. A. Mey.) Beck   Figure 412
Perennial. Saline lands, sandy deserts, fallow lands, banks of lakes and canals. Plain. Mirzachul, Kyzylkum. Medicinal, fodder, weed. Parasite. Host plants are species of *Haloxylon, Salsola, Anabasis, Kalidium, Atriplex, Calligonum*.

Figure 412 Cistanche salsa (photography by Natalya Beshko)

Genus 475. Leptorhabdos Schrenk ex Fisch. et C. A. Mey.

1544. Leptorhabdos parviflora (Benth.) Benth.
Annual. Banks of rivers, lakes and canals, sandy soils, fine-earth, gravelly and stony slopes, fallow lands, fields. Plain, foothills, montane zone. Nuratau, Aktau, Nuratau Relic Mountains, Malguzar, North Turkestan.

Genus 476. Euphrasia L.

1545. Euphrasia pectinata Ten.
Annual. Meadows, fine-earth slopes. Montane and alpine zones. Malguzar, North Turkestan. Medicinal, meliferous.

1546. Euphrasia regelii Wettst.
Annual. Meadows, fine-earth slopes. Montane and alpine zones. Malguzar, North Turkestan. Medicinal, meliferous.

Genus 477. Parentucellia Viv.

1547. Parentucellia flaviflora (Boiss.) Nevski   Figure 413
Annual. Fine-earth, gravelly and stony slopes, piedmont plains, clay desert. Plain, foothills, montane zone. Nuratau, Aktau, Nuratau Relic Mountains, Malguzar, North Turkestan, Mirzachul.

Figure 413 Parentucellia flaviflora (photography by Natalya Beshko)

Genus 478. Pedicularis L.

1548. Pedicularis dolichorrhiza Schrenk
Perennial. Fine-earth and stony slopes. Montane and alpine zones. North Turkestan. Medicinal.

1549. Pedicularis krylovii Bonati
Perennial. Stony slopes. Montane and alpine zones, North Turkestan.

1550. Pedicularis olgae Regel   Figure 414
Perennial. Fine-earth, gravelly and stony slopes. Montane and alpine zones. Nuratau, Nuratau Relic Mountains, Malguzar, North Turkestan.

Figure 414 Pedicularis olgae (photography by Natalya Beshko)

1551. Pedicularis peduncularis Popov
Perennial. Swamp meadows, swamps. Alpine zone. North Turkestan.

1552. Pedicularis popovii Vved.   Figure 415
Perennial. Fine-earth and stony slopes. Montane and alpine zones. North Turkestan.

Figure 415 Pedicularis popovii (photography by Natalya Beshko)

1553. Pedicularis rhinanthoides Schrenk
Perennial. Swamp meadows. Alpine zone. North Turkestan.

1554. Pedicularis sarawschanica Regel
Perennial. Wet stony slopes. Alpine zone. North Turkestan.

1555. Pedicularis waldheimii Bonati
Perennial. Screes. Alpine zone. North Turkestan.

Family 102. Campanulaceae

Genus 479. Campanula L.

1556. Campanula cashmeriana Royle
Perennial. Rocks. Montane and alpine zones. Malguzar, North Turkestan. Medicinal, meliferous,

ornamental.

1557. Campanula fedtschenkoana Trautv.
Perennial. Wet places near melting snow, stony slopes, rocks. Montane and alpine zones. North Turkestan.

1558. Campanula glomerata L.   Figure 416
Perennial. Fine-earth and stony slopes, river valleys, meadows. Montane and alpine zones. Malguzar, North Turkestan. Medicinal, meliferous, ornamental.

Figure 416 **Campanula glomerata** (photography by Natalya Beshko)

1559. Campanula lehmanniana Bunge
Perennial. Stony slopes, rocks. Montane and alpine zones. Malguzar, North Turkestan.

Genus 480. Asyneuma Griseb. & Schrenk

1560. Asyneuma argutum (Regel) Bornm.   Figure 417
Perennial. Fine-earth and stony slopes. Montane and alpine zones. Malguzar, North Turkestan. Medicinal.

Figure 417 Asyneuma argutum (photography by Natalya Beshko)

1561. Asyneuma trautvetteri (B. Fedtsch.) Bornm.
Perennial. Fine-earth and stony slopes. Montane and alpine zones. North Turkestan.

Genus 481. Sergia Al. Fed.

1562. Sergia regelii (Trautv.) Al. Fed.   Figure 418
Perennial. Rocks. Montane zone. Malguzar, North Turkestan.

Figure 418 Sergia regelii (photography by Natalya Beshko)

Genus 482. Cylindrocarpa Regel

1563. Cylindrocarpa sewerzowii (Regel) Regel
Perennial. Gravelly and stony slopes, rocks. Montane zone. Malguzar, North Turkestan.

Genus 483. Codonopsis Wall.

1564. Codonopsis clematidea (Schrenk) C. B. Clarke
Perennial. Banks of streams, stony slopes, wet places near springs, meadows. Montane zone. Malguzar, North Turkestan. Medicinal.

Family 103. Asteraceae [Compositae]

Genus 484. *Cichorium L.

1565. *Cichorium intybus L.
Perennial. Fallow lands, wastelands, river valleys, pebbles, fine-earth and stony slopes, fields, gardens, banks of canals. Plain, foothills, montane zone. Nuratau, Aktau, Nuratau Relic Mountains, Malguzar, North Turkestan, Mirzachul, Kyzylkum. Medicinal, fodder, food, weed, meliferous.

Genus 485. Picris L.

1566. Picris nuristanica Bornm.
Annual or biennial. Pebbles, river valleys, fine-earth, gravelly and stony slopes. Plain, foothills, montane zone. North Turkestan. Meliferous.

Genus 486. Garhadiolus Jaub. & Spach

1567. Garhadiolus hedypnois Jaub. & Spach [*Garhadiolus angulosus* Jaub. & Spach]
Annual. Sandy and clay deserts, fine-earth, gravelly and stony slopes, screes, pebbles, river valleys, fallow lands. Plain, foothills, montane zone. Nuratau, Aktau, Nuratau Relic Mountains, Malguzar, North Turkestan, Mirzachul, Kyzylkum. Weed.

1568. Garhadiolus papposus Boiss. & Buhse
Annual. Saline lands, sandy and clay deserts, fine-earth, gravelly and stony slopes, screes, pebbles, river valleys, fallow lands, wastelands. Plain, foothills, montane zone. Nuratau, Aktau, Nuratau Relic Mountains, Malguzar, North Turkestan, Mirzachul, Kyzylkum. Weed.

Genus 487. Launaea Cass.

1569. Launaea korovinii (Popov) Pavlov [*Rhabdotheca korovinii* (Popov) Kirp.]
Annual. Sandy desert. Plain. Kyzylkum.

1570. Launaea procumbens (Roxb.) Ramayya & Rajagopal [*Paramicrorhynchus procumbens* (Roxb.) Kipr.]
Perennial. Sandy and clay deserts, saline lands, banks of rivers, lakes and canals, fields, fallow lands, roadsides. Plain. Mirzachul, Kyzylkum. Weed.

Genus 488. Chondrilla L.

1571. Chondrilla aspera (Schrad. ex Will.) Poir.
Perennial or Biennial. Clay deserts, fine-earth, gravelly and stony slopes, rocks, screes, pebbles, banks of rivers and canals. Plain, foothills, montane and alpine zones. Nuratau, Nuratau Relic Mountains, Malguzar, North Turkestan, Mirzachul. Fodder.

1572. Chondrilla lejosperma Kar. & Kir.
Perennial. Fine-earth, gravelly and stony slopes, screes, pebbles, banks of rivers, lakes and canals. Plain, foothills, montane zone. Nuratau, Nuratau Relic Mountains, Malguzar, North Turkestan, Mirzachul. Fodder.

Genus 489. Heteroderis (Bunge) Boiss.

1573. Heteroderis pusilla (Boiss.) Boiss.
Annual. Sandy deserts, saline lands. Plain. Kyzylkum.

Genus 490. Lactuca L.

1574. Lactuca crambifolia (Bunge) Boiss. [*Steptorhamphus crambifolius* Bunge]
Perennial. Gravelly and stony slopes, screes, rocks. Foothills, montane zone. Nuratau, Aktau, Nuratau Relic Mountains, Malguzar, North Turkestan.

1575. Lactuca dissecta D. Don.
Annual. Fine-earth, gravelly and stony slopes, screes, rocks, shady wet places, banks of rivers, roadsides, fallow lands, wastelands. Plain, foothills, montane and alpine zones. North Turkestan.

1576. Lactuca glauciifolia Boiss.
Annual. Fine-earth, gravelly and stony slopes, rocks, pebbles, fallow lands, wastelands. Plain, foothills, montane and alpine zones. North Turkestan, Mirzachul.

1577. Lactuca orientalis (Boiss.) Boiss. [*Scariola orientalis* (Boiss.) Soják]
Semishrub. Fine-earth, gravelly and stony slopes, screes, rocks, pebbles, dry riverbeds. Plain,

foothills, montane and alpine zones. Nuratau, Aktau, Nuratau Relic Mountains, Malguzar, North Turkestan.

1578. Lactuca serriola Torner [*Lactuca altaica* Fisch. & C. A. Mey.]
Annual or biennial. Fine-earth, gravelly and stony slopes, pebbles, roadsides, banks of rivers, lakes and canals, fallow lands, wastelands, fields, gardens, settlements. Nuratau, Aktau, Nuratau Relic Mountains, Malguzar, North Turkestan, Mirzachul. Medicinal, food, fodder, oil, weed, rubber-bearing.

1579. Lactuca soongorica Regel [*Cephalorrhynchus soongoricus* (Regel) Kovalevsk.]
Perennial. Stony slopes, screes, rocks. Montane zone. Nuratau, Malguzar, North Turkestan.

1580. Lactuca spinidens Nevski
Annual or biennial. Stony slopes, fields, fallow lands. Nuratau.

1581. Lactuca tatarica (L.) C. A. Mey.
Perennial. Saline lands, pebbles, banks of rivers, lakes and canals, fields, fallow lands, roadsides. Plain, foothills, montane and alpine zones. Nuratau, Aktau, Nuratau Relic Mountains, Malguzar, North Turkestan, Mirzachul, Kyzylkum.

1582. Lactuca undulata Ledeb.
Annual. Sandy and clay deserts, banks of rivers, lakes and canals, saline lands, pebbles, fine-earth, gravelly and stony slopes, fields, gardens, fallow lands, roadsides. Plain, foothills. Nuratau Relic Mountains, Mirzachul, Kyzylkum.

Genus 491. Kovalevskiella Kamelin

1583. Kovalevskiella zeravschanica (Popov & Kovalevsk) Kamelin
Perennial. Fine-earth, gravelly and stony slopes. Montane and alpine zones. North Turkestan.

Genus 492. Crepidifolium Sennikov

1584. Crepidifolium serawschanicum (B. Fedtsch.) Sennikov [*Youngia serawschanica* (B. Fedtsch.) Babc. & Stebbins]
Perennial. Gravelly and stony slopes, screes, rocks, alpine meadows. Montane and alpine zones. North Turkestan.

Genus 493. *Sonchus L.

1585. *Sonchus arvensis L.
Perennial. Banks of rivers, lakes and canals, meadows, wastelands, fallow lands, fields, gardens, settlements. Nuratau, Aktau, Nuratau Relic Mountains, Malguzar, North Turkestan, Mirzachul. Medicinal, food, fodder, weed, meliferous.

1586. *Sonchus asper (L.) Hill

Annual. Fallow lands, fields, roadsides, wastelands, settlements. Nuratau, Nuratau Relic Mountains, Malguzar, North Turkestan, Mirzachul, Kyzylkum. Food, fodder, weed, meliferous.

1587. *Sonchus oleraceus (L.) L.

Annual. Fallow lands, fields, roadsides, banks of rivers, lakes and canals, wastelands, settlements. Nuratau, Aktau, Nuratau Relic Mountains, Malguzar, North Turkestan, Mirzachul, Kyzylkum. Medicinal, food, fodder, weed, meliferous.

1588. *Sonchus palustris L.

Perennial. Banks of rivers and canals, swamps, wet places, gardens. Plain, foothills, montane zone. Nuratau, Nuratau Relic Mountains, Malguzar, North Turkestan, Mirzachul, Kyzylkum. Weed, meliferous.

Genus 494. Taraxacum F. H. Wigg.

1589. Taraxacum bessarabicum (Hornem.) Hand.-Mazz.

Perennial. Meadows, swamps, pebbles, banks of rivers, lakes and canals, fallow lands, wet places. Plain, foothills, montane zone. Nuratau, Nuratau Relic Mountains, Mirzachul, Kyzylkum. Meliferous.

1590. Taraxacum bicorne Dahlst

Perennial. Meadows, saline lands, depressions, banks of rivers, lakes and canals, fallow lands, fields, roadsides. Plain, foothills. Nuratau Relic Mountains, Mirzachul, Kyzylkum. Meliferous, weed.

1591. Taraxacum brevirostre Hand.-Mazz.

Perennial. Fine-earth, gravelly and stony slopes, alpine meadows. Alpine zone. North Turkestan.

1592. Taraxacum contristans Kovalevsk.

Perennial. Wet places, saline lands, roadsides, banks of canals, fields, fallow lands. Plain, foothills. Malguzar, Mirzachul. Meliferous, weed.

1593. Taraxacum ecornutum Kovalevsk.

Perennial. Banks of rivers and canals, roadsides, fields, fallow lands, ravines. Plain, foothills, montane zone. Nuratau, Nuratau Relic Mountains, North Turkestan, Mirzachul. Meliferous, weed.

1594. Taraxacum elongatum Kovalevsk.

Perennial. Banks of rivers, meadows, swamps, fine-earth, gravelly and stony slopes. Montane zone. Malguzar, North Turkestan. Meliferous.

1595. Taraxacum erostre Zakirov

Perennial. Rocks, gravelly and stony slopes, screes, moraines. Montane and alpine zones. Malguzar, North Turkestan.

1596. Taraxacum glaucivirens Schischk.

Perennial. Fine-earth, gravelly and stony slopes, rocks, moraines. Montane and alpine zones. North Turkestan.

1597. Taraxacum maracandicum Kovalevsk.
Perennial. Wet places, banks of rivers and canals, meadows, fine-earth slopes, roadsides, settlements. Plain, foothills, Montane zone. Nuratau, Malguzar, North Turkestan.

1598. Taraxacum marginatum (Dahlst.) Raunk. [*Taraxacum reflexum* (Brenner) Hjelt]
Perennial. Stony slopes, screes, wet places, swamps, meadows. Montane zone. North Turkestan.

1599. Taraxacum minutilobium Popov ex Kovalevsk.
Perennial. Gravelly and stony slopes, moraines, rocks, screes, banks of rivers. Alpine zone. Malguzar, North Turkestan.

1600. Taraxacum modestum Schischk.
Perennial. Fine-earth, gravelly and stony slopes, wet places near melting snow. Montane and alpine zones. North Turkestan.

1601. Taraxacum monochlamydeum Hand.-Mazz.
Perennial. Fine-earth slopes, piedmont plains, gardens, fields, fallow lands, roadsides, wastelands, banks of rivers and canals. Plain, foothills, montane zone. Nuratau, Aktau, Nuratau Relic Mountains, Malguzar, North Turkestan, Mirzachul. Meliferous.

1602. Taraxacum nevskii Juz.
Perennial. Gravelly and stony slopes, screes. Montane zone. Malguzar, North Turkestan.

1603. Taraxacum nuratavicum Schischk.
Perennial. Wet places, swamps, banks of rivers. Montane zone. Nuratau. Meliferous.

1604. *Taraxacum officinale (L.) Weber ex F.H. Wigg.
Perennial. Banks of rivers, lakes and canals, fallow lands, wastelands, gardens, settlements. Plain, foothills, montane zone. Nuratau, Aktau, Nuratau Relic Mountains, Malguzar, North Turkestan, Mirzachul, Kyzylkum. Medicinal, food, fodder, meliferous.

1605. Taraxacum popovii Kovalevsk. ex Vainberg
Perennial. Gravelly and stony slopes, moraines, alpine meadows, screes, wet places near melting snow. Montane and alpine zones. Malguzar, North Turkestan.

1606. Taraxacum repandum Pavlov
Perennial. Fine-earth, gravelly and stony slopes. Foothills, montane zone. Malguzar, North Turkestan.

1607. Taraxacum sonchoides (D. Don) Sch. Bip. [*Taraxacum montanum* (C. A. Mey.) DC.]
Perennial. Fine-earth, gravelly and stony slopes, dry riverbeds, pebbles. Montane zone. Nuratau, Aktau, Malguzar, North Turkestan. Meliferous.

1608. Taraxacum strobilocephalum Kovalevsk.
Perennial. Wet places, banks of rivers and canals. Malguzar.

Genus 495. Heteracia Fisch. & C. A. Mey.

1609. Heteracia szovitzii Fisch. & C. A. Mey. [*Heteracia epapposa* (Regel & Schmalh.) Popov]
Annual. River valleys, sandy and clay deserts, saline lands, piedmont plains, fine-earth, gravelly and stony slopes, rocks, fields, fallow lands. Plain, foothills, montane zone. Nuratau, Aktau, Nuratau

Relic Mountains, Malguzar, Mirzachul, Kyzylkum.

Genus 496. Crepis L.

1610. Crepis kotschyana (Boiss.) Boiss. [*Barkhausia kotschyana* Boiss.]
Annual. Gravelly and stony slopes, rocks. Foothills, montane zone. Nuratau, Aktau, Nuratau Relic Mountains, Malguzar, North Turkestan.

1611. Crepis multicaulis Ledeb.
Perennial. Gravelly and stony slopes, screes, pebbles, moraines, river valleys, wet places. Montane and alpine zones. North Turkestan.

1612. Crepis pulchra L. [*Phaecasium pulchrum* (L.) Reich. f.]   Figure 419
Annual. Fine-earth, gravelly and stony slopes, screes, rocks, banks of rivers and canals, shady wet places, gardens, roadsides, fields, fallow lands. Plain, foothills, montane zone. Nuratau, Aktau, Nuratau Relic Mountains, Malguzar, North Turkestan, Mirzachul.

Figure 419 Crepis pulchra (photography by Natalya Beshko)

Genus 497. Pilosella Hill

1613. Pilosella echioides (Lumn.) F. W. Schultz & Sch. Bip. [*Hieracium echioides* Lumn.]
Perennial. Stony slopes. Montane and alpine zones. Malguzar, North Turkestan. Medicinal.

1614. Pilosella procera (Fr.) F. W. Schultz & Sch. Bip. [*Hieracium procerum* Fr.]
Perennial. Fine-earth, gravelly and stony slopes, meadows, ravines. Montane zone. North Turkestan.

Genus 498. Hieracium L.

1615. Hieracium raddeanum subsp. **regelianum** (Zahn) Greuter [*Hieracium regelianum* Zahn]
Perennial. Juniper forests, meadows, river valleys. Montane zone. North Turkestan.

1616. Hieracium robustum Fr.
Perennial. Meadows, fine-earth, gravelly and stony slopes, river valleys. Montane zone. North Turkestan. Weed.

1617. Hieracium virosum Pall.   Figure 420
Perennial. Meadows, fine-earth, gravelly and stony slopes, river valleys. Montane zone. North Turkestan. Weed.

Figure 420 Hieracium virosum (photography by Natalya Beshko)

Genus 499. Acanthocephalus Kar. & Kir.

1618. Acanthocephalus amplexifolius Kar. & Kir.
Annual. Saline lands, river valleys, dry riverbeds, ravines, fine-earth, gravelly and stony slopes. Plain, foothills, montane zone. Nuratau, Nuratau Relic Mountains, Malguzar, North Turkestan, Mirzachul. Weed.

1619. Acanthocephalus benthamianus Regel
Annual. River valleys, banks of canals, gardens, roadsides, ravines, fine-earth, gravelly and stony slopes. Plain, foothills, montane zone. Nuratau, Aktau, Nuratau Relic Mountains, Malguzar, North Turkestan, Mirzachul. Weed.

Genus 500. Scorzonera L.

1620. Scorzonera acanthoclada Franch.
Semishrub. Gravelly and stony slopes. Montane and alpine zones. North Turkestan. fodder, rubber-bearing.

1621. Scorzonera bracteosa C. Winkl.   Figure 421
Perennial. Fine-earth and stony slopes. Foothills, montane zone. Nuratau, Malguzar, North Turkestan. Rubber-bearing, meliferous.

Figure 421 Scorzonera bracteosa (photography by Natalya Beshko)

1622. Scorzonera circumflexa Krasch. & Lipsch.
Perennial. Fine-earth, gravelly and stony slopes, pebbles, clay deserts. Plain, foothills, montane and alpine zones. Nuratau, Aktau, Nuratau Relic Mountains, Malguzar, North Turkestan, Mirzachul. Rubber-bearing, meliferous.

1623. Scorzonera hissarica C. Winkl.
Perennial. Fine-earth, gravelly and stony slopes. Foothills, montane zone. Malguzar, North Turkestan.

1624. Scorzonera inconspicua Lipsch.
Perennial. Fine-earth, gravelly and stony slopes, pebbles, banks of rivers. Plain, foothills, montane zone. Malguzar, North Turkestan. Rubber-bearing, meliferous.

1625. Scorzonera litwinowii Krasch. & Lipsch.
Perennial. Fine-earth, gravelly and stony slopes. Foothills, montane zone. Nuratau. Rubber-bearing,

meliferous.

1626. Scorzonera ovata Trautv. Figure 422

Perennial. Fine-earth, gravelly and stony slopes. Foothills, montane zone. Nuratau, Nuratau Relic Mountains, Malguzar, North Turkestan. Rubber-bearing, meliferous.

Figure 422 Scorzonera ovata (photography by Natalya Beshko)

1627. Scorzonera songarica (Kar. & Kir.) Lipsch. & Vassilcz.
Perennial. Fine-earth, gravelly and stony slopes, river valleys. Montane zone. North Turkestan.

1628. Scorzonera tragopogonoides Regel & Schmalh.
Perennial. Fine-earth, gravelly and stony slopes. Foothills, montane and alpine zones, Nuratau, Nuratau Relic Mountains, Malguzar, North Turkestan, Mirzachul. Rubber-bearing, meliferous.

Genus 501. Takhtajaniantha Nazarova

1629. Takhtajaniantha pusilla (Pall.) Nazarova [*Scorzonera pusilla* Pall.]   Figure 423
Perennial. Saline lands, sandy and clay deserts, dry riverbeds, stony slopes. Plain, foothills. Nuratau Relic Mountains, Mirzachul, Kyzylkum. Rubber-bearing, meliferous.

Figure 423 **Takhtajaniantha pusilla** (photography by Natalya Beshko)

Genus 502. Epilasia (Bunge) Benth. & Hook. f.

1630. Epilasia acrolasia (Bunge) Lipsch.
Annual. Sandy and clay deserts, saline lands, fine-earth and stony slopes, river valleys, fallow lands. Plain, foothills. Nuratau, Nuratau Relic Mountains, Mirzachul, Kyzylkum.

1631. Epilasia hemilasia (Bunge) C. B. Clarke   Figure 424
Annual. Sandy and clay deserts, saline lands, fine-earth, gravelly and stony slopes, roadsides, fields, fallow lands. Plain, foothills. Nuratau, Aktau, Nuratau Relic Mountains, Malguzar, North Turkestan, Mirzachul, Kyzylkum.

Figure 424 Epilasia hemilasia (photography by Natalya Beshko)

1632. Epilasia mirabilis Lipsch.
Annual. Sandy and clay deserts, fields, fallow lands. Plain. Mirzachul.

Genus 503. Tragopogon L.

1633. Tragopogon capitatus S. A. Nikitin
Biennial. Fine-earth, gravelly and stony slopes, river valleys. Foothills, montane zone. Nuratau, Nuratau Relic Mountains, Malguzar, North Turkestan. Rubber-bearing, meliferous.

1634. Tragopogon conduplicatus S.A. Nikitin
Perennial. Gravelly and stony slopes. Montane zone. Nuratau, North Turkestan. Rubber-bearing, meliferous.

1635. Tragopogon kultiassovii Popov ex S. A. Nikitin
Perennial. Fine-earth, gravelly and stony slopes. Foothills, montane zone. North Turkestan. Rubber-

bearing, meliferous.

1636. Tragopogon malikus S. A. Nikitin

Perennial. Fine-earth, gravelly and stony slopes, river valleys. Foothills, montane zone. Nuratau, Aktau, Nuratau Relic Mountains, Malguzar, North Turkestan. Rubber-bearing, meliferous.

1637. Tragopogon porrifolius subsp. **longirostris** (Sch. Bip.) Greuter [*Tragopogon kraschennikovii* S. A. Nikitin]   Figure 425

Biennial. Fine-earth, gravelly and stony slopes, river valleys, fallow lands, fields. Foothills, montane zone. Nuratau, Aktau, Nuratau Relic Mountains, Malguzar, North Turkestan. Rubber-bearing, meliferous.

Figure 425 **Tragopogon porrifolius** subsp. **longirostris** (photography by Natalya Beshko)

1638. Tragopogon pseudomajor S. A. Nikitin

Biennial. Piedmont plains, fine-earth and stony slopes, river valleys. Plain, foothills, montane zone. Nuratau, Malguzar, North Turkestan. Rubber-bearing, meliferous.

1639. Tragopogon turkestanicus S. A. Nikitin ex Pavlov

Biennial. Gravelly and stony slopes. Foothills, montane and alpine zones. Nuratau, Malguzar, North Turkestan. Rubber-bearing, meliferous.

1640. Tragopogon vvedenskyi Popov ex Pavlov

Biennial. Fine-earth, gravelly and stony slopes, ravines. Foothills, montane and alpine zones. Nuratau, Malguzar, North Turkestan. Rubber-bearing, meliferous.

Genus 504. Koelpinia Pall.

1641. Koelpinia linearis Pall.

Annual. Sandy and clay deserts, piedmont plains, fine-earth, gravelly and stony slopes. Plain, foothills, montane and alpine zones. Nuratau, Aktau, Nuratau Relic Mountains, Malguzar, North

Turkestan, Mirzachul, Kyzylkum. Fodder, weed, meliferous.

1642. Koelpinia macrantha C. Winkl.

Annual. Sandy deserts, fine-earth, gravelly and stony slopes, river valleys. Plain, foothills. Nuratau, Kyzylkum. Fodder, weed, meliferous.

1643. Koelpinia tenuissima Pavlov & Lipsch.

Annual. Sandy and clay deserts, saline lands, fine-earth, gravelly and stony slopes. Plain, foothills, montane zone. Nuratau, Aktau, Nuratau Relic Mountains, Malguzar, North Turkestan, Mirzachul, Kyzylkum. Fodder, weed, meliferous.

1644. Koelpinia turanica Vassilcz.

Annual. Sandy and clay deserts, saline lands, fine-earth, gravelly and stony slopes. Plain, foothills, montane zone. Nuratau, Nuratau Relic Mountains, Mirzachul, Kyzylkum. Fodder, weed, meliferous.

Genus 505. Echinops L.

1645. Echinops knorringianus Iljin

Annual. Fine-earth, gravelly and stony slopes. Foothills. North Turkestan.

1646. Echinops leucographus Bunge

Perennial. Fine-earth, gravelly and stony slopes, piedmont plains. Plain, foothills. Nuratau, Nuratau Relic Mountains.

1647. Echinops maracandicus Bunge

Perennial. Fine-earth, gravelly and stony slopes, pebbles, screes. Foothills, montane and alpine zones. Nuratau, Aktau, Nuratau Relic Mountains, Malguzar, North Turkestan.

1648. Echinops nanus Bunge

Annual or biennial. Fine-earth, gravelly and stony slopes. Foothills, montane zone. North Turkestan.

1649. Echinops nuratavicus A. D. Li   Figure 426

Perennial. Gravelly and stony slopes, rocks. Montane zone. Nuratau.

Figure 426 Echinops nuratavicus (photography by Natalya Beshko)

Genus 506. Chardinia Desf.

1650. Chardinia orientalis (L.) Kuntze
Annual. Clay deserts, piedmont plains, fine-earth, gravelly and stony slopes, rocks. Foothills, montane zone. Nuratau, Aktau, Nuratau Relic Mountains, Malguzar, North Turkestan, Mirzachul.

Genus 507. Xeranthemum L.

1651. Xeranthemum longepapposum Fisch. & C. A. Mey.
Annual. Piedmont plains, fallow lands, pebbles, fine-earth, gravelly and stony slopes, rocks. Plain, foothills, montane zone. Nuratau, Aktau, Nuratau Relic Mountains, Malguzar, North Turkestan, Mirzachul.

Genus 508. Carduus L.

1652. Carduus pycnocephalus subsp. **albidus** (M. Bieb.) Kazmi [*Carduus albidus* M. Bieb.]
Annual. River valleys, dry riverbeds, banks of canals, gardens, roadsides, fields, fallow lands. Plain, foothills, montane zone. Nuratau, Aktau, Nuratau Relic Mountains, Malguzar, North Turkestan, Mirzachul. Weed, meliferous.

1653. Carduus nutans L.
Biennial or perennial. Fine-earth, gravelly and stony slopes, rocks, banks of rivers and canals, wet places, fallow lands, wastelands, roadsides. Plain, foothills, montane and alpine zones. Nuratau, Aktau, Nuratau Relic Mountains, Malguzar, North Turkestan, Mirzachul. Medicinal, food, weed, meliferous.

Genus 509. Picnomon Adans.

1654. Picnomon acarna (L.) Cass.
Annual or biennial. Gravelly and stony slopes, fallow lands, wastelands, roadsides, banks of canals. Plain, foothills, montane zone. Nuratau, Aktau, Nuratau Relic Mountains, Malguzar, North Turkestan, Mirzachul. Weed.

Genus 510. Onopordum L.

1655. Onopordum acantium L.
Biennial. Gravelly and stony slopes, ravines, wastelands, fallow lands, river valleys, banks of canals, fields, roadsides. Plain, foothills, montane and alpine zones. Nuratau, Aktau, Nuratau Relic Mountains, Malguzar, North Turkestan, Mirzachul. Medicinal, oil, weed, fodder, meliferous.

1656. Onopordum leptolepis DC.
Biennial. Fine-earth, gravelly and stony slopes, rocks, screes, river valleys, fallow lands, fields, roadsides, banks of canals. Plain, foothills, montane and alpine zones. Nuratau, Aktau, Nuratau Relic Mountains, Malguzar, North Turkestan, Mirzachul. Weed.

Genus 511. Cirsium Mill.

1657. Cirsium alatum (S. C. Gmel.) Bobrov
Perennial. Meadows, saline lands, wet places, banks of rivers and canals, piedmont plains, ravines, fields. Plain, foothills, montane zone. Malguzar. Weed.

1658. Cirsium arvense (L.) Scop. [*Cirsium ochrolepidium* Juz.]
Perennial. Sandy deserts, fallow lands, fields, banks of rivers and canals, wet places, meadows, riparian forests, fine-earth slopes, gardens. Plain, foothills, montane and alpine zones. Nuratau, Aktau, Nuratau Relic Mountains, Malguzar, North Turkestan, Mirzachul, Kyzylkum. Medicinal, poisonous, oil, weed, meliferous.

1659. Cirsium brevipapposum Czerniak.
Perennial. Swamps, wet places near springs. Montane and alpine zones. Malguzar, North Turkestan.

1660. Cirsium esculentum (Siev.) C. A. Mey.
Perennial. Wet places, banks of rivers, lakes and canals, meadows, fine-earth slopes, gardens. Plain, foothills, montane and alpine zones. Malguzar, North Turkestan. Medicinal, weed, meliferous.

1661. Cirsium semenovii Regel
Biennial or perennial. Stony slopes, pebbles, screes, moraines. Montane and alpine zones. North Turkestan.

1662. Cirsium straminispinum C. Jeffrey ex Cufod [*Cirsium polyacanthum* Hochst. ex Kar. & Kir.]
Biennial or perennial. Fine-earth slopes, ravines, wet places. Foothills, montane and alpine zones. North Turkestan. Weed, meliferous.

1663. Cirsium turkestanicum (Regel) Petr.
Perennial. Gravelly and stony slopes, rocks, screes, river valleys, swamps, banks of canals. Plain, foothills, montane and alpine zones. Nuratau, Nuratau Relic Mountains, Malguzar, North Turkestan, Mirzachul. Weed, meliferous.

1664. Cirsium vulgare (Savi) Ten.
Biennial. Wet places, banks of rivers and canals, fallow lands, fields, wastelands, gravelly and stony slopes. Plain, foothills, montane zone. Nuratau, Aktau, Nuratau Relic Mountains, Malguzar, North Turkestan, Mirzachul. Medicinal, food, weed, meliferous.

Genus 512. Arctium L.

1665. Arctium aureum (C. Winkl.) Kuntze [*Cousinia aurea* C. Winkl.]
Perennial. Gravelly and stony slopes, pebbles. Foothills, montane zone. Nuratau Relic Mountains,

Malguzar, North Turkestan. Fodder.

1666. Arctium chloranthum (Kult.) S. López, Romasch., Susanna & N.Garcia [*Cousinia chlorantha* Kult.]

Perennial. Gravelly and stony slopes. Foothills, montane zone. Nuratau, Nuratau Relic Mountains, Malguzar, North Turkestan.

1667. Arctium haesitabundum (Juz.) S. López, Romasch., Susanna & N.Garcia [*Cousinia haesitabunda* Juz.]

Perennial. Gravelly and stony slopes. Montane zone, Nuratau, Malguzar. UzbRDB 1.

1668. Arctium karatavicum (Regel & Schmalh.) Kuntze [*Cousinia karatavica* Regel & Schmalh.]   Figure 427

Perennial. Gravelly and stony slopes, rocks. Foothills, montane zone. Nuratau, Nuratau Relic Mountains, Malguzar.

Figure 427 Arctium karatavicum (photography by Natalya Beshko)

1669. Arctium korolkowii (Regel & Schmalh.) Kuntze [*Cousinia korolkowii* Regel & Schmalh.]

Perennial. Gravelly and stony slopes. Foothills, montane zone. Nuratau.

1670. Arctium leiospermum Juz. & Ye.V. Serg.

Biennial. River valleys, banks of canals, roadsides, wet places, gardens. Plain, foothills, montane zone. Nuratau, Malguzar, North Turkestan. Medicinal, food, weed, meliferous.

1671. Arctium pallidivirens (Kult.) S. Lopez, Romanschenko, Susanna & N. Garcia [*Anura pallidivirens* (Kult.) Tscherneva]   Figure 428
Perennial. Fine-earth, gravelly and stony slopes. Montane zone. Nuratau, Malguzar. UzbRDB 1. Endemic to the Nuratau Mountains.

Figure 428 **Arctium pallidivirens** (photography by Natalya Beshko)

Genus 513. Cousinia L.

1672. Cousinia alpina Bunge
Biennial. Fine-earth, gravelly and stony slopes, screes, rocks. Montane and alpine zones. North Turkestan.

1673. Cousinia ambigens Juz.
Perennial. Gravelly and stony slopes. Foothills. Nuratau.

1674. Cousinia botschantzevii Juz. ex Tscherneva   Figure 429
Biennial. Fine-earth, gravelly and stony slopes. Montane zone. Nuratau, Aktau. Endemic to the Nuratau Mountains.

Figure 429 Cousinia botschantzevii (photography by Natalya Beshko)

1675. Cousinia bungeana Regel & Schmalh.

Annual. Clay deserts. Plain. Mirzachul.

1676. Cousinia coronata Franch.

Biennial. Fine-earth, gravelly and stony slopes. Montane zone. Aktau, Malguzar, North Turkestan.

1677. Cousinia dissectifolia Kult.

Biennial. Fine-earth, gravelly and stony slopes. Montane zone. Nuratau.

1678. Cousinia dshisakensis Kult.   Figure 430

Biennial. Gravelly and stony slopes. Montane zone. Nuratau, Aktau, Malguzar, North Turkestan. UzbRDB 2.

Figure 430 **Cousinia dshisakensis** (photography by Natalya Beshko)

1679. Cousinia dubia Popov

Biennial. Gravelly and stony slopes. Montane zone. Nuratau, Malguzar, North Turkestan.

1680. Cousinia eriotricha Juz.   Figure 431

Perennial. Gravelly and stony slopes, screes. Montane and alpine zones. Nuratau, Malguzar, North Turkestan.

Figure 431 **Cousinia eriotricha** (photography by Natalya Beshko)

1681. Cousinia franchetii C. Winkl.

Perennial. Stony slopes. Montane zone. North Turkestan.

1682. Cousinia hamadae Juz.

Perennial. Gravelly and stony slopes. Foothills. Nuratau Relic Mountains.

1683. Cousinia horridula Juz.

Perennial. Fine-earth, gravelly and stony slopes. Foothills, montane zone. Nuratau, Malguzar, North Turkestan.

1684. Cousinia integrifolia Franch.

Biennial. Fine-earth and stony slopes. Montane zone. Malguzar, North Turkestan.

1685. Cousinia lappacea Schrenk

Perennial. Stony slopes, rocks. Montane zone. Malguzar, North Turkestan. Fodder, weed.

1686. Cousinia microcarpa Boiss.   Figure 432

Perennial. Fine-earth, gravelly and stony slopes, river valleys, banks of canals, fallow lands, fields, roadsides, wastelands. Plain, foothills, montane zone. Nuratau, Aktau, Nuratau Relic Mountains, Malguzar, North Turkestan, Mirzachul, Kyzylkum. Meliferous, weed.

Figure 432 Cousinia microcarpa (photography by Natalya Beshko)

1687. Cousinia mollis Schrenk

Biennial. Fine-earth, gravelly and stony slopes, screes, sandy and clay deserts, fallow lands, river valleys. Plain, foothills. Nuratau, Nuratau Relic Mountains, North Turkestan, Mirzachul, Kyzylkum.

1688. Cousinia olgae Regel & Schmalh.

Biennial. Fine-earth, gravelly and stony slopes, pebbles, piedmont plains. Plain, foothills. Nuratau, Aktau, Nuratau Relic Mountains, Malguzar, North Turkestan.

1689. Cousinia outichaschensis Franch.

Biennial. Gravelly and stony slopes, rocks. Montane and alpine zones. Malguzar, North Turkestan.

1690. Cousinia oxiana Tscherneva

Biennial. Sandy deserts. Plain. Kyzylkum.

1691. Cousinia polytimetica Tscherneva

Annual. Piedmont plain, river valleys. Plain, foothills. Nuratau, Nuratau Relic Mountains, Malguzar.

1692. Cousinia prolifera Jaub. & Spach.

Annual. Sandy and clay deserts, saline lands, piedmont plains, fields, fallow lands. Plain, foothills.

Nuratau Relic Mountains, Mirzachul, Kyzylkum.

1693. Cousinia pseudodshisakensis Tscherneva & Vved.
Biennial. Gravelly and stony slopes, screes, rocks. Foothills, montane zone. Nuratau, Nuratau Relic Mountains, Malguzar, North Turkestan.

1694. Cousinia pygmaea C. Winkl.
Biennial. Fine-earth and stony slopes, fallow lands. Montane zone. Malguzar, North Turkestan.

1695. Cousinia radians Bunge   Figure 433
Biennial. Fine-earth, gravelly and stony slopes, piedmont plains, fallow lands. Plain, foothills, montane zone. Nuratau, Aktau, Nuratau Relic Mountains, Malguzar, North Turkestan, Mirzachul. Weed, meliferous.

Figure 433 Cousinia radians (photography by Natalya Beshko)

1696. Cousinia regelii C. Winkl.
Biennial. Fine-earth, gravelly and stony slopes. Foothills. Aktau.

1697. Cousinia resinosa Juz.
Perennial. Fine-earth, gravelly and stony slopes, piedmont plains, sandy and clay deserts, roadsides, fallow lands, wastelands. Plain, foothills, montane zone. Nuratau, Aktau, Nuratau Relic Mountains, Malguzar, North Turkestan, Mirzachul, Kyzylkum. Fodder, weed, meliferous.

1698. Cousinia schtschurowskiana Regel & Schmalh.   Figure 434
Biennial. Gravelly and stony slopes. Montane zone. Nuratau, Aktau.

Figure 434 Cousinia schtschurowskiana (photography by Natalya Beshko)

1699. Cousinia simulatrix C. Winkl.
Biennial. Gravelly and stony slopes, ravines. Montane zone. Nuratau.

1700. Cousinia sogdiana Bornm.   Figure 435
Biennial. Sandy deserts. Plain. Kyzylkum.

Figure 435 Cousinia sogdiana (photography by Natalya Beshko)

1701. Cousinia spiridonovii Juz.
Perennial. Clay deserts, river valleys, banks of canals, fallow lands. Plain. Nuratau Relic Mountains, Mirzachul.

1702. Cousinia tenella Fisch. & C. A. Mey.
Annual. Sandy and clay deserts, saline lands, piedmont plains, river valleys, fields, roadsides, fallow lands. Plain, foothills. Nuratau, Nuratau Relic Mountains, Mirzachul, Kyzylkum.

1703. Cousinia umbrosa Bunge
Perennial. Fine-earth slopes, river valleys, shady wet places. Foothills, montane zone. Nuratau, Aktau, Malguzar, North Turkestan. Fodder, weed, meliferous.

1704. Cousinia verticillaris Bunge
Perennial. Gravelly and stony slopes. Alpine zone. Malguzar, North Turkestan.

1705. Cousinia xanthina Bornm.
Perennial. Fine-earth slopes. Alpine zone. North Turkestan.

Genus 514. Lipskyella Juz.

1706. Lipskyella annua (C. Winkl.) Juz.
Annual. Sandy deserts. Plain. Kyzylkum.

Genus 515. Saussurea DC.

1707. Saussurea elegans Ledeb.
Perennial. Fine-earth and stony slopes, meadows, river valleys. Foothills, montane zone. North Turkestan.

1708. Saussurea salsa (Pall. ex Pall.) Spreng.
Perennial. Saline lands, banks or rivers, lakes and canals. Plain. Mirzachul.

Genus 516. Jurinea Cass.

1709. Jurinea abramowii Regel & Herder
Perennial. Stony slopes, screes. Montane and alpine zones. North Turkestan.

1710. Jurinea androssovii Iljin
Perennial. Gravelly and stony slopes, rocks. Alpine zone. North Turkestan.

1711. Jurinea ferganica (Iljin) Iljin
Perennial. Gravelly and stony slopes, screes, rocks. Foothills, montane zone. Malguzar, North Turkestan.

1712. Jurinea helichrysifolia Popov ex Iljin   Figure 436
Perennial. Fine-earth, gravelly and stony slopes. Montane zone. Malguzar, NorthTurkestan.

Figure 436 Jurinea helichrysifolia (photography by Natalya Beshko)

1713. Jurinea kokanica Iljin

Perennial. Fine-earth, gravelly and stony slopes, river valleys. Foothills, montane zone. Nuratau, Nuratau Relic Mountains, Malguzar, North Turkestan. Ornamental, meliferous.

1714. Jurinea komarovii Iljin

Perennial. Gravelly and stony slopes, pebbles, rocks. Montane and alpine zones. North Turkestan.

1715. Jurinea lasiopoda Trautv.

Perennial. Stony slopes, rocks. Foothills, montane zone. Nuratau, Nuratau Relic Mountains.

1716. Jurinea maxima C. Winkl.   Figure 437
Perennial. Fine-earth and stony slopes. Montane and alpine zones. Malguzar, North Turkestan. Ornamental, meliferous.

Figure 437 Jurinea maxima (photography by Natalya Beshko)

1717. Jurinea olgae Regel & Schmalh.   Figure 438
Perennial. Rocks, Fine-earth, gravelly and stony slopes, screes. Foothills, montane zone. Nuratau, Aktau, Nuratau Relic Mountains, Malguzar, North Turkestan. Ornamental, meliferous.

Figure 438 Jurinea olgae (photography by Natalya Beshko)

1718. Jurinea suffruticosa Regel

Semishrub. Piedmont plains, gravelly and stony slopes, rocks. Plain, foothills, montane zone. Nuratau.

1719. Jurinea trautvetteriana Regel & Schmalh.   Figure 439

Perennial. Fine-earth, gravelly and stony slopes. Foothills, montane zone. Nuratau, Aktau, Malguzar, North Turkestan. Ornamental, meliferous.

Figure 439 Jurinea trautvetteriana (photography by Natalya Beshko)

1720. Jurinea zakirovii Iljin   Figure 440

Perennial. Fine-earth, gravelly and stony slopes. Montane zone. Nuratau. UzbRDB 1. Endemic to the Nuratau Mountains.

Figure 440 Jurinea zakirovii (photography by N. Yu. Beshko)

Genus 517. Amberboa (Pers.) Less.

1721. Amberboa turanica Iljin
Annual. Sandy and clay deserts, piedmont plains, fallow lands. Plain, foothills. Nuratau, Nuratau Relic Mountains, Malguzar, Mirzachul, Kyzylkum.

Genus 518. Centaurea L.

1722. *Centaurea benedicta (L.) L. [*Cnicus benedictus* L.]
Annual. Clay deserts, wastelands, fallow lands, fields, wet places, banks of rivers and canals. Plain, foothills. Nuratau, Aktau, Nuratau Relic Mountains, Malguzar, North Turkestan, Mirzachul. Medicinal, oil, weed.

1723. Centaurea bruguierana subsp. **belangeriana** (DC.) Bornm. [*Centaurea belangeriana* (DC.) Stapf]
Annual. Sandy, clayey and skeleton soils, fields, fallow lands, wastelands, roadsides. Plain, foothills, montane zone. Nuratau, Aktau, Nuratau Relic Mountains, Malguzar, Mirzachul, Kyzylkum. Weed.

1724. *Centaurea iberica Trevir. ex Spreng.
Biennial. Wastelands, banks of canals, river valleys, roadsides, surroundings of settlements, fine-earth, gravelly and stony slopes. Foothills, montane zone. Nuratau, Aktau, Nuratau Relic Mountains, Malguzar, North Turkestan, Mirzachul, Kyzylkum. Medicinal, weed.

1725. *Centaurea solstitialis L.
Biennial. Fallow lands, wastelands, fields, banks of canals, surroundings of settlements, fine-earth and gravelly slopes. Plain, foothills. Nuratau, Malguzar, North Turkestan, Mirzachul. Medicinal, weed.

1726. Centaurea virgata subsp. **squarrosa** (Boiss.) Gugler [*Centaurea squarrosa* Willd.]
Perennial. Fine-earth, gravelly and stony slopes, piedmont plains, sandy and clay deserts, fields, fallow lands. Plain, foothills, montane zone. Nuratau, Aktau, Nuratau Relic Mountains, Malguzar, North Turkestan, Mirzachul, Kyzylkum. Medicinal, weed, meliferous.

Genus 519. Cyanus Mill.

1727. Cyanus depressus (M. Bieb.) Soják [*Centaurea depressa* M. Bieb.]
Annual. Fallow lands, wastelands, fields, banks of canals, fine-earth and gravelly slopes. Plain, foothills, montane zone. Nuratau, Aktau, Nuratau Relic Mountains, Malguzar, North Turkestan, Mirzachul. Medicinal, ornamental, meliferous, weed.

1728. *Cyanus segetum Hill [*Centaurea cyanus* L.]
Annual. Fallow lands, fields, wastelands, surroundings of settlements, banks of canals. Plain, foothills, montane zone. Nuratau, Aktau, Nuratau Relic Mountains, Malguzar, North Turkestan. Medicinal, poisonous, dye, ornamental, meliferous, weed.

Genus 520. Rhaponticum Ludw.

1729. Rhaponticum repens (L.) Hidalgo [*Acroptilon repens* (L.) DC.]
Perennial. Sandy and clay desets, saline lands, banks of rivers, lakes and canals, fine-earth, gravelly and stony slopes, fields, fallow lands, roadsides, wastelands, gardens. Plain, foothills, montane zone. Nuratau, Aktau, Nuratau Relic Mountains, Malguzar, North Turkestan, Mirzachul, Kyzylkum. Medicinal, poisonous, meliferous, weed. Quarantine invasive species.

Genus 521. Hyalea (DC.) Jaub. & Spach

1730. Hyalea pulchella (Ledeb.) Koch   Figure 441
Annual. Sandy desert. Plain. Kyzylkum.

Figure 441 Hyalea pulchella (photography by Natalya Beshko)

Genus 522. *Stizolophus Cass.

1731. *Stizolophus balsamita (Lam.) K. Koch
Annual. Fields, fallow lands, roadsides, gravelly and stony slopes, banks of rivers, lakes and canals. Plain, foothills, montane zone. Nuratau, Aktau, Nuratau Relic Mountains, Malguzar, North

Turkestan, Mirzachul. Medicinal, dye, weed.

Genus 523. Crupina Cass.

1732. Crupina vulgaris Pers. ex Cass. [*Crupina oligantha* Tscherneva]
Annual. Gravelly and stony slopes, rocks, screes, pebbles, banks of rivers, gardens, vineyards, fallow lands, fields. Plain, foothills, montane zone. Nuratau, Aktau, Nuratau Relic Mountains, Malguzar, North Turkestan. Weed.

Genus 524. Serratula L.

1733. Serratula algida Iljin
Perennial. Rocks, gravelly and stony slopes, moraines. Alpine zone. North Turkestan. Medicinal.

1734. Serratula lancifolia Zakirov
Perennial. Fine-earth, gravelly and stony slopes. Montane zone. Malguzar. UzbRDB 1.

Genus 525. Carthamus L.

1735. Carthamus oxyacanthus M. Bieb.
Annual. Fallow lands, wastelands, pebbles, fields, banks of rivers and canals, surroundings of settlements, stony slopes. Plain, foothills, montane zone. Nuratau, Aktau, Nuratau Relic Mountains, Malguzar, North Turkestan, Mirzachul. Weed.

1736. *Carthamus tinctorius L.
Annual. Fields, fallow lands, wastelands, roadsides, fine-earth slopes (escaped from culture). Plain, foothills, montane zone. Nuratau, Aktau, Malguzar, North Turkestan, Mirzachul. Dye, oil, food, industrial.

1737. Carthamus turkestanicus Popov
Annual. Fine-earth, gravelly and stony slopes, wastelands, fields, fallow lands, roadsides, banks of canals, surroundings of settlements. Foothills, montane zone. Nuratau, Aktau, Nuratau Relic Mountains, Malguzar, North Turkestan, Mirzachul. Weed.

Genus 526. *Eclipta L.

1738. *Eclipta prostrata (L.) L.
Annual. Fields, fallow lands, banks of canals. Plain. Mirzachul. Weed.

Genus 527. Bidens L.

1739. Bidens tripartita L.
Annual. Wet places, banks of rivers, lakes and canals, fields, fallow lands. Plain, foothills, montane zone. Nuratau, Malguzar, North Turkestan, Mirzachul. Medicinal, essential oil, dye, weed.

Genus 528. Senecio L.

1740. Senecio franchetii C. Winkl.
Perennial. Fine-earth and stony slopes. Montane zone. North Turkestan.

1741. Senecio olgae Regel & Schmalh.
Perennial. Stony slopes. Montane zone. Malguzar, North Turkestan.

1742. Senecio subdentatus (Bunge) Ledeb.   Figure 442
Annual. Sandy deserts, saline lands, banks of rivers and canals, pebbles, fine-earth, gravelly and stony slopes. Plain, foothills, montane zone. Nuratau, Nuratau Relic Mountains, Malguzar, North Turkestan, Mirzachul, Kyzylkum. Fodder.

Figure 442 Senecio subdentatus (photography by Natalya Beshko)

Genus 529. Ligularia Cass.

1743. Ligularia alpigena Pojark.
Perennial. Fine-earth, gravelly and stony slopes, screes, banks of rivers, swamps. Montane and alpine zones. North Turkestan. Meliferous.

1744. Ligularia heterophylla Rupr.
Perennial. Fine-earth, gravelly and stony slopes, banks of rivers. Montane zone. Nuratau, Malguzar, North Turkestan. Meliferous.

1745. Ligularia thomsonii (C. B. Clarke) Pojark.   Figure 443
Perennial. Fine-earth, gravelly and stony slopes, screes, pebbles, moraines, subalpine meadows, swamps, banks of rivers. Montane and alpine zones. Malguzar, North Turkestan. Poisonous, weed, meliferous.

Figure 443 **Ligularia thomsonii** (photography by Natalya Beshko)

Genus 530. Tussilago L.

1746. Tussilago farfara L.
Perennial. River valleys, wet places, pebbles, fine-earth slopes. Montane zone. North Turkestan. Medicinal, essential oil, meliferous.

Genus 531. Karelinia Less.

1747. Karelinia caspia (Pall.) Less.   Figure 444
Perennial. Saline lands, banks of rivers, lakes and canals, fields, fallow lands. Plain, foothills. Nuratau, Aktau, Nuratau Relic Mountains, Malguzar, Mirzachul, Kyzylkum. Weed, meliferous.

Figure 444 Karelinia caspia (photography by Natalya Beshko)

Genus 532. Inula L.

1748. Inula britannica L.
Perennial. Fine-earth slopes, wet places, banks of rivers and canals, fields, fallow lands. Plain, foothills, montane zone. Malguzar, North Turkestan, Mirzachul. Medicinal, fodder, meliferous.

1749. Inula caspica F.K. Blum. ex Ledeb.
Annual or biennial. Saline lands, banks of rivers and canals, fields. Plain, foothills. Malguzar, Mirzachul. Medicinal, meliferous.

1750. Inula glauca C. Winkl.
Perennial. Gravelly and stony slopes, rocks. Montane zone. Malguzar, North Turkestan.

1751. Inula helenium L.
Perennial. Fine-earth slopes, banks of rivers, wet places. Foothills, montane zone. Malguzar, North Turkestan. Medicinal, ornamental, dye, meliferous.

1752. Inula orientalis Lam. [*Inula macrophylla* Kar. & Kir.]
Perennial. Fine-earth, gravelly slopes. Foothills, montane zone. Malguzar, North Turkestan. Medicinal, dye, meliferous.

1753. Inula rhizocephala Schrenk
Perennial. Fine-earth, gravelly and stony slopes, alpine meadows. Montane and alpine zones. Malguzar, North Turkestan. Medicinal, fodder.

Genus 533. Pulicaria Gaerth.

1754. Pulicaria dysenterica subsp. **uliginosa** Nyman [*Pulicaria uliginosa* Steven ex DC.]
Annual. Banks of rivers and canals, wet places. North Turkestan. Poisonous, insecticide.
1755. Pulicaria gnaphalodes (Vent) Boiss.
Perennial. Rocks, gravelly and stony slopes, pebbles, dry riverbeds. Plain, foothills, montane zone. Nuratau, Aktau, Nuratau Relic Mountains, Malguzar. Essential oil.
1756. Pulicaria salviifolia Bunge
Perennial. Gravelly and stony slopes, pebbles, dry riverbeds, banks of rivers and canals. Foothills, montane zone. Nuratau, Aktau, Nuratau Relic Mountains, Malguzar, North Turkestan. Essential oil.

Genus 534. Pentanema Cass.

1757. Pentanema albertoregelia (C. Winkl.) Gorschk. [*Vicoa albertoregelia* C. Winkl.]
Perennial. Gravelly and stony slopes, rocks. Montane zone. Nuratau, Malguzar, North Turkestan.
1758. Pentanema divaricatum Cass.
Annual. Gravelly and stony slopes, dry riverbeds, fields, wastelands. Plain, foothills. Nuratau.

Genus 535. Helichrysum Mill.

1759. Helichrysum maracandicum Popov   Figure 445
Perennial. Fine-earth, gravelly and stony slopes. Montane zone. Nuratau, Malguzar, North Turkestan. Medicinal, ornamental.

Figure 445 Helichrysum maracandicum (photography by Natalya Beshko)

1760. Helichrysum mussae Nevski

Perennial. Gravelly and stony slopes, rocks. Montane zone. Nuratau, Malguzar, North Turkestan.

1761. Helichrysum nuratavicum Krasch.   Figure 446

Perennial. Gravelly and stony slopes, rocks. Montane zone. Nuratau. UzbRDB 2. Endemic to the Nuratau Mountains.

Figure 446 **Helichrysum nuratavicum** (photography by N. Yu. Beshko)

1762. Helichrysum tianschanicum Regel

Perennial. Gravelly and stony slopes, pebbles, banks of rivers. Foothills, montane and alpine zones. North Turkestan.

Genus 536. Laphangium (Hilliard & B. L. Burtt) Tzvelev

1763. Laphangium luteoalbum (L.) Tzvelev [*Gnaphalium luteoalbum* L.]

Annual. River valleys, wet places, banks of canals, fallow lands fields. Nuratau, Malguzar, North Turkestan. Medicinal, fodder, ornamental, weed.

Genus 537. Gnaphalium L.

1764. Gnaphalium supinum L.

Perennial. Fine-earth, gravelly and stony slopes. Montane and alpine zones. Malguzar, North Turkestan.

Genus 538. Leontopodium R. Br. ex Cass.

1765. Leontopodium fedtschenkoanum Beauverd

Perennial. Fine-earth slopes, meadows, swamps. Montane and alpine zones. North Turkestan. Ornamental.

Genus 539. Filago L.

1766. Filago arvensis Lam.
Annual. Fine-earth, gravelly and stony slopes, piedmont plains, fields, fallow lands, roadsides, wastelands. Plain, foothills, montane and alpine zones. Nuratau, Aktau, Nuratau Relic Mountains, Malguzar, North Turkestan. Medicinal, ornamental, weed.

1767. Filago filaginoides (Kar. & Kir.) Wagenitz [*Evax filaginoides* Kar. & Kir.]
Annual. Sandy deserts, saline lands, banks of lakes and canals. Plain. Kyzylkum.

1768. Filago germanica (L.) Huds. (*Filago vulgaris* Lam.)
Annual. Gravelly and stony slopes, fields, wastelands. Foothills, montane zone. Nuratau, Aktau, Nuratau Relic Mountains, Mirzachul. Medicinal, weed.

1769. Filago pyramidata L.
Annual. Fine-earth, gravelly and stony slopes, piedmont plains, fields, fallow lands, roadsides, wastelands. Foothills, montane zone. Nuratau, Nuratau Relic Mountains, Malguzar, North Turkestan. Medicinal, weed.

Genus 540. Cymbolaena Smoljan.

1770. Cymbolaena griffithii (A. Gray) Wagenitz
Annual. Fine-earth, gravelly and stony slopes, piedmont plains, fallow lands, wastelands. Plain, foothills, montane zone. Nuratau, Aktau, Nuratau Relic Mountains, Malguzar, North Turkestan.

Genus 541. Solidago L.

1771. Solidago kuhistanica Juz.
Perennial. Fine-earth and stony slopes, rocks, banks of rivers, meadows. Montane and alpine zones. Nuratau, North Turkestan. Ornamental, meliferous.

Genus 542. Pseudolinosyris Novopokr.

1772. Pseudolinosyris grimmii (Regel & Schmalh.) Novopokr.
Semishrub. Fine-earth, gravelly and stony slopes, rocks, screes. Foothills, montane zone. Nuratau, Aktau, Malguzar, North Turkestan.

Genus 543. Galatella Cass.

1773. Galatella coriacea Novopokr.
Perennial. Fine-earth, gravelly and stony slopes. Foothills, montane and alpine zones. Nuratau,

Malguzar, North Turkestan. Weed.

Genus 544. Kalimeris (Cass.) Cass.

1774. Kalimeris altaica (Willd.) Nees ex Fisch. Mey. & Avé-Lall. [*Aster canescens* (Ness) Fisjun]
Perennial. Fine-earth, gravelly and stony slopes, roadsides, banks of canals, fallow lands, fields. Plain, foothills, montane zone. Nuratau, Aktau, Nuratau Relic Mountains, Malguzar, North Turkestan, Mirzachul.

Genus 545. *Tripolium L.

1775. *Tripolium pannonicum subsp. **tripolium** (L.) Greuter
Annual. Saline lands, banks of canals, roadsides, fields, fallow lands. Plain. Nuratau Relic Mountains, Mirzachul, Kyzylkum. Fodder, ornamental, meliferous.

Genus 546. Rhinactinidia Novopokr.

1776. Rhinactinidia popovii (Botsch.) Botsch.
Perennial. Rocks, stony slopes. Montane and alpine zones. Nuratau, Malguzar, North Turkestan.

Genus 547. Erigeron L.

1777. Erigeron acris L.
Biennial. Meadows, fallow lands, saline lands, banks of rivers, lakes and canals, rocks, fine-earth, gravelly and stony slopes. Plain, foothills, montane and alpine zones. Nuratau, Aktau, Nuratau Relic Mountains, Malguzar, North Turkestan, Mirzachul. Medicinal, weed.

1778. Erigeron bellidiformis Popov
Perennial. Fine-earth, gravelly and stony slopes. Alpine zone. North Turkestan

1779. Erigeron cabulicus (Boiss.) Botsch. [*Psychrogeton cabulicus* Boiss.]
Perennial. Fine-earth, gravelly and stony slopes. Montane and alpine zones. Malguzar, North Turkestan. Ornamental.

1780. *Erigeron canadensis L. [*Conyza canadensis* (L.) Cronquist]
Annual. Gardens, fields, fallow lands, wastelands, banks of rivers, lakes and canals, roadsides. Plain, foothills, montane zone. Nuratau, Aktau, Nuratau Relic Mountains, Malguzar, North Turkestan, Mirzachul, Kyzylkum. Medicinal, weed, meliferous.

1781. Erigeron khorossanicus Boiss. [*Psychrogeton aucherii* (DC.) Grierson]
Biennial. Fine-earth slopes, banks of rivers and canals, fallow lands. Foothills, montane and alpine zones. Nuratau, Aktau, Nuratau Relic Mountains, Malguzar, North Turkestan.

1782. Erigeron pallidus Popov

Perennial. Fine-earth, gravelly and stony slopes, pebbles, moraines. Montane and alpine zones. Malguzar, NorthTurkestan.

1783. Erigeron petiolaris Vierh. [*Erigeron pseudoneglectus* Popov]

Perennial. Fine-earth slopes, subalpine and alpine meadows. Montane and alpine zones. North Turkestan.

1784. Erigeron popovii Botsch.

Biennial or perennial. Meadows, banks of rivers. Montane zone. North Turkestan.

1785. Erigeron pseudoseravschanicus Botsch.

Perennial. Fine-earth and stony slopes, banks of rivers, meadows. Foothills, montane and alpine zones. North Turkestan.

1786. Erigeron seravschanicus Popov

Perennial. Gravelly and stony slopes, shady wet places, subalpine meadows, moraines. Montane and alpine zones. North Turkestan.

1787. Erigeron sogdianus Popov

Perennial. Subalpine and alpine meadows, rocks, screes, moraines. Montane and alpine zones. North Turkestan.

1788. Erigeron umbrosus (Kar. & Kir.) Boiss.

Annual. Gravelly and stony slopes, screes, rocks, shady wet places. Montane and alpine zones. North Turkestan.

Genus 548. Psychrogeton Boiss.

1789. Psychrogeton amorphoglossus (Boiss.) Novopokr. [*Psychrogeton leucophyllus* (Bunge) Novopokr.]

Perennial. Gravelly and stony slopes, rocks. Montane and alpine zones. North Turkestan.

1790. Psychrogeton pseuderigeron (Bunge) Novopokr. ex Nevski

Perennial. Fine-earth, gravelly and stony slopes, rocks, screes, pebbles. Montane and alpine zones. North Turkestan.

Genus 549. Symphyotrichum Nees

1791. Symphyotrichum ciliatum (Ledeb.) G. L. Nesom [*Brachyactis ciliata* (Ledeb.) Ledeb.]

Annual. Saline lands, banks of rivers, lakes and canals. Plain. Mirzachul, Kyzylkum.

Genus 550. Lachnophyllum Bunge

1792. Lachnophyllum gossypinum Bunge

Annual. Fine-earth, gravelly and stony slopes, piedmont plains, clay deserts, fallow lands, pebbles, fields. Plain, foothills, montane zone. Nuratau, Aktau, Nuratau Relic Mountains, Malguzar, North Turkestan, Mirzachul. Medicinal, essential oil, fodder.

Genus 551. Anthemis L.

1793. Anthemis deserticola Krasch. & Popov
Annual. Saline lands. Plain. Mirzachul, Kyzylkum.

Genus 552. Achillea L.

1794. Achillea arabica Kotschy [*Achillea biebersteinii* Afan.]   Figure 447
Perennial. Fine-earth, gravelly and stony slopes, pebbles, river valleys, piedmont plains, fallow lands, wastelands, banks of canals. Plain, foothills, montane zone. Nuratau, Aktau, Nuratau Relic Mountains, Malguzar, North Turkestan, Mirzachul. Medicinal.

Figure 447 *Achillea arabica* (photography by Natalya Beshko)

1795. Achillea filipendulina Lam.   Figure 448
Perennial. Fine-earth, gravelly and stony slopes, screes, pebbles, river valleys, fallow lands, banks of canals, gardens. Foothills, montane and alpine zones. Nuratau, Aktau, Nuratau Relic Mountains, Malguzar, North Turkestan. Medicinal, essential oil, ornamental.

Figure 448 Achillea filipendulina (photography by Natalya Beshko)

1796. Achillea millefolium L.   Figure 449

Perennial. River valleys, fine-earth slopes, meadows. Montane and alpine zones. Malguzar, North Turkestan. Medicinal, ornamental.

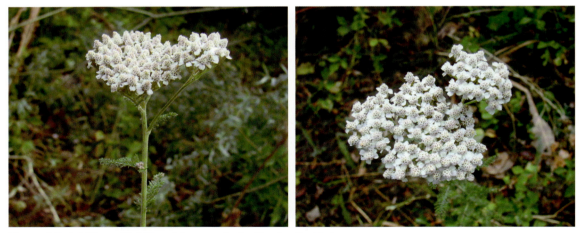

Figure 449 Achillea millefolium (photography by Natalya Beshko)

1797. Achillea santolinoides subsp. wilhelmsii (K. Koch) Greuter [*Achillea kermanica* Gand.]   Figure 450

Perennial. Sandy and clay deserts, piedmont plains, fine-earth, gravelly and stony slopes, river valleys, banks of canals, fallow lands, fields. Plain, foothills, montane zone. Nuratau, Aktau,

Nuratau Relic Mountains, Malguzar, Mirzachul, Kyzylkum. Medicinal.

Figure 450 Achillea santolinoides subsp. **wilhelmsii** (photography by Natalya Beshko)

Genus 553. Handelia Heimerl

1798. Handelia trichophylla Heimerl   Figure 451
Perennial. Fine-earth, gravelly slopes, screes, pebbles, river valleys, fields, fallow lands. Plain, foothills, montane zone. Nuratau, Nuratau Relic Mountains, Malguzar, North Turkestan. Medicinal, ornamental.

Figure 451 Handelia trichophylla (photography by Natalya Beshko)

Genus 554. Pseudohandelia Tzvelev

1799. Pseudohandelia umbellifera (Boiss.) Tzvelev   Figure 452
Biennial or perennial. Sandy and clay deserts, fine-earth and gravelly slopes, fallow lands. Plain, foothills. Nuratau, Aktau, Nuratau Relic Mountains, Malguzar, North Turkestan, Mirzachul, Kyzylkum. Medicinal, ornamental.

Figure 452 Pseudohandelia umbellifera (photography by Natalya Beshko)

Genus 555. Tanacetopsis (Tzvelev) Kovalevsk.

1800. Tanacetopsis karataviensis (Kovalevsk.) Kovalevsk. Figure 453
Perennial. Stony slopes, rocks. Montane zone. Nuratau.

Figure 453 Tanacetopsis karataviensis (photography by Natalya Beshko)

1801. Tanacetopsis mucronata (Regel & Schmalh.) Kovalevsk.
Perennial. Stony slopes, rocks, pebbles. Montane and alpine zones. North Turkestan.

1802. Tanacetopsis urgutensis (Popov ex Tzvel.) Kovalevsk.
Perennial. Gravelly and stony slopes, rocks. Montane and alpine zones. Malguzar, North Turkestan.

Genus 556. Lepidolopsis Poljakov

1803. Lepidolopsis turkestanica (Regel & Schmalh.) Poljakov
Perennial. Fine-earth, gravelly and stony slopes, fallow lands, pebbles, banks of rivers. Foothills, montane zone. Nuratau, Aktau, Nuratau Relic Mountains, Malguzar, North Turkestan.

Genus 557. Mausolea Bunge

1804. Mausolea eriocarpa (Bunge) Poljakov ex Podlech.   Figure 454
Semishrub. Sandy deserts. Plain. Kyzylkum. Essential oil.

Figure 454 Mausolea eriocarpa (photography by Natalya Beshko)

Genus 558. Artemisia L.

1805. Artemisia absinthium L.
Annual, perennial. River valleys, banks of canals, roadsides, fallow lands, gardens, meadows, wastelands, fine-earth, gravelly and stony slopes. Plain, foothills, montane zone. Nuratau, Nuratau Relic Mountains, Malguzar, North Turkestan, Mirzachul. Medicinal, food (spice-aromatic), essential oil, poisonous.

1806. *Artemisia annua L.
Annual. River valleys, banks of canals, roadsides, fields, fallow lands, wastelands, gardens, surroundings of settlements. Plain, foothills, montane zone. Nuratau, Aktau, Nuratau Relic Mountains, Malguzar, North Turkestan, Mirzachul, Kyzylkum. Medicinal, essential oil, insecticide, dye, weed.

1807. Artemisia cina Berg ex Poljakov
Semishrub. Saline lands, dry riverbeds, sandy soils. Plain. Mirzachul. Medicinal, essential oil, poisonous.

1808. Artemisia diffusa Krasch. ex Poljakov
Semishrub. Sandy and clay deserts, pebbles, saline lands, fine-earth, gravelly and stony slopes. Plain, foothills. Nuratau Relic Mountains, Mirzachul, Kyzylkum. Medicinal, essential oil, fodder.

1809. Artemisia dracunculus L.
Perennial. Fine-earth, gravelly and stony slopes, river valleys, fallow lands, wastelands. Plain, foothills, montane zone. Nuratau, Malguzar, North Turkestan. Medicinal, essential oil, food (spice-aromatic), fodder, meliferous, weed, ornamental.

1810. Artemisia ferganensis Krasch. ex Poljakov
Semishrub. Sandy and clay deserts, fallow lands, fine-earth, gravelly and stony slopes, river valleys, dry riverbeds, pebbles, screes. Plain, foothills, montane and alpine zones. Nuratau, Malguzar, North Turkestan, Mirzachul. Medicinal, essential oil, fodder.

1811. Artemisia eremophila Krasch. & Butkov
Semishrub. Fine-earth, gravelly and stony slopes, pebbles, piedmont plains. Plain, foothills. Nuratau, North Turkestan.

1812. Artemisia glanduligera Krasch. ex Poljakov
Semishrub. Pebbles, cliffs, fine-earth, gravelly and stony slopes. Foothills, montane zone. Nuratau, Malguzar, North Turkestan.

1813. Artemisia glaucina Krasch. ex Poljakov
Semishrub. Pebbles, banks of rivers and canals, fine-earth, gravelly and stony slopes, ravines. Foothills, montane zone. Nuratau, Malguzar, North Turkestan.

1814. Artemisia juncea Kar. & Kir.   Figure 455
Semishrub. Dry riverbeds, river valleys, pebbles, fine-earth, gravelly and stony slopes, screes, rocks. Foothills, montane zone. Nuratau. Medicinal, essential oil.

Figure 455 Artemisia juncea (photography by Natalya Beshko)

1815. Artemisia lehmanniana Bunge
Semishrub. Fine-earth, gravelly and stony slopes, rocks. Montane and alpine zones. Malguzar, North Turkestan.

1816. Artemisia leucodes Schrenk   Figure 456

Annual or biennial. Sandy and clay deserts, fine-earth, gravelly and stony slopes, river valleys, banks of canals, roadsides. Plain, foothills, montane zone. Nuratau, Aktau, Nuratau Relic Mountains, Malguzar, North Turkestan, Mirzachul, Kyzylkum. Medicinal, essential oil, fodder, weed.

Figure 456 Artemisia leucodes (photography by Natalya Beshko)

1817. Artemisia persica Boiss.
Semishrub. Fine-earth, gravelly and stony slopes, pebbles, screes. Montane and alpine zones. Malguzar, North Turkestan. Medicinal, essential oil.

1818. Artemisia porrecta Krasch. ex Poljakov
Semishrub. Banks of canals, roadsides, fallow lands, screes, dry riverbeds, pebbles, fine-earth, gravelly and stony slopes. Plain, foothills, montane zone. Nuratau, Nuratau Relic Mountains, Malguzar, North Turkestan. Medicinal, essential oil.

1819. Artemisia prolixa Krasch. ex Poljakov
Semishrub. Fine-earth, gravelly and stony slopes, screes. Foothills, montane zone. NorthTurkestan.

1820. Artemisia rutifolia Stephan ex Spreng.
Semishrub. Gravelly and stony slopes, rocks, pebbles, screes. Montane and alpine zones. Nuratau, Malguzar, North Turkestan. Medicinal, essential oil.

1821. Artemisia santolinifolia Turcz. ex Krasch.
Perennial. Dry riverbeds, banks of rivers, piedmont plains, fine-earth, gravelly and stony slopes. Plain, foothills. Nuratau, North Turkestan. Medicinal, essential oil, fodder.

1822. Artemisia scoparia Waldst. & Kitam.   Figure 457
Annual or biennial. Sandy and clay deserts, piedmont plains, river valleys, fine-earth, gravelly and stony slopes, fields, fallow lands, roadsides, wastelands. Plain, foothills, montane and alpine zones. Nuratau, Aktau, Nuratau Relic Mountains, Malguzar, North Turkestan, Mirzachul, Kyzylkum. Medicinal, essential oil, insecticide, fodder, weed.

Figure 457 Artemisia scoparia (photography by Natalya Beshko)

1823. Artemisia scotina Nevski
Semishrub. Fine-earth, gravelly and stony slopes, piedmont plains. Plain, foothills, montane zone. Nuratau, Malguzar. Essential oil.

1824. Artemisia serotina Bunge
Semishrub. Dry riverbeds, piedmont plains, fine-earth, gravelly and stony slopes. Foothills, montane

zone. Nuratau, Malguzar, North Turkestan. Medicinal, essential oil, fodder.

1825. Artemisia sieversiana Ehrh.

Annual or biennial. Saline lands, banks of rivers and canals, roadsides, gardens, fields, fallow lands, wastelands. Plain. Mirzachul. Medicinal, essential oil, fodder.

1826. Artemisia sogdiana Bunge   Figure 458

Semishrub. Piedmont plains, fine-earth, gravelly and stony slopes, dry riverbeds. Foothills, montane zone. Nuratau, Aktau, Nuratau Relic Mountains, Malguzar, North Turkestan, Mirzachul. Medicinal, essential oil, fodder.

Figure 458 **Artemisia sogdiana** (photography by Natalya Beshko)

1827. Artemisia subsalsa Filatova

Semishrub. Saline lands, banks of rivers, lakes ands canals. Plain. Mirzachul.

1828. Artemisia tenuisecta Nevski   Figure 459

Semishrub. Fine-earth, gravelly and stony slopes, screes. Montane zone. Nuratau, Aktau, Malguzar, North Turkestan. Essential oil, fodder.

Figure 459 **Artemisia tenuisecta** (photography by Natalya Beshko)

1829. Artemisia tournefortiana Rchb.
Annual or biennial. River valleys, banks of canals, roadsides, fine-earth slopes, fields, pebbles, wastelands, gardens, surroundings of settlements. Plain, foothills, montane zone. Nuratau, Aktau, Malguzar, North Turkestan, Mirzachul. Weed.

1830. Artemisia turanica Krasch.   Figure 460
Semishrub. Sandy and clay deserts, saline lands, pebbles, fine-earth, gravelly and stone slopes. Plain, foothills. Nuratau Relic Mountains, Mirzachul, Kyzylkum. Essential oil, fodder.

Figure 460 **Artemisia turanica** (photography by Natalya Beshko)

1831. Artemisia vulgaris L.

Perennial. River valleys, ravines, banks of canals, roadsides, fields, fine-earth and stony slopes, fallow lands, surroundings of settlements. Plain, foothills, montane and alpine zones. Nuratau, Nuratau Relic Mountains, Malguzar, North Turkestan, Mirzachul. Medicinal, essential oil, insecticide, fodder, weed.

Genus 559. Trichantemis Regel & Schmalh.

1832. Trichanthemis karataviensis Regel & Schmalh.
Perennial. Fine-earth, gravelly and stony slopes. Montane zone. Nuratau.

Genus 560. Waldheimia Kar. & Kir.

1833. Waldheimia glabra (Decne) Regel
Perennial. Alpine meadows, moraines, rocks. Alpine zone. North Turkestan.

1834. Waldheimia tomentosa (Decne) Regel
Perennial. Gravelly and stony slopes, screes, rocks, pebbles, banks of rivers. Alpine zone. North Turkestan.

Genus 561. Pyrethrum Zinn

1835. Pyrethrum galae Popov
Perennial. Fine-earth, gravelly and stony slopes, rocks. Montane zone. Nuratau.

1836. Pyrethrum pyrethroides (Kar. & Kir.) B. Fedtsch. ex Krasch.   Figure 461
Perennial. Fine-earth, gravelly and stony slopes, screes, pebbles, moraines, rocks. Montane and alpine zones. Malguzar, North Turkestan.

Figure 461 Pyrethrum pyrethroides (photography by Natalya Beshko)

Genus 562. Tanacetum L.

1837. Tanacetum parthenifolium (Willd.) Sch. Bip. [*Pyrethrum parthenifolium* Willd.]
Perennial. Ravines, banks of rivers, pebbles, rocks, stony slopes. Montane zone. Malguzar, North Turkestan. Medicinal, essential oil, insecticide, ornamental.

1838. Tanacetum pseudachillea C. Winkl.
Perennial. Banks of rivers, fine-earth, gravelly and stony slopes. Montane and alpine zones. North Turkestan. Medicinal, essential oil.

Genus 563. Lepidolopha C. Winkl.

1839. Lepidolopha komarowii C. Winkl.
Shrub. Fine-earth and stony slopes, rocks. Montane zone. Nuratau, Malguzar. Essential oil.

1840. Lepidolopha nuratavica Krasch.   Figure 462
Shrub. Fine-earth and stony slopes, rocks. Montane zone. Nuratau. UzbRDB 1.

Figure 462 Lepidolopha nuratavica (photography by N. Yu. Beshko)

Genus 564. Microcephala Pobed.

1841. Microcephala lamellata (Bunge) Pobed.
Annual. Sandy and clay deserts, saline lands, fine-earth, gravelly and stony slopes, rocks. Plain, foothills, montane zone. Nuratau, Nuratau Relic Mountains, Malguzar, North Turkestan, Mirzachul, Kyzylkum. Essential oil.

1842. Microcephala turcomanica (C. Winkl.) Pobed.
Annual. Sandy and clay deserts, wastelands. Plain. Mirzachul. Essential oil.

Genus 565. *Tripleurospermum Sch. Bip.

1843. *Tripleurospermum disciforme (C. A. Mey.) Sch. Bip.
Biennial. Fine-earth, gravelly and stony slopes, pebbles, river valleys, banks of canals, meadows, roadsides, fallow lands, fields. Plain, foothills, montane zone. Malguzar, North Turkestan. Weed.

Genus 566. *Xanthium L.

1844. *Xanthium spinosum L.
Annual. Wastelands, roadsides, fallow lands, banks of canals, surroundings of settlements. Plain, foothills, montane zone. Nuratau, Aktau, Nuratau Relic Mountains, Malguzar, North Turkestan, Mirzachul, Kyzylkum. Medicinal, oil, poisonous, weed.

1845. *Xanthium strumarium L.
Annual. River valleys, banks of canals, wastelands, fields, fallow lands, roadsides, surroundings of settlements. Plain, foothills, montane zone. Nuratau, Aktau, Nuratau Relic Mountains, Malguzar, North Turkestan, Mirzachul, Kyzylkum. Medicinal, oil, poisonous, weed.

Family 104. Caprifoliaceae

Genus 567. Lonicera L.

1846. Lonicera bracteolaris Boiss. & Buhse.   Figure 463
Shrub. Fine-earth, gravelly and stony slopes, river valleys, pebbles, ravines. Montane zone. Nuratau, North Turkestan. Ornamental, meliferous.

Figure 463 **Lonicera bracteolaris** (photography by Natalya Beshko)

1847. Lonicera humilis Kar. & Kir. [*Lonicera altmannii* Regel & Schmalh.] Figure 464
Shrub. Fine-earth, gravelly and stony slopes, river valleys, ravines. Montane and alpine zones. Malguzar, North Turkestan. Ornamental, meliferous.

Figure 464 **Lonicera humilis** (photography by Natalya Beshko)

1848. Lonicera korolkowii Stapf.
Shrub. Fine-earth, gravelly and stony slopes, river valleys, screes, rocks. Foothills, montane zone. North Turkestan. Ornamental, meliferous.

1849. Lonicera microphylla Willd. ex Schult.
Shrub. Fine-earth and stony slopes, rocks, river valleys, subalpine meadows. Montane zone. Malguzar, North Turkestan. Medicinal, meliferous, ornamental.

1850. Lonicera nummulariifolia Jaub. & Spach   Figure 465
Shrub. Fine-earth, gravelly and stony slopes, river valleys, ravines, rocks. Montane and alpine zones. Nuratau, Aktau, Malguzar, North Turkestan. Medicinal, meliferous, ornamental.

Figure 465 **Lonicera nummulariifolia** (photography by Natalya Beshko)

1851. Lonicera olgae Regel & Schmalh.
Shrub. Stony slopes, rocks. Montane and alpine zones. North Turkestan. Ornamental, meliferous.

1852. Lonicera paradoxa Pojark.
Shrub. Stony slopes, rocks. Montane and alpine zones. North Turkestan. Ornamental, meliferous. Threatened species.

1853. Lonicera semenovii Regel
Shrub. Stony slopes, rocks, moraines. Montane and alpine zones. North Turkestan. Ornamental, meliferous.

1854. Lonicera simulatrix Pojark.
Shrub. Fine-earth, gravelly and stony slopes. Montane and alpine zones. North Turkestan.

1855. Lonicera zeravshanica Pojark.
Shrub. Fine-earth, gravelly and stony slopes, river valleys. Montane and alpine zones. North Turkestan. Ornamental, meliferous.

Genus 568. Dipsacus L.

1856. Dipsacus dipsacoides (Kar. & Kir.) Botsch.   Figure 466
Perennial. Fine-earth, gravelly and stony slopes. Montane zone. Nuratau, Malguzar, North Turkestan, Mirzachul. Medicinal, ornamental, meliferous.

Figure 466 Dipsacus dipsacoides (photography by Natalya Beshko)

1857. Dipsacus laciniatus L.
Biennial. Fine-earth slopes, river valleys, fields, gardens, fallow lands. Plain, foothills, montane

zone. Nuratau, Malguzar, North Turkestan, Mirzachul. Medicinal, ornamental, meliferous.

Genus 569. *Cephalaria Schrad.

1858. Cephalaria syriaca (L.) Schrad. ex Roem. & Schult.
Annual. Piedmont plains, fallow lands, fields. Plain, foothills, montane zone. Nuratau, Malguzar, North Turkestan, Mirzachul. Medicinal, oil.

Genus 570. Scabiosa L.

1859. Scabiosa flavida Boiss. & Hausskn.
Annual. Fine-earth, gravelly and stony slopes. Foothills. Nuratau.
1860. Scabiosa olivierii Coult.
Annual. Sandy and clay deserts, piedmont plains, gravelly and stony slopes. Plain, foothills, montane zone. Nuratau, Aktau, Nuratau Relic Mountains, Malguzar, North Turkestan, Mirzachul, Kyzylkum.
1861. Scabiosa rhodantha Kar. & Kir.
Annual. Fine-earth, gravelly and stony slopes, banks of rivers, piedmont plains, sandy and clay deserts, fields, fallow lands. Foothills, montane zone. Nuratau, Nuratau Relic Mountains, Kyzylkum.
1862. Scabiosa songarica Schrenk   Figure 467
Perennial. Fine-earth, gravelly and stony slopes. Foothills, montane and alpine zones. Nuratau, Malguzar, North Turkestan. Medicinal, ornamental, meliferous.

Figure 467 Scabiosa songarica (photography by Natalya Beshko)

Genus 571. Valeriana L.

1863. Valeriana chionophila Popov & Kult.   Figure 468
Perennial. Fine-earth and stony slopes, rocks, wet places near melting snow. Foothills, montane and alpine zones. Nuratau, Aktau, Malguzar, North Turkestan. Medicinal.

Figure 468 Valeriana chionophila (photography by Natalya Beshko)

1864. Valeriana fedtschenkoi Coincy
Perennial. Rocks, stony slopes, screes, banks of rivers, subalpine and alpine meadows. Montane and alpine zones. North Turkestan. Medicinal.

1865. Valeriana ficarifolia Boiss.
Perennial. Shady wet places, banks of rivers, ravines, fine-earth, gravelly and stony slopes, screes. Foothills, montane and alpine zones. Nuratau, Malguzar, North Turkestan.

Genus 572. Valerianella Mill.

1866. Valerianella coronata (L.) DC.
Annual. Fine-earth, gravelly and stony slopes, fallow lands, pebbles. Plain, Foothills. North Turkestan, Mirzachul.

1867. Valerianella cymbocarpa C. A. Mey.
Annual. Fine-earth, gravelly and stony slopes, dry riverbeds. Plain, foothills, montane zone. Nuratau, Malguzar, North Turkestan.

1868. Valerianella dactylophylla Boiss. & Hohen.
Annual. Fine-earth and stony slopes, shady places. Montane zone. Nuratau, Malguzar.

1869. Valerianella muricata (Steven ex M. Bieb.) W. H. Baxter
Annual. Fine-earth and stony slopes, screes, rocks, pebbles, fallow lands. Plain, foothills, montane zone. Nuratau, Malguzar, Mirzachul.

1870. Valerianella oxyrrhyncha Fisch. & C. A. Mey. [*Valerianella diodon* Boiss.]
Annual. Fine-earth and stony slopes, pebbles, banks of rivers, fields, fallow lands. Plain, foothills, montane and alpine zones. Nuratau, Aktau, Malguzar, North Turkestan, Mirzachul.

1871. Valerianella plagiostephana Fisch. & C. A. Mey.
Annual. Fine-earth and stony slopes, screes, rocks, pebbles, river valleys. Plain, foothills, montane zone. Malguzar, North Turkestan.

1872. Valerianella szovitsiana Fisch. & C. A. Mey.
Annual. Sandy and clay deserts, saline lands, fine-earth and stony slopes, rocks, screes. Plain, foothills, montane zone. Nuratau, Aktau, Malguzar, North Turkestan, Mirzachul, Kyzylkum.

1873. Valerianella turkestanica Regel & Schmalh.
Annual. Fine-earth, gravelly and stony slopes, pebbles, banks of rivers, clay deserts, saline lands, fields, fallow lands. Plain, foothills, montane zone. Nuratau, Aktau, Malguzar, North Turkestan, Mirzachul.

1874. Valerianella vvedenskyi Lincz.
Annual. Stony slopes, river valleys, pebbles. Foothills, montane zone. North Turkestan.

Family 105. Apiaceae [Umbelliferae]

Genus 573. Eryngium L.

1875. Eryngium caeruleum M. Bieb.
Perennial, monocarpic. Fine-earth, gravelly and stony slopes, river valleys, banks of canals, gardens, fallow lands, roadsides, wastelands, fields. Plain, foothills, montane zone. Nuratau, Aktau, Nuratau Relic Mountains, Malguzar, North Turkestan, Mirzachul. Medicinal, weed.

1876. Eryngium macrocalyx Schrenk
Perennial, monocarpic. Stony and gravelly slopes, screes, dry riverbeds, fallow lands, roadsides. Foothills, montane zone. Nuratau, Malguzar, North Turkestan. Medicinal, essential oil.

1877. Eryngium octophyllum Korovin
Perennial, polycarpic. Stony slopes, rocks, dry riverbeds. Foothills, montane zone. Nuratau, Malguzar. Medicinal, essential oil.

Genus 574. Echinophora L.

1878. Echinophora tenuifolia subsp. **sibthorpiana** (Guss.) Tutin [*Echinophora sibthorpiana* Guss.]
Perennial, monocarpic. Fine-earth, gravelly and stony slopes, piedmont plains, wastelands, fallow lands, fields, pebbles, gardens, roadsides, banks of canals. Plain, foothills. Nuratau, Aktau, Nuratau Relic Mountains, Malguzar. Medicinal, essential oil, food (spice-aromatic), fodder.

Genus 575. Anthriscus Pers.

1879. Anthriscus glacialis Lipsky
Perennial, polycarpic. Stony slopes, stream banks. Montane and alpine zones. Nuratau. Fodder.

Genus 576. Scandix L.

1880. Scandix nodosa L. [*Chaerophyllum nodosum* (L.) Crantz; *Myrrhoides nodosa* (L.) Cannon; *Physocaulis nodosus* (L.) W. D. J. Koch]
Annual. River valleys, shady wet places. Foothills, montane zone. Nuratau. Food, essential oil.
1881. Scandix pecten-veneris L.
Annual. Fine-earth, gravelly and stony slopes, river valleys, banks of canals, fields, gardens, fallow lands, wastelands, roadsides. Plain, foothills, montane zone. Nuratau, Aktau, Nuratau Relic Mountains, Malguzar, North Turkestan, Mirzachul, Kyzylkum. Weed, medicinal.
1882. Scandix stellata Banks & Sol.
Annual. Fine-earth, gravelly and stony slopes, screes, rocks, pebbles, dry riverbeds, roadsides, fields, fallow lands. Plain, foothills, montane zone. Nuratau, Aktau, Nuratau Relic Mountains, Malguzar, North Turkestan. Weed.

Genus 577. Kozlovia Lipsky

1883. Kozlovia paleacea Lipsky
Perennial, monocarpic. Fine-earth slopes, fallow lands. Foothills, montane zone. Nuratau, Malguzar, North Turkestan. Weed.

Genus 578.*Torilis Adans.

1884. *Torilis arvensis (Huds.) Link.
Annual. Fallow lands, fields, gardens, roadsides, wastelands, banks of rivers and canals, fine-earth, gravelly and stony slopes. Plain, foothills, montane zone. Nuratau, Aktau, Nuratau Relic Mountains,

Malguzar, North Turkestan, Mirzachul, Kyzylkum. Essential oil, weed.

1885. *Torilis leptophylla (L.) Rchb. f.
Annual. Fine-earth, gravelly and stony slopes, wastelands, fallow lands, fields, roadsides, banks of rivers and canals, gardens, surroundings of settlements. Foothills, montane zone. Nuratau, Aktau, Nuratau Relic Mountains, Malguzar, North Turkestan, Mirzachul, Kyzylkum. Weed, food.

Genus 579. Turgenia Hoffm.

1886. Turgenia latifolia (L.) Hoffm.
Annual. Fields, fallow lands, sandy and clay deserts, river valleys, fine-earth, gravelly and stony slopes, dry riverbeds. Plain, foothills, montane zone. Nuratau, Aktau, Nuratau Relic Mountains, Malguzar, North Turkestan, Mirzachul, Kyzylkum. Essential oil, weed.

Genus 580. *Orlaya Hoffm.

1887. *Orlaya grandiflora (L.) Hoffm.
Annual. Fields, fallow lands. Plain. Mirzachul. Essential oil, weed.

Genus 581. Daucus L.

1888. Daucus carota L.
Biennial. Roadsides, wastelands, fallow lands, fields, gardens, banks of canals, river valleys, meadows. Foothills, montane zone. Nuratau, Aktau, Nuratau Relic Mountains, Malguzar, North Turkestan, Mirzachul, Kyzylkum. Food, essential oil, medicinal, weed.

Genus 582. *Coriandrum L.

1889. *Coriandrum sativum L.
Annual. Surrounds of settlements, roadsides, fallow lands, fields, river valleys (escaped from culture). Plain, foothills, montane zone. Nuratau, Aktau, Nuratau Relic Mountains, Malguzar, North Turkestan, Mirzachul. Food (spice-aromatic), essential oil, meliferous.

Genus 583. Schrenkia Fisch. & C. A. Mey.

1890. Schrenkia golickeana B. Fedtsch.
Perennial, monocarpic. Fine-earth and stony slopes, pebbles, rocks. Foothills, montane and alpine zones. Nuratau, Malguzar, North Turkestan. Medicinal.

1891. Schrenkia pungens Regel & Schmalh.
Perennial, polycarpic. Stony slopes, rocks, screes, dry riverbeds. Foothills, montane zone. Nuratau, Malguzar.

1892. Schrenkia vaginata (Ledeb.) Fisch. & C. A. Mey.
Perennial, monocarpic. Stony slopes, rocks, screes, pebbles. Montane and alpine zones. Nuratau, North Turkestan.

Genus 584. Lipskya (K.-Pol.) Nevski

1893. Lipskya insignis Nevski
Perennial, monocarpic. Stony slopes, rocks. Foothills, montane zone. Malguzar. Essential oil, food (spice-aromatic). Threatened species, UzbRDB 3.

Genus 585. Schtschurowskia Regel & Schmalh.

1894. Schtschurowskia meifolia Regel & Schmalh.
Perennial, polycarpic. Stony slopes, screes, pebbles. Montane and alpine zones. Malguzar, North Turkestan.

Genus 586. Korshinskya Lipsky

1895. Korshinskya olgae Lipsky
Perennial, monocarpic. Fine-earth, gravelly and stony slopes. Foothills, montane zone. Nuratau, Malguzar, North Turkestan.

Genus 587. Aulacospermum Ledeb.

1896. Aulacospermum roseum Korovin
Perennial, monocarpic. Gravelly and stony slopes, screes, alpine meadows, stream banks, wet places near melting snow. Montane and alpine zones. Malguzar, North Turkestan.

1897. Aulacospermum simplex Rupr.
Perennial, monocarpic. Fine-earth, gravelly and stony slopes, screes, rocks. Montane and alpine zones. Malguzar, North Turkestan.

1898. Aulacospermum tenuisectum Korovin
Perennial, monocarpic. Fine-earth, gravelly and stony slopes, screes, rocks. Montane and alpine zones. Malguzar, North Turkestan.

Genus 588. Eremodaucus Bunge

1899. Eremodaucus lehmannii Bunge
Annual. Fields, fallow lands, wastelands, roadsides, piedmont plains, fine-earth slopes. Plain, foothills, montane zone. Nuratau, Aktau, Nuratau Relic Mountains, Malguzar, North Turkestan, Mirzachul, Kyzylkum. Fodder, weed.

Genus 589. *Conium L.

1900. *Conium maculatum L.
Biennial. Wastelands, gardens, fields, banks of rivers and canals. Foothills, montane zone. Nuratau, Aktau, Malguzar, North Turkestan. Medicinal, poisonous, essential oil, weed.

Genus 590. Prangos Lindl.

1901. Prangos didyma (Regel) Pimenov & V. N. Tikhom.
Perennial, monocarpic. Sandy and clay deserts, piedmont plains, dry riverbeds. Plain, foothills. Nuratau Relic Mountains, Mirzachul, Kyzylkum. Fodder, essential oil, meliferous.

1902. Prangos fedtschenkoi (Regel & Schmalh.) Korovin   Figure 469
Perennial, monocarpic. Fine-earth, gravelly and stony slopes, dry riverbeds, pebbles, roadsides, fields. Foothills, montane zone. Nuratau, Aktau, Malguzar, North Turkestan. Fodder, essential oil, meliferous.

Figure 469 Prangos fedtschenkoi (photography by Natalya Beshko)

1903. Prangos ornata Kuzmina   Figure 470
Perennial, polycarpic. Fine-earth, gravelly and stony slopes, river valleys, dry riverbeds. Foothills, montane zone. Nuratau. Fodder, essential oil, meliferous.

Figure 470 **Prangos ornata** (photography by Natalya Beshko)

1904. Prangos pabularia Lindl.   Figure 471

Perennial, polycarpic. Fine-earth, gravelly and stony slopes, river valleys, dry riverbeds. Montane and alpine zones. Nuratau, Malguzar, North Turkestan. Fodder, essential oil, medicinal, meliferous.

Figure 471 **Prangos pabularia** (photography by Natalya Beshko)

Genus 591. Bupleurum L.

1905. Bupleurum falcatum L. subsp. **exaltatum** (M. Bieb.) Rouy & Camus [*Bupleurum exaltatum* M. Bieb.]
Perennial, polycarpic. Stony slopes, screes, rocks, dry riverbeds, pebbles. Montane and alpine zones. Nuratau, Malguzar, North Turkestan. Medicinal, fodder.

Genus 592. Elaeosticta Fenzl

1906. Elaeosticta allioides (Regel & Schmalh.) Kljuykov, Pimenov & V. N. Tikhom.
Perennial, monocarpic. Fine-earth and stony slopes, dry riverbeds. Foothills, montane zone. Nuratau, Aktau, Malguzar, North Turkestan. Essential oil, fodder.
1907. Elaeosticta hirtula (Regel & Schmalh.) Kljuykov, Pimenov & V. N. Tikhom.
Perennial, monocarpic. Fine-earth, gravelly and stony slopes, screes. Foothills, montane and alpine zones. Nuratau, Malguzar, North Turkestan. Essential oil.
1908. Elaeosticta polycarpa (Korovin) Kljuykov, Pimenov & V. N. Tikhom.
Perennial, polycarpic. Fine-earth slopes. Foothills, montane zone. Nuratau Relic Mountains. Essential oil.
1909. Elaeosticta vvedenskyi (Kamelin) Kljuykov, Pimenov & V. N. Tikhom.
Perennial, monocarpic. Fine-earth, gravelly and stony slopes. Foothills, montane zone. Nuratau. Essential oil.

Genus 593. Galagania Lipsky

1910. Galagania fragrantissima Lipsky
Perennial, monocarpic. Fine-earth, gravelly and stony slopes, fallow lands. Foothills, montane zone. Nuratau, Malguzar, North Turkestan. Essential oil, food (spice-aromatic).
1911. Galagania tenuisecta (Regel & Schmalh.) M. G. Vassiljeva & Pimenov
Perennial, monocarpic. Fine-earth, gravelly slopes, dry riverbeds, fallow lands. Foothills, montane zone. Nuratau, Malguzar, North Turkestan. Essential oil, weed.

Genus 594. Hyalolaena Bunge

1912. Hyalolaena depauperata Korovin
Perennial, monocarpic. Gravelly and stony slopes, rocks. Foothills, montane zone. Nuratau. Essential oil.
1913. Hyalolaena jaxartica Bunge
Perennial, monocarpic. Fine-earth, gravelly and stony slopes, sandy and clay deserts, dry riverbeds, pebbles, rocks. Plain, foothills. Nuratau, Mirzachul. Essential oil.

Genus 595. Oedibasis Koso.-Pol.

1914. Oedibasis apiculata (Kar. & Kir.) Koso.-Pol.
Perennial, monocarpic. Sandy deserts, piedmont plains, fine-earth, gravelly and stony slopes, rocks, fallow lands. Plain, foothills, montane and alpine zones. Nuratau, Malguzar, North Turkestan.

1915. Oedibasis tamerlanii (Lipsky) Korovin ex Nevski
Perennial, monocarpic. Fine-earth, gravelly and stony slopes, rocks. Foothills, montane zone. Nuratau, Malguzar, North Turkestan.

Genus 596. Elwendia Boiss.

1916. Elwendia capusii (Franch.) Pimenov & Kljuykov [*Bunium capusii* (Franch.) Korovin]
Perennial, polycarpic. Piedmont plains, fine-earth slopes, river valleys. Foothills, montane zone. Nuratau, Aktau, Nuratau Relic Mountains, Malguzar, North Turkestan.

1917. Elwendia chaerophylloides (Regel & Schmalh.) Pimenov & Kljuykov [*Bunium chaerophylloides* (Regel & Schmalh.) Drude]
Perennial, polycarpic. Fine-earth, gravelly and stony slopes. Foothills, montane zone. Nuratau, Aktau, Nuratau Relic Mountains, Malguzar, North Turkestan. Medicinal, food (spice-aromatic), essential oil.

1918. Elwendia intermedia (Korovin) Pimenov & Kljuykov [*Bunium intermedium* Korovin]
Perennial, polycarpic. Fine-earth slopes, river valleys, shady wet places. Montane zone. Nuratau, Malguzar, North Turkestan.

1919. Elwendia persica (Boiss.) Pimenov & Kljuykov [*Bunium persicum* (Boiss.) B. Fedtsch.]
Perennial, polycarpic. Fine-earth, gravelly and stony slopes, screes. Foothills, montane zone. Nuratau, Malguzar, North Turkestan. Medicinal, food (spice-aromatic), essential oil.

Genus 597. Falcaria Fabr.

1920. Falcaria vulgaris Bernh.
Perennial, monocarpic. River valleys, banks of canals, fine-earth, gravelly slopes, wet places, gardens, roadsides. Plain, foothills, montane zone. Nuratau, Nuratau Relic Mountains, Malguzar, North Turkestan. Medicinal, food, essential oil.

Genus 598. Carum L.

1921. Carum carvi L.
Biennial. Banks of rivers, wet places. Montane and alpine zones. Nuratau, Malguzar, North Turkestan. Medicinal, essential oil, food (spice-aromatic).

Genus 599. Apium L.

1922. Apium nodiflorum (L.) Lag.
Perennial, polycarpic. Banks of rivers and canals. Nuratau. Medicinal, food.

Genus 600. Cuminum L.

1923. Cuminum setifolium (Boiss.) Koso-Pol.
Annual. Sandy and clay deserts, fine-earth slopes. Nuratau, Kyzylkum. Essential oil, weed.

Genus 601. Aphanopleura Boiss.

1924. Aphanopleura capillifolia (Regel & Schmalh.) Lipsky
Annual. Sandy and clay deserts, dry riverbeds, fine-earth, gravelly and stony slopes, riparian forests, fallow lands, roadsides. Plain, foothills, montane zone. Nuratau, Aktau, Nuratau Relic Mountains, Malguzar, North Turkestan, Mirzachul, Kyzylkum. Essential oil, weed.

Genus 602. Pimpinella L.

1925. Pimpinella peregrina L.
Biennial. Banks of rivers and canals, gardens, roadsides, fields, fallow lands, riparian forests. Foothills, montane zone. Nuratau, Malguzar, North Turkestan. Essential oil, medicinal, fodder, meliferous.

Genus 603. Sium L.

1926. Sium sisaroideum DC.
Perennial, polycarpic. Banks of rivers and canals, swamps. Plain, foothills, montane zone. Nuratau. Medicinal, oil.

Genus 604. Berula W. D. J. Koch

1927. Berula erecta (Huds.) Coville
Perennial, polycarpic. Banks of rivers and canals. Foothills, montane zone. Nuratau, Malguzar, North Turkestan. Medicinal, poisonous.

Genus 605. Seseli L.

1928. Seseli calycinum (Korovin) Pimenov & Sdobnina
Perennial, polycarpic. Rocks. Montane zone. Nuratau, Malguzar, North Turkestan.

1929. Seseli korovinii Schischk.   Figure 472
Perennial, polycarpic. Screes, rocks, gravey and stony slopes. Montane and alpine zones. Nuratau, Malguzar, North Turkestan.

Figure 472 Seseli korovinii (photography by Natalya Beshko)

1930. Seseli lehmannianum Boiss.
Perennial, polycarpic. Fine-earth, gravelly and stony slopes, rocks, pebbles. Montane and alpine zones. Malguzar, North Turkestan.

1931. Seseli mucronatum (Schrenk) Pimenov & Sdobnina
Perennial, polycarpic. Meadows, river valleys, fine-earth and stony slopes, rocks. Montane and alpine zones. Malguzar, North Turkestan.

1932. Seseli seravschanicum Pimenov & Sdobnina   Figure 473
Perennial, monocarpic. Stony slopes, rocks. Montane and alpine zones. North Turkestan.

Figure 473 **Seseli seravschanicum** (photography by Natalya Beshko)

1933. Seseli tenuisectum Regel & Schmalh.

Perennial, polycarpic. Rocks, screes, pebbles. Montane zone. North Turkestan.

1934. Seseli turbinatum Korovin

Perennial, polycarpic. Stony slopes. Montane zone. Nuratau. UzbRDB 2.

Genus 606. Libanotis Haller ex Zinn

1935. Libanotis schrenkiana C. A. Mey. ex Schischk. [*Seseli schrenkianum* (C. A. Mey. ex Schischk.) Pimenov & Sdobnina]

Perennial, polycarpic. Fine-earth and stony slopes, pebbles, rocks. Montane and alpine zones. North Turkestan.

Genus 607. Mediasia Pimenov

1936. Mediasia macrophylla (Regel & Schmalh.) Pimenov   Figure 474
Perennial, polycarpic. Fine-earth, gravelly and stony slopes, screes. Montane and alpine zones. Nuratau, Malguzar, North Turkestan. Medicinal, food (spice-aromatic), essential oil.

Figure 474 Mediasia macrophylla (photography by Natalya Beshko)

Genus 608. Alposelinum Pimenov

1937. Alposelinum albomarginatum (Schrenk) Pimenov
Perennial, monocarpic. Stony slopes, moraines, screes, pebbles, alpine meadows. Montane and alpine zones. North Turkestan.

Genus 609. Conioselinum Hoffm.

1938. Conioselinum vaginatum (Spreng.) Thell. [*Conioselinum tataricum* Fisch. ex Hoffm.]
Perennial, polycarpic. Fine-earth, gravelly and stony slopes, ravines, river valleys, pebbles, screes, shady wet places, swamp meadows. Montane and alpine zones. North Turkestan.

Genus 610. Angelica L.

1939. Angelica archangelica subsp. **decurrens** (Ledeb.) Kuvaev
Perennial, monocarpic. Stream banks, wet places near melting snow, springs, wet gravelly and fine-earth slopes, meadows, pebbles. Montane and alpine zones. Nuratau, North Turkestan. Medicinal, essential oil, meliferous.

1940. Angelica brevicaulis (Rupr.) B. Fedtsch.   Figure 475
Perennial, monocarpic. Banks of rivers, pebbles, moraines, wet gravelly slopes, screes, swamp meadows. Montane and alpine zones. North Turkestan. Medicinal, essential oil.

Figure 475 Angelica brevicaulis (photography by Natalya Beshko)

1941. Angelica ternata Regel & Schmalh.
Perennial, polycarpic. Gravelly slopes, screes, moraines, rocks, pebbles. Montane and alpine zones. North Turkestan. Essential oil, food (spice-aromatic).

Genus 611. Ferula L.

1942. Ferula angreni Korovin   Figure 476
Perennial, polycarpic. Fine-earth and gravelly slopes, river valleys, screes. Montane zone. Nuratau, Malguzar. Essential oil, fodder, meliferous.

Figure 476 Ferula angreni (photography by Natalya Beshko)

1943. Ferula bucharica (Lipsky) Koso-Pol. [*Ladyginia bucharica* Lipsky]
Perennial, polycarpic. Fine-earth, stony and gravelly slopes. Montane zone. Malguzar. Essential oil, saponin-bearing, meliferous.

1944. Ferula diversivittata Regel & Schmalh.   Figure 477
Perennial, monocarpic. Fine-earth slopes, dry riverbeds. Foothills, montane zone. Nuratau. Medicinal, essential oil, meliferous.

Figure 477 Ferula diversivittata (photography by Natalya Beshko)

1945. Ferula dshizakensis Korovin
Perennial, polycarpic. Stony and gravelly slopes, rocks. Montane zone. Nuratau, Nuratau Relic Mountains, Malguzar. Essential oil, fodder, meliferous.

1946. Ferula fedtschenkoana Koso-Pol.
Perennial, polycarpic. Fine-earth, gravelly slopes. Montane and alpine zones. Malguzar, North Turkestan. Essential oil, melliferous. Threatened species, UzbRDB 2.

1947. Ferula ferganensis Lipsky ex Korovin
Perennial, polycarpic. Stony slopes, rocks, screes. Montane zone. Nuratau. Essential oil, meliferous.

1948. Ferula foetida (Bunge) Regel   Figure 478
Perennial, monocarpic. Sandy and clay deserts, piedmont plains, fine-earth slopes. Plain, foothills. Nuratau, Nuratau Relic Mountains, Kyzylkum. Medicinal, essential oil, fodder, meliferous.

Figure 478 Ferula foetida (photography by Natalya Beshko)

1949. Ferula helenae Rakhm. & Melibaev

Perennial, monocarpic. Stony slopes. Foothills. Nuratau Relic Mountains. Essential oil, fodder, meliferous. Threatened species. Endemic to the Nuratau Relic Mountains.

1950. Ferula kokanica Regel & Schmalh.　　Figure 479

Perennial, monocarpic. Fine-earth, gravelly and stony slopes, screes, pebbles, rocks. Foothills, montane and alpine zones. Nuratau, Nuratau Relic Mountains, Malguzar, North Turkestan. Medicinal, essential oil, fodder, meliferous.

Figure 479 Ferula kokanica (photography by Natalya Beshko)

1951. Ferula kuhistanica Korovin   Figure 480
Perennial, monocarpic. Fine-earth, gravelly and stony slopes. Montane and alpine zones. Aktau, Malguzar, North Turkestan. Medicinal, essential oil, fodder, meliferous.

Figure 480 Ferula kuhistanica (photography by Natalya Beshko)

1952. Ferula lehmannii Boiss.

Perennial, monocarpic. Sandy deserts. Plain. Kyzylkum. Essential oil, meliferous.

1953. Ferula mollis Korovin

Perennial, polycarpic. Fine-earth, gravelly and stony slopes. Foothills, montane zone. North Turkestan. Essentialoil, fodder, meliferous.

1954. Ferula moschata (H. Reinsch) Koso-Pol. [*Ferula sumbul* (Kauffm.) Hook. f.]   Figure 481

Perennial, polycarpic. Fine-earth and stony slopes, screes, rocks, ravines. Montane zone. Nuratau, Malguzar, North Turkestan. Medicinal, essential oil, meliferous. Threatened species, UzbRDB 2.

Figure 481 Ferula moschata (photography by N. Yu. Beshko)

1955. Ferula ovczinnikovii Pimenov

Perennial, monocarpic. Stony and gravelly slopes, screes. Montane zone. Malguzar. Essential oil, meliferous.

1956. Ferula ovina (Boiss.) Boiss.   Figure 482

Perennial, polycarpic. Fine-earth, gravelly and stony slopes, rocks. Montane and alpine zones. Nuratau, Aktau, Malguzar, North Turkestan. Essential oil, fodder, meliferous.

Figure 482 **Ferula ovina** (photography by Natalya Beshko)

1957. Ferula penninervis Regel & Schmalh.   Figure 483
Perennial, polycarpic. Stony slopes, screes, pebbles, dry riverbeds, rocks. Montane and alpine zones. Nuratau, Malguzar, North Turkestan. Essential oil, medicinal, fodder, meliferous.

Figure 483 **Ferula penninervis** (photography by Natalya Beshko)

1958. Ferula samarkandica Korovin
Perennial, monocarpic. Fine-earth, gravelly and stony slopes, dry riverbeds, rocks. Montane and alpine zones. Nuratau, Malguzar, North Turkestan. Medicinal, essential oil, fodder, meliferous.

1959. Ferula schtschurowskiana Regel & Schmalh.

Perennial, monocarpic. Sandy and clay, deserts, saline lands, gravelly slopes. Plain, foothills, montane zone. Nuratau, Malguzar, North Turkestan, Mirzachul. Essential oil, fodder, meliferous.

1960. Ferula varia (Schrenk) Trautv.

Perennial, monocarpic. Sandy deserts, fine-earth and stony slopes. Plain, foothills. Nuratau Relic Mountains. Medicinal, essential oil, fodder, meliferous.

Genus 612. Heracleum L.

1961. Heracleum lehmannianum Bunge   Figure 484

Perennial or biennial, monocarpic. Subalpine meadows, river valleys, ravines, pebbles, wet screes. Montane and alpine zones. Nuratau, Malguzar, North Turkestan. Medicinal, essential oil, fodder, meliferous.

Figure 484 **Heracleum lehmannianum** (photography by Natalya Beshko)

1962. Heracleum olgae Regel & Schmalh. [*Tetrataenium olgae* (Regel & Schmalh.) Manden]

Perennial, monocarpic. Screes, pebbles, fine-earth slopes. Montane and alpine zones. North Turkestan. Essential oil.

Genus 613. Semenovia Regel & Herd.

1963. Semenovia dasycarpa (Regel & Schmalh.) Korovin   Figure 485

Perennial, polycarpic. Fine-earth and stony slopes, ravines, stream banks, shady wet places, rocks. Montane and alpine zones. North Turkestan.

Figure 485 Semenovia dasycarpa (photography by Natalya Beshko)

1964. Semenovia heterodonta Manden

Perennial, polycarpic. Rocks, screes, stony slopes. Alpine zone. North Turkestan.

1965. Semenovia pimpinelloides (Nevski) Manden

Perennial, polycarpic. Stony slopes, rocks. Montane zone. Nuratau.

REFERENCES

Adylov T. A. 1983. Conspectus Florae Asiae Mediae. Vol. 7. Tashkent: Fan Publishers, 415. (In Russian)

Adylov T. A. 1987. Conspectus Florae Asiae Mediae. Vol. 9. Tashkent: Fan Publishers, 400. (In Russian)

Adylov T. A., Zuckerwanik T. I. 1993. Conspectus Florae Asiae Mediae. Vol. 10. Tashkent: Fan Publishers, 692. (In Russian)

Akhani H., Edwards G. & Roalson E. H. 2007. Diversification of the world Salsoleae s.l. (Chenopodiaceae): molecular phylogenetic analysis of nuclear and chloroplast datasets and a revised classification. Int. J. Pl. Sci., 168: 931–956.

Alibekov L. A., Nishanov S. A. 1978. Natural conditions and resources of Dzhizak region. Tashkent: Uzbekistan Publishers, 254. (In Russian)

Al-Shehbaz I. A., German D.A., Mummenhoff K., Moazzeni H. 2014. Systematics, tribal placements, and synopses of the *Malcolmia* s.l. segregates (Brassicaceae). Harvard Papers in Botany, London, 19 (1): 53–71.

Aramov, S. A. 2012. Geographical Atlas of Uzbekistan. Tashkent: Goskomgeodeskadastr, 119. [In Russian]

Azimova D. E. 2017. Flora of the Malguzar Ridge. Tashkent: Abstr. PhD Diss., 44. (In Uzbek and Russian)

Batoshov A. R. 2016. Flora of South-East Kyzylkum Relic Mountains. Tashkent: Abstr. Doct. Diss., 75. (In Uzbek, Russian and English)

Batoshov A. R., Beshko N. Yu. 2013. Characteristics of the flora and plant cover of the relic mountains of the South Kyzylkum. Arid Ecosystems, 19 (3): 73–78. (In Russian)

Batoshov A. R., Beshko N. Yu. 2015. Comparative analysis of geophytes of the flora of relic mountains of South-eastern Kyzylkum and Nuratau Mountains. Uzbek. Biol. Journ., 5: 29–33. (In Russian)

Belolipov I. V. 1973. Interesting floristic findings in the vegetation of Nuratau Mountains. Uzbek. Biol. Journ., 2: 48–49. (In Russian)

Beshko N. Yu. 2000a. Flora of the Nuratau Nature Reserve. Tashkent: Abstr. Cand. Diss., 26 (In Russian)

Beshko N. Yu. 2000b. Flora of the planned biosphere reserve Nuratau-Kyzylkum. In: Biodiversity conservation on the protected areas of Uzbekistan. Tashkent: Chinor ENK, 21–43. (In Russian)

Beshko N. Yu. 2011. Flora of vascular plants of the Nuratau nature reserve. In: Proceedings of nature reserves of Uzbekistan. Issue 7. Tashkent: Chinor ENK, 19–78. (In Russian)

Beshko N. Yu., Azimova D. E. 2013. New floristic findings on the North-West Pamir-Alay

(Uzbekistan). Turczaninowia, (16) 1: 197–203. (In Russian)

Beshko N. Yu., Azimova D. E. 2014. The genus Astragalus L. in the flora of Nuratau and Malguzar mountains: the comparative analysis. Uzb. Biol. Journ., Special issue: 20-21. (In Russian)

Beshko N. Yu., Tojibaev K. Sh. & Batoshov A. R. 2013a. Tulips of the Nuratau Mountains and South-Eastern Kyzylkum (Uzbekistan). Stapfia Reports, 99: 198–204.

Beshko N. Yu., Tojibaev K. Sh., Batoshov A. R. & Azimova D. E. 2014. Botanical-geographical regionalization of Uzbekistan. Kuhistan and Nuratau districts. Uzbek. Biol. Journ., 3: 30–34. (In Russian)

Beshko, N. Yu., Zagrebin, V., Popov, A., Khasanov, F., Mitropolskaya, O., Magdiev, Kh. 2013. Recommendations for Protected Areas System Development in Uzbekistan. UNDP, GEF, Main Department of Forestry of Ministry of Agriculture and Water Resources of the Republic of Uzbekistan. Tashkent: Baktria Press, 256. [In Russian].

Bondarenko O. N., Nabiev M. M. 1972. Conspectus Florae Asiae Mediae. Vol. 3. Tashkent: Fan Publishers, 266. (In Russian)

Botirova L. A. 2012. Vegetation of the Zaaminsu River basin. Tashkent: Abstr. Cand. Diss., 24 (In Uzbek)

Botschantzev V. P., Butkov A. Ya., Vvedensky A. I., Nikiforova N. B. & Pazij V. K. 1961. An identification guide of wild plants of the Hungry Steppe. Tashkent:Tashkent State University, 216. (In Russian)

Brummitt R. K., Powell C. E. 1992. Authors of plant names. Royal Botanic Gardens, Kew. 732 p.

Budantsev A.L. 1996. Plant resources of Russia and adjacent states: Flowering plants, their chemical composition and use. Part I. Families Lycopodiaceae – Ephedraceae. Part II. Addenda. St.-Petersburg: World and Family -95, 571. (In Russian)

CABI. 2017. Invasive Species Compendium. CAB International, Wallingford, UK. Available from: www.cabi.org/isc.

Christenhusz M. J. M., Reveal J. L., Farjon A., Gardner M. F., Mill R. R. & Chase M. W. 2011a. A linear sequence of extant families and genera of lycophytes and ferns. Phytotaxa, 19: 7–54.

Christenhusz M. J. M., Reveal J. L., Farjon A., Gardner M. F., Mill R. R. & Chase M. W. 2011b. A new classification and linear sequence of extant gymnosperms. Phytotaxa, 19: 55-70.

Clayton W.D., Vorontsova M.S., Harman K.T. & Williamson H. 2006. GrassBase - The Online World Grass Flora. Available from: http://www.kew.org/data/grasses-db.html.

Czukavina A. P. 1984. Flora of the Tajik SSR. Vol. 7. Academy of Sciences of USSR, Leningrad, 563. (In Russian)

Degtjareva G. V., Kljuykov E. V., Samigullin T. H., Valiejo-Roman C. M. & Pimenov M. G. 2013. ITS phylogeny of Middle Asian geophilic Umbelliferae-Apioideae genera with comments on their morphology and utility of psbA-trnH sequences. Plant Syst. Evol., 299: 985–1010.

Demurina E. M. 1975. Vegetation of the western part of Turkestan ridge and its spurs. Tashkent: Fan Publishers, 189. (In Russian)

Downie S. R., Spalik K., Katz-Downie D. S. & Reduron J.-P. 2010. Major clades within Apiaceae subfamily Apioideae as inferred by phylogenetic analysis of nrDNA ITS sequences. Plant Div. Evol.,

128: 111–136.

Egorova T. V. 1999. Sedges (Carex L.) of Russia and adjanced states (within the limits of the former USSR). St. Petersburg: State Chemical-Pharmaceutical Academy, 771.

eMonocot: an online resource for monocot plants. Available from:http://e-monocot.org/.

Esankulov A. S. 2012. Flora of the Zaamin Nature Reserve. Tashkent: Abstr. Cand. Diss., 19 (In Russian)

Fedorov A. A. 1984. Plant resources of the U.S.S.R.: Flowering plants, their chemical composition and use. Families Magnoliaceae–Limoniaceae. Leningrad: Science Publishers, 460. (In Russian)

Gammerman A. F., Grom I. I. 1976. Wild medicinal plants of the USSR. Moscow: Medicine Publishers, 288. (In Russian)

Geltman D. V. 2013. Spurges (Euphorbia L., Euphorbiaceae) of East Europe and the Caucasus in the mirrow of new system of the genus. Turczaninowia, 16 (2): 30–40. (In Russian)

Hill L. Fritillaria: a list of published names. 2017. Available from: http://www.fritillariaicones.com/info/names/frit.names.pdf.

Horandl E., Emadzade K. 2012. Evolutionary classification: A case study on the diverse plant genus Ranunculus L. (Ranunculaceae). Perspectives in Plant Ecology, Evolution and Systematics, 14: 310–324.

ILDIS World Database of Legumes (version 10, 2005). Available from: www.ildis.com.

IUCN. 2020 The IUCN Red List of Threatened Species. Available from: http://www.iucnredlist.org.

IUCN/ISSC Invasive Species Specialist Group (ISSG). 2014. Global Invasive Species Database version 2014-2. Available from: http://193.206.192.138/gisd/

Jizzax Viloyat Hokimligi. Official web-site. Available from: http://jizzax.uz.

Kamelin R V. 1973a. Florogenetic analysis of the native flora of the Mountain Middle Asia. Leningrad: Science Publishers, 133-138. (in Russian)

Kamelin R V. 1973b. To the knowledge of the flora of Nuratau Mountains. Botan. Journ., 58 (5): 625–637. (In Russian).

Kamelin R V. 1990. Flora of Syrdarya Karatau: Materials for floristic regionalization of Middle Asia. Leningrad: Science Publishers 70–107. (in Russian)

Kamelin R. V. 1979. The Kuhistan district of mountainous Middle Asia: Botanical-geographical analysis. Leningrad: Science Publishers, 166. (In Russian)

Kamelin R. V., Kovalevskaja S. S., Nabiev M. M. 1981. Conspectus Florae Asiae Mediae. Vol. 6. Tashkent: Fan Publishers, 395. (In Russian)

Khassanov F. O. 2015. Conspectus Florae Asiae Mediae. Vol. 11. Tashkent: Fan Publishers, 456. (In Russian)

Khassanov F. O., Esankulov A. S. & Tirkasheva M. B. 2013. Flora of the Zaamin Nature Reserve. Tashkent: [s.n.], 119. (In Russian)

Khassanov F. O., Rakhimova N. 2012. Taxonomical revision of genus Iris L. (Iridaceae Juss.) in the Flora of Central Asia. Stapfia: reports, 97: 121-126.

Kinzikaeva G. K. 1988. Flora of the Tajik SSR. Vol. 9. Academy of Sciences of USSR, Leningrad, 568. (In Russian)

Kochkareva T. F. 1986. Flora of the Tajik SSR. Vol. 8. Academy of Sciences of USSR, Leningrad,

519. (In Russian)

Konnov A. A. 1973. Flora of juniper forests of the Shakhristan. Dushanbe: Donish Publishers, 176. (In Russian)

Konnov A. A. 1990. Juniper formations of Central Asia and adjacent areas. Novosibirsk: Abstr. Doct. Diss., 34. (In Russian)

Korovin E. P. 1923. Plant formations of the Nurata Valley. In: Proceedings of the Turkestan Scientific Society. Vol. 1. Tashkent: Turkestan Publishing House, 2–5. (In Russian)

Korovin E. P. 1934. Vegetation of Middle Asia and South Kazakhstan. Moscow & Tashkent: [s.n.], 479. (In Russian)

Korovin E. P. 1961. Vegetation of Middle Asia and South Kazakhstan. Vol. 1. Tashkent: [s.n.], 452. (In Russian)

Korovin E. P. 1962. Vegetation of Middle Asia and South Kazakhstan. Vol. 2. Tashkent: [s.n.], 547. (In Russian).

Kovalevskaja S. S. 1963. Conspectus Florae Asiae Mediae. Vol. 1. Tashkent: Fan Publishers, 226. (In Russian)

Kovalevskaja S. S. 1971. Conspectus Florae Asiae Mediae. Vol. 2. Tashkent: Fan Publishers, 360. (In Russian)

Kudrjashev, S. N. 1941. Flora of Uzbekistan (Vol.1). Tashkent: Uzbek Department of the Academy of Sciences of the USSR. [In Russian]

Kudryashev S. N. 1941. Flora of Uzbekistan. Vol. 1. Uzbek Department of the Academy of Sciences of the USSR, Tashkent, 568. (In Russian)

Kultiassov M. V. 1923. Essay on the vegetation of the Pistalitau Mountains. In: Proceedings of the Turkestan Scientific Society. Tashkent: Turkestan Publishing House, 89–102. (In Russian)

Kurmikov A. G., Belolipov I. V. 2012. Wild medicinal plants of Uzbekistan. Tashkent: Extremum Press, 278. (In Russian)

Lammers T. G. 2007. World checklist and bibliography of Campanulaceae. Kew, UK: Royal Botanic Gardens, 675.

Larin I. V. 1956. Forage plants of hayfields and pastures of the USSR. (In 3 vol). Moskow, Leningrad: State agricultural publishing house. (In Russian)

Lazkov G. A. 2016. Family Labiatae Juss. in flora of Kyrgyzstan. Pocheon, Republic of Korea: Geobook, 384.

Lopez-Vinyallonga S., Mehregan I., Garcia-Jacas N., Tscherneva O., Susanna A. & Kadereit J.W. 2009. Phylogeny and evolution of the Arctium-Cousinia complex (Compositae, Cardueae-Carduinae). Taxon, 58 (1): 153–171. (In Russian)

Lopez-Vinyallonga S., Romaschenko K., Susanna A. & Garcia-Jacas N. 2011. Systematics of the Arctioid group: Disentangling Arctium and Cousinia (Cardueae, Carduinae). Taxon, 60(2): 539–554.

Nabiev M. M. 1986. Conspectus Florae Asiae Mediae. Vol. 8. Tashkent: Fan Publishers, 191. (In Russian)

Nikitin V. V. 1983. Weeds of the flora of the USSR. Leningrad: Science Publishers, 454. (In Russian)

Olson D. M., Dinerstein E. 2002. The Global 200: Priority ecoregions for global conservation.

Annals of the Missouri Botanical Garden, 89(2): 199–224.

Ovczinnikov P. N. 1957. Flora of the Tajik SSR. Vol. 1. Academy of Sciences of the USSR, Moscow & Leningrad, 547. (In Russian)

Ovczinnikov P. N. 1963. Flora of the Tajik SSR. Vol. 2. Academy of Sciences of the USSR, Moscow & Leningrad, 456. (In Russian)

Ovczinnikov P. N. 1968. Flora of the Tajik SSR. Vol. 3. Academy of Sciences of the USSR, Leningrad, 711. (In Russian)

Ovczinnikov P. N. 1975. Flora of the Tajik SSR. Vol. 4. Academy of Sciences of the USSR, Leningrad, 576. (In Russian)

Ovczinnikov P. N. 1978. Flora of the Tajik SSR. Vol. 5. Academy of Sciences of the USSR, Leningrad, 678. (In Russian).

Ovczinnikov P. N. 1981. Flora of the Tajik SSR. Vol. 6. Academy of Sciences of the USSR, Leningrad, 727. (In Russian)

Pachomova M. G. 1974. Conspectus Florae Asiae Mediae. Vol. 4. Tashkent: Fan Publishers, 274. (In Russian)

Pachomova M. G. 1976. Conspectus Florae Asiae Mediae. Vol. 5. Tashkent: Fan Publishers, 375. (In Russian)

Podlech D. & Zarre S. (with collaboration of Ekici M., Maassoumi A. A. R. & Sytin A.). 2013. A taxonomic revision of the genus Astragalus L. (Leguminosae) in the Old World (vols. 1–3). Wien: Naturhistorisches Museum, 2439.

Popov M. G., Androsov N. V. 1937. Vegetation of the Guralash Nature Reserve and Zaamin forestry. Tashkent: Department of Sciences, 40. (In Russian).

Pratov U. P. 1998. The Red Data Book of the Republic of Uzbekistan. Vol. 1. Plants. Tashkent: Chinor ENK. 335. (in Uzbek and Russian)

Pratov U. P. 2006. The Red Data Book of the Republic of Uzbekistan. Vol. 1. Plants and Fungi. Tashkent: Chinor ENK. 250. (in Uzbek, Russian and English)

Pratov U. P., Khassanov F. O. The Red Data Book of the Republic of Uzbekistan. Vol. 1. Plants and Fungi. 2009. Tashkent: Chinor ENK. 360. (in Uzbek, Russian and English)

Rassulova M. R. 1991. Flora of the Tajik SSR. Vol. 10. Academy of Sciences of USSR, Leningrad, 624. (In Russian)

Sadykov A.S. 1984. The Red Data Book of the Uzbek SSR. Vol. 2. Plants. Tashkent. 150. (in Uzbek and Russian)

Salmaki Y., Zarre S., Ryding O., Lindqvist C., Schneunert A., Brauchler C. & Heubl G. 2012. Phylogeny of the tribe Phlomideae (Lamioideae: Lamiaceae) with special focus on Eremostachys and Phlomoides: New insights from nuclear and chloroplast sequences. Taxon, 61 (1): 161–179.

Sennikov A. N., Tojibaev K. Sh, Khassanov F. O., Beshko N. Yu. 2016. The Flora of Uzbekistan Project. Phytotaxa, 282 (2): 107–118.

Sennikov, A. N. 2016. Flora of Uzbekistan (Vol.1). Toshkent: Navroz Publishers, xviii + 173. [In Russian]

Sennikov, A. N. 2017. Flora of Uzbekistan (Vol. 2). Toshkent: Navroz Publishers, xii + 200. [In

Russian]

Sokolov P. D. 1986. Plant resources of the U. S. S. R.: Flowering plants, their chemical composition and use. Families Paeoniaceae – Thymelaeaceae. Leningrad: Science Publishers, 336. (In Russian)

Sokolov P. D. 1987. Plant resources of the U. S. S. R.: Flowering plants, their chemical composition and use. Families Hydrangeaceae – Haloragaceae. Leningrad: Science Publishers, 326. (In Russian)

Sokolov P. D. 1988. Plant resources of the U. S. S. R.: Flowering plants, their chemical composition and use. Families Rutaceae – Elaeagnaceae. Leningrad: Science Publishers, 357. (In Russian)

Sokolov P. D. 1990. Plant resources of the U. S. S. R.: Flowering plants, their chemical composition and use. Families Caprifoliaceae – Plantaginaceae. Leningrad: Science Publishers, 328. (In Russian)

Sokolov P. D. 1991. Plant resources of the U.S.S.R.: Flowering plants, their chemical composition and use. Families Hippuridaceae – Lobeliaceae. St.-Petersburg: Science Publishers, 200. (In Russian)

Sokolov P. D. 1993. Plant resources of the U. S. S. R.: Flowering plants, their chemical composition and use. Familia Asteraceae (Compositae). St.-Petersburg: Science Publishers, 352. (In Russian)

Sokolov P. D. 1994. Plant resources of Russia and adjacent states: Flowering plants, their chemical composition and use. Families Butomaceae – Typhaceae. St.-Petersburg: Science Publishers, 271. (In Russian)

Takhtajan A. 1986. Floristic regions of the World. Los Angeles, London: Berkeley, 522.

The Angiosperm Phylogeny Group. 2016. An update of the Angiosperm Phylogeny Group classification for the orders and families of flowering plants: APG IV. Botanical Journal of the Linnean Society, 181 (1): 1–20.

The Global Compositae Checklist. 2009. Available from: http://www.compositae.org/checklist.

The Gymnosperm Database. 2017. Available from: http://www.conifers.org/.

The International Plant Names Index. 2012. Available from: http://www.ipni.org.

The Plant List (Version 1.1). 2017. Available from: http://www.theplantlist.org/.

The Republic of Uzbekistan: Biodiversity Conservation, National Strategy and Action Plan. 1998. Tashkent. Available from: https://www.cbd.int/doc/world/uz/uz-nr-01-en.pdf.

The State Committee of the Republic of Uzbekistan on Statistics. Official web-site. 2017. Available from: http://www.stat.uz/.

The World Umbellifer Database. 1999. Available from: http://www.umbellifers.com/frames.html?http://rbg-web2.rbge.org.uk/URC/urchomepage.html.

Tirkasheva M. B. 2011. Vegetation of the Sanzar River basin. Tashkent: Abstr. Cand. Diss., 21. (In Uzbek)

Tojibaev K. Sh., Beshko N. Yu. & Popov V. A. 2016. Botanical-geographical regionalization of Uzbekistan. Bot. Journ., 101 (10): 1105–1132. (In Russian)

Tojibaev K. Sh., Beshko N. Yu. 2007. The cadaster of rare and endemic plants of Dzhizak and Navoi provinces of the Republic of Uzbekistan. In: Biodiversity of Uzbekistan — monitoring and use. Tashkent: [s.n.], 200-208. (In Russian).

Tojibaev K. Sh., Beshko N. Yu., Karimov F. I., Batoshov A. R., Turginov O. T. & Azimova D. 2014a. The Data Base of Flora of Uzbekistan. Journal for Arid Land Studies, 24 (1): 157–160.

Tojibaev K. Sh., Beshko N.Yu., Popov V. A., Jang C. G. & Chang K. S. 2017. Botanical Geography of Uzbekistan. Pocheon, Republic of Korea: Korea National Arboretum, 250.

Tojibaev K.Sh., Beshko N.Yu. 2015. Reassessment of diversity and analysis of distribution in Tulipa (Liliaceae) in Uzbekistan. Nordic Journal of Botany, 33: 324–334.

Tojibaev K.Sh., Beshko N.Yu., Azimova D. E. & Turginov O. T. 2015. Distribution patterns of species of the genus Astralalus L. (sect. Macrocystis, Laguropsis and Chaetodon) in the territory of Mountain Middle Asian province. Turczaninowia, 18 (2): 17–38. (In Russian)

Tsvelev N. N. 1983. Grasses of the Soviet Union. New Delhi: Oxonian Press, 1196.

UNDP, State Committee for Nature Protection, Academy Sciences of the Repubic of Uzbekistan. 2015. The Fifth National Report of the Republic of Uzbekistan on the Biodiversity Conservation. Tashkent: [s.n.], 58. [In Russian]

Vvedensky A. I. 1953. Flora of Uzbekistan. Vol. 2. Academy of Sciences of the Uzbek SSR, Tashkent, 549. (In Russian)

Vvedensky A. I. 1955. Flora of Uzbekistan. Vol. 3. Academy of Sciences of the Uzbek SSR, Tashkent, 825. (In Russian)

Vvedensky A. I. 1959. Flora of Uzbekistan. Vol. 4. Academy of Sciences of the Uzbek SSR, Tashkent, 507. (In Russian)

Vvedensky A. I. 1961. Flora of Uzbekistan. Vol. 5. Academy of Sciences of the Uzbek SSR, Tashkent, 667. (In Russian)

Vvedensky A. I. 1962. Flora of Uzbekistan. Vol. 6. Academy of Sciences of the Uzbek SSR, Tashkent, 630. (In Russian)

Wiegleb G., Bobrov A. A. & Zalewska-Galosz J. 2017. A taxonomic account of Ranunculus section Batrachium (Ranunculaceae). Phytotaxa, 319 (1): 001–055.

Williams M. W., Konovalov V. G. 2008. Central Asia temperature and precipitation data, 1879-2003: USA National Snow and Ice Data Center. Available from: http://nsidc.org/data/docs/noaa/g02174_central_asia_data/index.html.

Zakirov K. Z. 1955. Flora and vegetation of the Zeravschan river basin. (Vol. 1.) Tashkent: [s.n.], 205 p. (In Russian)

Zakirov K. Z. 1962. Flora and vegetation of the Zeravschan river basin. (Vol. 2.) Tashkent: [s.n.], 446 p. (In Russian)

Zakirov K. Z., Burygin V. A. 1956. About some relict plants of the Nuratau ridge. Botan Journ., 41 (9): 1331–1332. (In Russian).

Zakirov P. K. 1969. Vegetation cover of the Nuratau Mountains. Tashkent: Fan Publishers, 142. (In Russian)

Zakirov P. K. 1971. Botanical geography of low mountains of Kyzylkum and Nuratau ridge. Tashkent: Fan Publishers. 203. (In Russian)

Zakirov, K. Z. 1971. Vegetation cover of Uzbekistan and the ways of its practical use (I-vol). Tashkent: Fan Publishers. [In Russian]

Zakirov, K. Z. 1976. Vegetation cover of Uzbekistan and the ways of its practical use (III-vol). Tashkent: Fan Publishers. [In Russian]

Zakirov, K. Z. 1984. Vegetation cover of Uzbekistan and the ways of its practical use (IV-vol).

Tashkent: Fan Publishers. [In Russian]

Zakirov, K. Z., 1973. Vegetation cover of Uzbekistan and the ways of its practical use (II-vol). Tashkent: Fan Publishers. [In Russian]

INDEX OF LATIN NAMES

A

Abutilon   271
Abutilon therophrasti   271
Acanthocephalus   428
Acanthocephalus amplexifolius   428
Acanthocephalus benthamianus   429
Acantholimon   297
Acantholimon erythraeum   297
Acantholimon nuratavicum   298
Acantholimon subavenaceum   299
Acantholimon tataricum   299
Acanthophyllum   326
Acanthophyllum aculeatum   326
Acanthophyllum elatius   326
Acanthophyllum gypsophiloides   326
Acanthophyllum pungens   326
Acantholimon zakirovii   299
Acer   265
Acer pubescens   265
Acer semenovii   266
Acer turkestanicum   266
Achillea   459
Achillea arabica   459
Achillea biebersteinii   459
Achillea filipendulina   459
Achillea kermanica   460
Achillea millefolium   460
Achillea santolinoides subsp. wilhelmsii   460
Achnatherum splendens   107
Achnatherum turkomanicum   107
Acinos rotundiflorus   411
Aconitum   149
Aconitum talassicum   149
Aconitum zeravschanicum   149
Aconogonon coriarum   311
Aconogonon hissaricum   312
Acroptilon repens   449
Adiantum   36
Adiantum capillus-veneris   36
Adonis   147
Adonis aestivalis   147
Adonis turkestanica   148
Aegilops   128
Aegilops crassa   128
Aegilops cylindrica   128
Aegilops juvenalis   128
Aegilops squarrosa   128
Aegilops triuncialis   128
Aellenia glauca   340
Aellenia iliensis   340
Aellenia subaphylla   340
Aeluropus   117
Aeluropus intermedius   117

Aeluropus litoralis   117
Aethionema   293
Aethionema carneum   293
Agrimonia   229
Agrimonia asiatica   229
Agriophyllum   334
Agriophyllum lateriflorum   334
Agropyron   126
Agropyron badamense   126
Agropyron drobovii   132
Agropyron ferganense   131
Agropyron intermedium   132
Agropyron lachnophyllum   131
Agropyron lolioides   132
Agropyron repens   132
Agropyron setuliferum   126
Agropyron tianschanicum   132
Agropyron trichophorum   132
Agropyron ugamicum   131
Agrostemma   319
Agrostemma githago   319
Agrostis   111
Agrostis gigantea   111
Agrostis hissarica   112
Agrostis olympica   112
Agrostis stolonifera   112
Ailanthus   269
Ailantus altissima   269
Alcea   270
Alcea litvinovii   270
Alcea nudiflora   271
Alcea rhyticarpa   271
Alchemilla   228
Alchemilla fontinalis   228
Alchemilla krylovii   228
Alchemilla sibirica   228
Alhagi   211
Alhagi canescens   211
Alhagi kirghisorum   211
Alhagi pseudalhagi   211
Alisma   49
Alisma plantago-aguatica   49
Alismataceae   49
Alliaria   274
Alliaria petiolata   274
Allium   76
Allium alexeianum   76
Allium altissimum   77
Allium barsczewskii   77
Allium caesium   77
Allium carolinianum   78
Allium caspium   78
Allium clausum   79
Allium cupuliferum subsp. cupuliferum   80

Allium cupuliferum subsp. nuratavicum   80
Allium drepanophyllum   81
Allium filidens   81
Allium griffithianum   81
Allium gusaricum   82
Allium inconspicuum   82
Allium isakulii   82
Allium jodanthum   83
Allium karataviense   83
Allium kaufmannii   84
Allium kokanicum   84
Allium komarowii   85
Allium levichevii   85
Allium longicuspis   85
Allium oreodictyum   85
Allium oreophiloides   85
Allium oreophilum   85
Allium oschaninii   86
Allium praemixtum   86
Allium protensum   87
Allium sarawschanicum   87
Allium stephanophorum   87
Allium stipitatum   87
Allium suworowii   88
Allium taeniopetalum   88
Allium talassicum   88
Allium turkestanicum   89
Allium verticillatum   89
Allium xiphopetalum   90
Allochrusa   326
Allochrusa gypsophiloides   326
Alopecurus   111
Alopecurus arundinaceus   111
Alopecurus himalaicus   111
Alopecurus myosuroides   111
Alopecurus pratensis   111
Alposelinum   490
Alposelinum albomarginatum   490
Althaea   271
Althaea armeniaca   271
Althaea officinalis   271
Alyssum   285
Alyssum alyssoides   285
Alyssum campestre   285
Alyssum dasycarpum   285
Alyssum desertorum   286
Alyssum linifolium   286
Alyssum marginatum   286
Alyssum szovitsianum   286
Amaranthaceae   329
Amaranthus   329
Amaranthus albus   329
Amaranthus blitoides   329
Amaranthus deflexus   329

Amaranthus retroflexus 329
Amaryllidaceae 76
Amberboa 448
Amberboa turanica 448
Ammodendron 174
Ammodendron conollyi 174
Ammodendron lehmannii 174
Ammothamnus lehmanni 173
Amoria bonannii 177
Amoria fragifera 177
Amygdalus bucharica 233
Amygdalus communis 235
Amygdalus spinosissima 237
Anabasis 340
Anabasis eriopoda 340
Anabasis jaxartica 341
Anabasis salsa 341
Anacaridaceae 265
Anagallis 349
Anagallis arvensis 349
Anagallis foemina 349
Anchusa 363
Anchusa arvensis subsp. orientalis 363
Anchusa azurea 363
Anchusa italica 363
Andrachne 251
Andrachne fedtschenkoi 251
Andrachne telephioides 251
Androsace 349
Androsace dasyphylla 349
Androsace maxima 349
Androsace sericea 349
Anemone 157
Anemone biflora var. petiolulosa 157
Anemone narcissiflora subsp. protracta 157
Anemone petiolulosa 157
Anemone protracta 157
Anemone tschernaewii 158
Angelica 490
Angelica archangelica subsp. decurrens 490
Angelica brevicaulis 490
Angelica ternata 491
Angiospermae 47
Angiosperms 47
Anthemis 459
Anthemis deserticola 459
Anthriscus 480
Anthriscus glacialis 480
Antonia debilis 411
Anura pallidivirens 438
Apera 112
Apera interrupta 112
Aphanopleura 487
Aphanopleura capillifolia 487
Apiaceae 479
Apium 487
Apium nodiflorum 487
Apocynaceae 358
Aquilegia 145
Aquilegia lactiflora 145

Aquilegia vicaria 123 145
Arabidopsis 275
Arabidopsis parvula 274
Arabidopsis pumila 275
Arabidopsis thaliana 275
Arabis 277
Arabis auriculata 277
Arabis kokanica 276
Arabis montbretiana 277
Araceae 48
Arceuthobium 294
Arceuthobium oxycedri 294
Arctium 436
Arctium aureum 436
Arctium chloranthum 437
Arctium haesitabundum 437
Arctium karatavicum 437
Arctium korolkowii 437
Arctium leiospermum 437
Arctium pallidivirens 438
Arenaria 318
Arenaria griffithii 318
Arenaria leptoclados 319
Arenaria rotundifolia 318
Arenaria serpyllifolia 319
Arenaria serpyllifolia subsp. leptoclados 319
Aristida karelinii 105
Armeniaca vulgaris 233
Arnebia 360
Arnebia coerulea 360
Arnebia decumbens 361
Arnebia euchroma 362
Arnebia obovata 362
Arnebia transcaspica 362
Artemisia 465
Artemisia absinthium 465
Artemisia annua 465
Artemisia cina 465
Artemisia diffusa 465
Artemisia dracunculus 465
Artemisia eremophila 465
Artemisia ferganensis 465
Artemisia glanduligera 465
Artemisia glaucina 465
Artemisia juncea 465
Artemisia lehmanniana 466
Artemisia leucodes 467
Artemisia persica 468
Artemisia porrecta 468
Artemisia prolixa 468
Artemisia rutifolia 468
Artemisia santolinifolia 468
Artemisia scoparia 468
Artemisia scotina 468
Artemisia serotina 468
Artemisia sieversiana 469
Artemisia sogdiana 469
Artemisia subsalsa 469
Artemisia tenuisecta 469
Artemisia tournefortiana 470
Artemisia turanica 470

Artemisia vulgaris 471
Arum 48
Arum korolkowii 48
Asparagaceae 90
Asparagus 91
Asparagus brachyphyllus 91
Asparagus persicus 91
Asperuginoides 286
Asperuginoides axillaris 286
Asperugo 367
Asperugo procumbens 367
Asperula 351
Asperula aparine 353
Asperula glabrata 351
Asperula humifusa 352
Asperula oppositifolia 351
Asperula setosa 351
Asperula trichodes 351
Asphodelaceae 70
Aspleniaceae 36
Asplenium 37
Asplenium ruta-muraria 37
Asplenium trichomanes 37
Aster canescens 457
Asteraceae 422
Astragalus 182
Astragalus adpressipilosus 182
Astragalus alabugensis 182
Astragalus alopecias 182
Astragalus alpinus 182
Astragalus ammophilus 194
Astragalus ammotrophus 182
Astragalus angreni 183
Astragalus aphanassjievii 183
Astragalus aschuturi 183
Astragalus bactrianus 183
Astragalus bakaliensis 183
Astragalus belolipovii 183
Astragalus camptoceras 183
Astragalus campylorrhynchus 184
Astragalus campylotrichus 184
Astragalus chodshenticus 184
Astragalus commixtus 184
Astragalus compositus 184
Astragalus contortuplicatus 184
Astragalus cornu-bovis 185
Astragalus corydalinus 194
Astragalus dendroides 185
Astragalus dictamnoides 185
Astragalus dipelta 185
Astragalus dolichocarpus 185
Astragalus eximius 185
Astragalus falcigerus 185
Astragalus farctissimus 185
Astragalus ferganensis 185
Astragalus filicaulis 185
Astragalus flexus 186
Astragalus floccosifolius 186
Astragalus globiceps 187
Astragalus guttatus 187
Astragalus harpilobus 187
Astragalus inaequalifolius 187

INDEX OF LATIN NAMES

Astragalus intarrensis 187
Astragalus isfahanicus 187
Astragalus iskanderi 187
Astragalus jagnobicus 187
Astragalus juratzkanus 187
Astragalus kelleri 187
Astragalus knorringianus 188
Astragalus korovinianus 194
Astragalus kurdaicus 192
Astragalus lasiosemius 188
Astragalus lasiostylus 188
Astragalus leiophysa 188
Astragalus lentilobus 188
Astragalus leptophysus 189
Astragalus leptostachys 190
Astragalus lipskyi 190
Astragalus lithophilus 190
Astragalus macrocladus 190
Astragalus macronyx 191
Astragalus macropetalus 192
Astragalus macrotropis 192
Astragalus marguzaricus 192
Astragalus masenderanus 192
Astragalus maverranagri 187
Astragalus mucidus 193
Astragalus nematodes 193
Astragalus nivalis 193
Astragalus nobilis 193
Astragalus nuciferus 194
Astragalus ophiocarpus 194
Astragalus orbiculatus 194
Astragalus oxyglottis 194
Astragalus patentipilosus 194
Astragalus patentivillosus 194
Astragalus paucijugus 194
Astragalus pauper 194
Astragalus peduncularis 194
Astragalus persipolitanus 194
Astragalus petunnikovii 194
Astragalus platyphyllus 195
Astragalus pterocephalus 195
Astragalus pulcher 195
Astragalus quisqualis 195
Astragalus retamocarpus 195
Astragalus rubromarginatus 195
Astragalus russanovii 195
Astragalus saratagius 195
Astragalus sarytavicus 195
Astragalus schachimardanus 195
Astragalus schmalhausenii 195
Astragalus schrenkianus 195
Astragalus schugnanicus 196
Astragalus sesamoides 196
Astragalus sewerzowii 196
Astragalus sieversianus 196
Astragalus skorniakovii 192
Astragalus sogdianus 197
Astragalus stalinskyi 197
Astragalus stenanthus 197
Astragalus stenocystis 197
Astragalus stipulosus 195
Astragalus striatellus 187

Astragalus subbarbellatus 196
Astragalus subinduratus 197
Astragalus subscaposus 197
Astragalus substipitatus 197
Astragalus subverticillatus 198
Astragalus tibetanus 198
Astragalus titovii 198
Astragalus transoxanus 198
Astragalus turbinatus 198
Astragalus turczaninovii 198
Astragalus turkestanus 199
Astragalus unifoliolatus 200
Astragalus uninodus 200
Astragalus urgutinus 200
Astragalus variegatus 200
Astragalus vicarius 201
Astragalus villosissimus 201
Astragalus vladimiri 188
Astragalus xanthomeloides 201
Asyneuma 419
Asyneuma argutum 419
Asyneuma trautvetteri 420
Atraphaxis 306
Atraphaxis compacta 306
Atraphaxis karataviensis 306
Atraphaxis pyrifolia 306
Atraphaxis seravschanica 307
Atraphaxis spinosa 307
Atraphaxis virgata 307
Atriplex 332
Atriplex aucherii 332
Atriplex dimorphostegia 332
Atriplex flabellum 332
Atriplex hastata 333
Atriplex micrantha 332
Atriplex ornata 332
Atriplex prostrata subsp. calotheca 333
Atriplex tatarica 333
Aulacospermum 482
Aulacospermum roseum 482
Aulacospermum simplex 482
Aulacospermum tenuisectum 482
Avena 113
Avena barbata 113
Avena fatua 113
Avena persica 113
Avena sterilis subsp. ludoviciana 113
Avena trichophylla 113

B

Baeothryon 97
Baeothryon pumilum 97
Balsaminaceae 346
Barbarea 276
Barbarea brachycarpa subsp. minor 276
Barbarea minor 276
Barbarea vulgaris 276
Barkhausia kotschyana 427
Bassia 334
Bassia crassifolia 334
Bassia eriophora 334

Bassia hyssopifolia 334
Batrachium divaricatum 164
Batrachium pachycaulon 164
Batrachium rionii 162
Batrachium trichophyllum 164
Berberidaceae 143
Berberis 143
Berberis integerrima 143
Berberis nummularia 143
Berberis oblonga 144
Berula 487
Berula erecta 487
Betula 243
Betula pendula 243
Betulaceae 243
Bidens 451
Bidens tripartita 451
Biebersteinia 263
Biebersteinia multifida 263
Biebersteiniaceae 263
Bistortia elliptica 313
Blysmus 97
Blysmus compressus 97
Boissiera 125
Boissiera squarrosa 125
Bolbosaponaria 325
Bolbosaponaria sewerzowii 325
Bolboschoenus 95
Bolboschoenus maritimus 95
Bolboschoenus popovii 96
Bongardia 145
Bongardia chrysogonum 145
Boraginaceae 359
Botriochloa 104
Botriochloa ischaemum 104
Botrychium 34
Botrychium lunaria 34
Brachyactis ciliata 458
Brachypodium 125
Brachypodium sylvaticum 125
Brassica 287
Brassica campestris 287
Brassica elongata 287
Brassica rapa 287
Brassicaceae 274
Bromus 123
Bromus danthoniae 123
Bromus inermis 124
Bromus japonicus 124
Bromus lanceolatus 124
Bromus oxyodon 124
Bromus paulsenii 124
Bromus popovii 124
Bromus scoparius 124
Bromus scoparius 124
Bromus sericeus 124
Bromus sewerzowii 124
Bromus sterilis 124
Bromus tectorum 124
Bryonia 243
Bryonia cretica subsp. dioica 243
Bryonia dioica 243

Bryonia melanocarpa 244
Buchingera axillaris 286
Bufonia 317
Bufonia oliveriana 317
Buglossoides 360
Buglossoides arvensis 360
Bunium capusii 486
Bunium chaerophylloides 486
Bunium intermedium 486
Bunium persicum 486
Bupleurum 485
Bupleurum exaltatum 485
Bupleurum falcatum subsp. exaltatum 485
Butomaceae 49
Butomus 49
Butomus umbellatus 49

C

Calamagrostis 112
Calamagrostis anthoxanthoides 112
Calamagrostis dubia 112
Calamagrostis epigeios 112
Calamagrostis holciformis 112
Calamagrostis laguroides 112
Calamagrostis pseudophragmites 112
Callianthemum 157
Callianthemum alatavicum 157
Calligonum 307
Calligonum aphyllum 307
Calligonum caput-medusae 307
Calligonum leucocladum 308
Calligonum macrocarpum 309
Calligonum microcarpum 310
Callipeltis 354
Callipeltis cucularis 354
Calystegia 373
Calystegia sepium 373
Camelina 294
Camelina microcarpa 294
Camelina rumelica 294
Camelina sylvestris 294
Campanula 418
Campanula cashmeriana 418
Campanula fedtschenkoana 419
Campanula glomerata 419
Campanula lehmanniana 419
Campanulaceae 418
Camphorosma 334
Camphorosma monspeliaca 334
Campyloptera carnea 293
Cannabaceae 241
Cannabis 241
Cannabis sativa 241
Capparaceae 273
Capparis 273
Capparis spinosa 273
Caprifoliaceae 473
Capsella 294
Capsella bursa-pastoris 294
Caragana 181
Caragana alaica 181

Caragana turkestanica 181
Cardaria repens 290
Carduus 435
Carduus albidus 435
Carduus nutans 435
Carduus pycnocephalus subsp. albidus 435
Carex 100
Carex diluta 100
Carex divisa 100
Carex enervis 100
Carex karoi 100
Carex melanantha 100
Carex melanostachya 101
Carex microglochin 101
Carex orbicularis 101
Carex pachystilis 101
Carex parva 101
Carex physodes 101
Carex pseudofoetida 101
Carex serotina 101
Carex songorica 101
Carex stenocarpa 102
Carex stenophylla 102
Carex subphysodes 102
Carex turkestanica 102
Carthamus 450
Carthamus oxyacanthus 450
Carthamus tinctorius 450
Carthamus turkestanicus 450
Carum 486
Carum carvi 486
Caryophyllaceae 314
Catabrosa 116
Catabrosa aquatica 116
Catabrosa capusii 116
Celastraceae 245
Celtis 241
Celtis australis subsp. caucasica 241
Centaurea 448
Centaurea belangeriana 448
Centaurea benedicta 448
Centaurea bruguierana subsp. belangeriana 448
Centaurea cyanus 448
Centaurea depressa 448
Centaurea iberica 448
Centaurea solstitialis 448
Centaurea squarrosa 448
Centaurea virgata subsp. squarrosa 448
Centaurium 355
Centaurium erythraea subsp. turcicum 355
Centaurium pulchellum 355
Centaurium spicatum 355
Centaurium turcicum 355
Cephalaria 477
Cephalaria syriaca 477
Cephalorrhynchus soongoricus 424
Cerastium 315
Cerastium arvense 315
Cerastium bungeanum 316

Cerastium cerastoides 316
Cerastium dentatum 316
Cerastium dichotomum 316
Cerastium dichotomum subsp. inflatum 316
Cerastium falcatum 316
Cerastium fontanum subsp. vulgare 316
Cerastium glomeratum 316
Cerastium holosteoides 316
Cerastium inflatum 316
Cerastium lithospermifolium 316
Cerastium pentandrum 316
Cerastium perfoliatum 316
Cerastium pusillum 316
Cerastium semidecandrum 316
Cerasus amygdaliflora 238
Cerasus erythrocarpa 235
Cerasus mahaleb 236
Cerasus verrucosa 238
Cerasus vulgaris 234
Ceratocarpus 333
Ceratocarpus arenarius 333
Ceratocarpus utriculosus 333
Ceratocephala 159
Ceratocephala falcata 159
Ceratocephala testiculata 159
Ceratoides ewersmanniana 333
Ceratoides latens 333
Ceratophyllaceae 136
Ceratophyllum 136
Ceratophyllum demersum 136
Ceterach 37
Ceterach officinarum 37
Chaerophyllum nodosum 480
Chaetolimon 300
Chaetolimon setiferum 300
Chalcanthus 289
Chalcanthus renifolius 289
Chamaenerion angustifolium 261
Chamaesphacos 406
Chamaesphacos ilicifolius 406
Chamaesyce canescens 252
Chamaesyce turcomanica 254
Chardinia 435
Chardinia orientalis 435
Cheilanthes 35
Cheilanthes persica 35
Chenopodium 330
Chenopodium album 330
Chenopodium botrys 331
Chenopodium chenopodioides 331
Chenopodium foliosum 331
Chenopodium glaucum 332
Chenopodium murale 332
Chenopodium rubrum 332
Chenopodium vulvaria 332
Chesneya 181
Chesneya ternata 181
Chesneya turkestanica 182
Chondrilla 423
Chondrilla aspera 423

Chondrilla lejosperma 423
Chorispora 283
Chorispora bungeana 283
Chorispora elegans 284
Chorispora macropoda 283
Chorispora sabulosa 284
Chorispora tenella 284
Chrozophora 251
Chrozophora gracilis 252
Chrozophora hierosolymitana 251
Chrozophora obliqua 251
Chrozophora sabulosa 252
Chrozophora tinctoria 251
Chrysaspis campestris 177
Cicer 211
Cicer flexuosum 211
Cicer grande 211
Cicer macracanthum 213
Cicer paucijugum 213
Cicer pungens 213
Cicer songaricum 213
Cichorium 422
Cichorium intybus 422
Cirsium 436
Cirsium alatum 436
Cirsium arvense 436
Cirsium brevipapposum 436
Cirsium esculentum 436
Cirsium ochrolepidium 436
Cirsium polyacanthum 436
Cirsium semenovii 436
Cirsium straminispinum 436
Cirsium turkestanicum 436
Cirsium vulgare 436
Cistaceae 272
Cistanche 415
Cistanche mongolica 415
Cistanche salsa 415
Cithareloma 284
Cithareloma lehmannii 284
Clematis 158
Clematis orientalis 158
Clemensia 166
Clementsia semenovii 166
Cleomaceae 273
Cleome 273
Cleome fimbriata 273
Climacoptera 339
Climacoptera lanata 339
Climacoptera longistylosa 339
Climacoptera minkvitziae 340
Climacoptera obtusifolia 340
Climacoptera olgae 340
Climacoptera transoxana 340
Clinopodium 411
Clinopodium alpinum 411
Clinopodium debile 411
Clypeola 286
Clypeola jonthlaspi 286
Cnicus benedictus 448
Codonopsis 422
Codonopsis clematidea 422

Colchicaceae 51
Colchicum 51
Colchicum kesselringii 51
Colchicum luteum 53
Colpodium 120
Colpodium humile 120
Colpodium parviflorum 120
Colutea 179
Colutea paulsenii 179
Compositae 422
Conioselinum 490
Conioselinum tataricum 490
Conioselinum vaginatum 490
Conium 483
Conium maculatum 483
Conringia 288
Conringia clavata 288
Conringia orientalis 288
Conringia persica 288
Conringia planisiliqua 288
Consolida 150
Consolida camptocarpa 150
Consolida leptocarpa 151
Consolida rugulosa 152
Convolvulaceae 370
Convolvulus 371
Convolvulus arvensis 371
Convolvulus fruticosus 371
Convolvulus hamadae 372
Convolvulus korolkowii 372
Convolvulus lineatus 372
Convolvulus pilosellifolius 373
Convolvulus pseudocantabricus 373
Convolvulus subhirsutus 373
Conyza canadensis 457
Coriandrum 481
Coriandrum sativum 481
Corispermum 334
Corispermum lehmanninum 334
Cortusa turkestanica 347
Corydalis 141
Corydalis aitchisonii 141
Corydalis glaucescens 142
Corydalis gortschakovii 142
Corydalis ledebouriana 142
Corydalis nudicaulis 143
Corydalis schelesnowiana 143
Corydalis sewerzowii 141
Cotoneaster 218
Cotoneaster goloskokovii 218
Cotoneaster multiflorus 218
Cotoneaster nummarioides 219
Cotoneaster nummularius 219
Cotoneaster oliganthus 219
Cotoneaster songaricus 220
Cotoneaster suavis 221
Cousinia 438
Cousinia alpina 438
Cousinia ambigens 438
Cousinia aurea 436
Cousinia botschantzevii 438
Cousinia bungeana 439

Cousinia chlorantha 437
Cousinia coronata 439
Cousinia dissectifolia 439
Cousinia dshisakensis 439
Cousinia dubia 440
Cousinia eriotricha 440
Cousinia franchetii 440
Cousinia haesitabunda 437
Cousinia hamadae 440
Cousinia horridula 440
Cousinia integrifolia 440
Cousinia karatavica 437
Cousinia korolkowii 437
Cousinia lappacea 440
Cousinia microcarpa 441
Cousinia mollis 441
Cousinia olgae 441
Cousinia outichaschensis 441
Cousinia oxiana 441
Cousinia polytimetica 441
Cousinia prolifera 441
Cousinia pseudodshisakensis 442
Cousinia pygmaea 442
Cousinia radians 442
Cousinia regelii 442
Cousinia resinosa 442
Cousinia schtschurowskiana 442
Cousinia simulatrix 443
Cousinia sogdiana 443
Cousinia spiridonovii 444
Cousinia tenella 444
Cousinia umbrosa 444
Cousinia verticillaris 444
Cousinia xanthina 444
Crambe 288
Crambe cordifolia subsp. kotschyana 288
Crambe kotschyana 288
Crassulaceae 166
Crataegus 223
Crataegus pontica 223
Crataegus songarica 224
Crataegus turkestanica 224
Crepidifolium 424
Crepidifolium serawschanicum 424
Crepis 427
Crepis kotschyana 427
Crepis multicaulis 427
Crepis pulchra 427
Crocus 67
Crocus korolkowii 67
Crucianella 351
Crucianella chlorostachys 351
Crucianella exasperata 351
Crucianella filifolia 351
Cruciata 354
Cruciata pedemontana 354
Crupina 450
Crupina oligantha 450
Crupina vulgaris 450
Crypsis 108
Crypsis schoeoides 108

Cryptospora 282
Cryptospora falcata 282
Cryptospora inconspicua 282
Cryptospora omissa 282
Cucurbitaceae 243
Cullen 178
Cullen drupaceum 178
Cuminum 487
Cuminum setifolium 487
Cupressaceae 44
Cuscuta 370
Cuscuta approximata 370
Cuscuta brevistyla 370
Cuscuta campestris 370
Cuscuta cupulata 370
Cuscuta europaea 371
Cuscuta lehmanniana 371
Cuscuta monogyna 371
Cuscuta pedicellata 371
Cuscuta pellucida 371
Cutandia rigescens 116
Cutandia Willk 116
Cyanus 448
Cyanus depressus 448
Cyanus segetum 448
Cylindrocarpa 422
Cylindrocarpa sewerzowii 422
Cymbolaena 456
Cymbolaena griffithii 456
Cynanchum 358
Cynanchum acutum subsp. sibiricum 358
Cynanchum sibiricum 358
Cynodon 113
Cynodon dactylon 113
Cynoglossum 370
Cynoglossum creticum 370
Cyperaceae 95
Cyperus 98
Cyperus glaber 98
Cyperus longus 98
Cyperus rotundus 98
Cystidospermum cheirolepis 252
Cystopteris 36
Cystopteris fragilis 36

D

Dactylis 117
Dactylis glomerata 117
Dactylorhiza 65
Dactylorhiza umbrosa 65
Datisca 244
Datisca cannabina 244
Datiscaceae 244
Datura 376
Datura stramonium 376
Daucus 481
Daucus carota 481
Delphinium 152
Delphinium barbatum 152
Delphinium batalinii 153
Delphinium biternatum 154

Delphinium confusum 154
Delphinium leptocarpum 151
Delphinium longipedunculatum 154
Delphinium oreophilum 155
Delphinium poltoratzkii 155
Delphinium propinquum 155
Delphinium rugulosum 152
Delphinium semibarbatum 155
Dendrostellera arenaria 272
Deschampsia 112
Deschampsia cespitosa 112
Deschampsia koelerioides 112
Descurainia 275
Descurainia sophia 275
Dianthus 327
Dianthus baldzhuanicus 327
Dianthus brevipetalus 327
Dianthus crinitus subsp. tetralepis 327
Dianthus darvazicus 327
Dianthus helenae 328
Dianthus subscabridus 329
Dianthus tetralepis 327
Diarthron 272
Diarthron arenaria 272
Diarthron vesiculosum 272
Dictamnus 269
Dictamnus angustifolius 269
Didymophysa 292
Didymophysa fedtschenkoana 292
Digitaria 104
Digitaria ischaemum 104
Digitaria sanguinalis 104
Dipsacus 476
Dipsacus dipsacoides 476
Dipsacus laciniatus 476
Diptychocarpus 283
Diptychocarpus strictus 283
Dodartia 413
Dodartia orientalis 413
Draba 286
Draba lanceolata 286
Draba melanopus 286
Draba nemorosa 286
Draba nuda 286
Draba stenocarpa 287
Drabopsis nuda 286
Dracocephalum 388
Dracocephalum diversifolium 388
Dracocephalum imberbe 388
Dracocephalum nodulosum 389
Dracocephalum nuratavicum 389
Dracocephalum scrobiculatum 389
Drepanocaryum 386
Drepanocaryum sewerzowii 386
Dryopteris 38
Dryopteris filix-mas 38

E

Echinochloa 104
Echinochloa crus-galli 104
Echinophora 480
Echinophora sibthorpiana 480

Echinophora tenuifolia subsp. sibthorpiana 480
Echinops 434
Echinops knorringianus 434
Echinops leucographus 434
Echinops maracandicus 434
Echinops nanus 434
Echinops nuratavicus 434
Echium 363
Echium biebersteinii 363
Echium vulgare 363
Eclipta 450
Eclipta prostrata 450
Elaeagnaceae 238
Elaeagnus 239
Elaeagnus angustifolia 239
Elaeosticta 485
Elaeosticta allioides 485
Elaeosticta hirtula 485
Elaeosticta polycarpa 485
Elaeosticta vvedenskyi 485
Eleocharis 97
Eleocharis argyrolepis 97
Eleocharis mitracarpa 98
Eleocharis quinqueflora 98
Eleocharis uniglumis 98
Elwendia 486
Elwendia capusii 486
Elwendia chaerophylloides 486
Elwendia intermedia 486
Elwendia persica 486
Elymus 131
Elymus bungeanus 131
Elymus dentatus 131
Elymus drobovii 132
Elymus hispidus 132
Elymus lolioides 132
Elymus repens 132
Elymus uralensis 132
Eminium 49
Eminium lehmanii 49
Enneapogon 114
Enneapogon persicus 114
Ephedra 40
Ephedra botschantzevii 40
Ephedra equisetina 40
Ephedra fedtschenkoae 40
Ephedra foliata 41
Ephedra intermedia 42
Ephedra regeliana 43
Ephedra strobilacea 44
Ephedraceae 40
Epilasia 431
Epilasia acrolasia 431
Epilasia hemilasia 432
Epilasia mirabilis 432
Epilobium 261
Epilobium angustifolium 261
Epilobium hirsutum 261
Epilobium komarovii 262
Epilobium minutiflorum 262
Epilobium nervosum 263

Epilobium palustre   263
Epilobium roseum subsp.
   subsessile   263
Epilobium tetragonum   263
Epilobium tianschanicum   263
Epilobium turkestanicum   263
Epilobium velutinum   261
Epipactis   65
Epipactis latifolia   65
Equisetaceae   34
Equisetum   34
Equisetum arvense   34
Equisetum ramosissimum   34
Eragrostis   115
Eragrostis minor   115
Eragrostis pilosa   115
Eranthis   156
Eranthis longistipitata   156
Eremodaucus   483
Eremodaucus lehmannii   483
Eremogone   318
Eremogone griffithii   318
Eremolimon   303
Eremolimon sogdianum   303
Eremopyrum   126
Eremopyrum bonaepartis   126
Eremopyrum distans   127
Eremopyrum orientale   128
Eremopyrum triticeum   128
Eremurus   70
Eremurus anisopterus   70
Eremurus chloranthus   70
Eremurus fuscus   70
Eremurus inderiensis   70
Eremurus kaufmannii   70
Eremurus lactiflorus   71
Eremurus nuratavicus   71
Eremurus olgae   72
Eremurus regelii   73
Eremurus robustus   74
Eremurus sogdianus   74
Eremurus tianschanicus   75
Eremurus turkestanicus   75
Erianthus   103
Erianthus ravennae   103
Ericaceae   351
Erigeron   457
Erigeron acris   457
Erigeron bellidiformis   457
Erigeron canadensis   457
Erigeron khorossanicus   457
Erigeron pallidus   458
Erigeron petiolaris   458
Erigeron popovii   458
Erigeron pseudoneglectus   458
Erigeron pseudoseravschanicus   458
Erigeron seravschanicus   458
Erigeron sogdianus   458
Erigeron umbrosus   458
Eriocaulaceae   93
Eriocaulon   93
Eriocaulon sieboldianum   93

Erodium   260
Erodium ciconium   260
Erodium cicutarium   260
Erodium hoefftianum   260
Erodium oxyrrhynchum   260
Erophila   287
Erophila minima   287
Erophila verna   287
Eruca   287
Eruca sativa   287
Eruca vesicaria   287
Eryngium   479
Eryngium caeruleum   479
Eryngium macrocalyx   479
Eryngium octophyllum   479
Erysimum   275
Erysimum cyaneum   275
Erysimum marschallianum   275
Erysimum nuratense   276
Erysimum violascens   276
Euclidium   284
Euclidium syriacum   284
Eudicots   135
Euphorbia   252
Euphorbia chamaesyce   252
Euphorbia cheirolepis   252
Euphorbia densa   252
Euphorbia esula   253
Euphorbia falcata   253
Euphorbia franchetii   253
Euphorbia glomerulans   253
Euphorbia granulata   254
Euphorbia helioscopia   254
Euphorbia humilis   254
Euphorbia inderiensis   254
Euphorbia jaxartica   253
Euphorbia kanaorica   255
Euphorbia rapulum   255
Euphorbia rosularis   255
Euphorbia sarawschanica   255
Euphorbia sororia   255
Euphorbia szowitsii   255
Euphorbia turczaninowii   256
Euphorbiaceae   251
Euphrasia   416
Euphrasia pectinata   416
Euphrasia regelii   416
Eurotia ceratoides   333
Eurotia ewersmanniana   333
Evax filaginoides   456

F

Fabaceae   172
Falcaria   486
Falcaria vulgaris   486
Fallopia   314
Fallopia convolvulus   314
Ferns   33
Ferula   491
Ferula angreni   491
Ferula bucharica   491
Ferula diversivittata   492

Ferula dshizakensis   492
Ferula fedtschenkoana   492
Ferula ferganensis   492
Ferula foetida   492
Ferula helenae   493
Ferula kokanica   493
Ferula kuhistanica   494
Ferula lehmannii   495
Ferula mollis   495
Ferula moschata   495
Ferula ovczinnikovii   495
Ferula ovina   495
Ferula penninervis   496
Ferula samarkandica   496
Ferula schtschurowskiana   497
Ferula sumbul   495
Ferula varia   497
Fessia   92
Fessia puschkinioides   92
Festuca   121
Festuca alaica   121
Festuca amblyodes   121
Festuca arundinacea   121
Festuca griffithiana -   121
Festuca olgae   121
Festuca pratensis   121
Festuca rubra   121
Festuca rupicola   121
Festuca valesiaca   121
Filago   456
Filago arvensis   456
Filago filaginoides   456
Filago germanica   456
Filago pyramidata   456
Filago vulgaris   456
Fimbristylis   98
Fimbristylis dichotoma   98
Frankenia   294
Frankenia bucharica   294
Frankenia bucharica subsp.
   vvedenskyi   295
Frankenia hirsuta   295
Frankenia pulverulenta   295
Frankenia vvedenskyi   295
Frankeniaceae   294
Fraxinus   377
Fraxinus angustifolia subsp.
   syriaca   377
Fraxinus sogdiana   377
Fraxinus syriaca   377
Fritillaria   64
Fritillaria olgae   64
Fumaria   143
Fumaria vaillantii   143

G

Gagea   54
Gagea afghanica   54
Gagea bergii   54
Gagea calyptrifolia   54
Gagea capillifolia   54
Gagea capusii   54

Gagea chomutovae 54
Gagea divaricata 55
Gagea dschungarica 55
Gagea filiformis 55
Gagea gageoides 55
Gagea graminifolia 56
Gagea hissarica 57
Gagea kamelinii 57
Gagea liotardii 58
Gagea nabievii 58
Gagea olgae 58
Gagea ova 57
Gagea reinhardii 59
Gagea reticulata 59
Gagea stipitata 59
Gagea stolonifera 60
Gagea subtilis 60
Gagea taschkentica 60
Gagea tenera 61
Gagea turkestanica 61
Gagea vegeta 61
Gagea vvedenskyi 61
Galagania 485
Galagania fragrantissima 485
Galagania tenuisecta 485
Galatella 456
Galatella coriacea 456
Galium 352
Galium aparine 352
Galium ceratopodum 352
Galium decaisnei 353
Galium humifusum 352
Galium ibicinum 353
Galium karakulense 352
Galium pamiroalaicum 352
Galium pseudorivale 353
Galium ruthenicum 353
Galium setaceum 353
Galium songaricum 353
Galium spurium 353
Galium spurium subsp. ibicinum 353
Galium tenuissimum 353
Galium tianschanicum 353
Galium tricornutum 353
Galium turkestanicum 353
Galium verticellatum 353
Galium verum 353
Gamanthus 340
Gamanthus gamocarpus 340
Garhadiolus 422
Garhadiolus angulosus 422
Garhadiolus hedypnois 422
Garhadiolus papposus 422
Gastrolychnis 325
Gastrolychnis apetala 324
Gastrolychnis longicarpophora 325
Gegea pseudoreticulata 59
Gentiana 356
Gentiana barbata 357
Gentiana leucomelaena 356
Gentiana olivieri 355 356
Gentiana prostrata 357

Gentiana riparia 357
Gentiana sibirica 357
Gentiana turkestanorum 357
Gentianaceae 355
Gentianella 357
Gentianella sibirica 357
Gentianella turkestanorum 357
Gentianopsis 357
Gentianopsis barbata 357
Geraniaceae 257
Geranium 257
Geranium collinum 257
Geranium divaricatum 257
Geranium kotschyi subsp. charlesii 258
Geranium linearilobum 258
Geranium pusillum 258
Geranium regelii 257
Geranium robertianum 258
Geranium rotundifolium 258
Geranium saxatile 259
Geum 228
Geum heterocarpum 228
Girgensohnia 340
Girgensohnia diptera 340
Girgensohnia oppositiflora 340
Gladiolus 69
Gladiolus italicus 69
Glaucium 139
Glaucium elegans 139
Glaucium fimbrilligerum 140
Glaucium squamigerum 140
Glaux maritima 349
Gleditsia 172
Gleditsia caspia 172
Glyceria 120
Glyceria plicata 120
Glycyrrhiza 206
Glycyrrhiza aspera 206
Glycyrrhiza glabra 206
Glycyrrhiza triphylla 206
Gnaphalium 455
Gnaphalium luteoalbum 455
Gnaphalium supinum 455
Goldbachia 277
Goldbachia laevigata 277
Goldbachia sabulosa 277
Goldbachia torulosa 278
Goldbachia verrucosa 278
Gramineae 103
Grossulariaceae 165
GYMNOSPERMAE 39
Gymnospermium 144
Gymnospermium albertii 144
GYMNOSPERMS 39
Gypsophila 325
Gypsophila cephalotes 325
Gypsophila herniarioides 325
Gypsophila paniculata 325

H

Hackelia 368
Hackelia uncinata 368

Halerpestes 159
Halerpestes sarmentosa 159
Halimocnemis 345
Halimocnemis mollissima 345
Halimocnemis villosa 345
Halimodendron 180
Halimodendron halodendron 180
Halocharis 344
Halocharis hispida 344
Halocnemum 336
Halocnemum strobilaceum 336
Halogeton 345
Halogeton glomeratus 345
Halopeplis 336
Halopeplis pygmaea 336
Haloragaceae 169
Halostachys 336
Halostachys belangeriana 336
Halothamnus 340
Halothamnus glaucus 340
Halothamnus iliensis 340
Halothamnus subaphyllus 340
Haloxylon 342
Haloxylon ammodendron 342
Haloxylon aphyllum 342
Haloxylon persicum 343
Handelia 462
Handelia trichophylla 462
Haplophyllum 266
Haplophyllum acutifolium 266
Haplophyllum bungei 267
Haplophyllum lasianthum 267
Haplophyllum latifolium 267
Haplophyllum pedicellatum 267
Haplophyllum perforatum 266
Haplophyllum ramosissimum 269
Haplophyllum robustum 269
Haplophyllum versicolor 267
Hedysarum 206
Hedysarum flavescens 206
Hedysarum jaxarticum 206
Hedysarum mogianicum 206
Hedysarum montanum 206
Hedysarum nuratense 207
Hedysarum plumosum 208
Helianthemum 272
Helianthemum songaricum 272
Helichrysum 454
Helichrysum maracandicum 454
Helichrysum mussae 455
Helichrysum nuratavicum 455
Helichrysum tianschanicum 455
Helictotrichon 113
Helictotrichon asiaticum 113
Helictotrichon desertorum 113
Heliotropium 359
Heliotropium arguzioides 359
Heliotropium bogdanii 359
Heliotropium dasycarpum 359
Heliotropium ellipticum 359
Heliotropium lasiocarpum 359
Heliotropium olgae 360

Henrardia 126
Henrardia persica 126
Heracleum 497
Heracleum lehmannianum 497
Heracleum olgae 497
Herniaria 319
Herniaria glabra 319
Herniaria hirsuta 319
Hesperis 278
Hesperis sibirica 278
Heteracia 426
Heteracia epapposa 426
Heteracia szovitzii 426
Heteranthelium 129
Heteranthelium piliferum 129
Heterocaryum 367
Heterocaryum laevigatum 367
Heterocaryum macrocarpum 367
Heterocaryum subsessile 367
Heterocaryum szovitsianum 367
Heteroderis 423
Heteroderis pusilla 423
Hibiscus 272
Hibiscus trionum 272
Hieracium 428
Hieracium echioides 427
Hieracium procerum 428
Hieracium raddeanum subsp. regelianum 428
Hieracium regelianum 428
Hieracium robustum 428
Hieracium virosum 428
Hippophae 238
Hippophae rhamnoides 238
Hippuris 380
Hippuris vulgaris 380
Holoschoenus 97
Holoschoenus vulgaris 97
Holosteum 316
Holosteum glutinosum 317
Holosteum umbellatum 316
Holosteum umbellatum subsp. glutinosum 317
Hordeum 130
Hordeum bogdanii 130
Hordeum brevisubulatum 130
Hordeum bulbosum 130
Hordeum geniculatum 131
Hordeum leporinum 131
Hordeum marinum subsp. gussoneanum 131
Hordeum murinum subsp. leporinum 131
Hordeum spontaneum 131
Hornungia 291
Hornungia procumbens 291
Hulthemia 233
Hulthemia persica 233
Hyalea 449
Hyalea pulchella 449
Hyalolaena 485
Hyalolaena depauperata 485

Hyalolaena jaxartica 485
Hydrocharitaceae 49
Hymenolobus procumbens 291
Hyoscyamus 375
Hyoscyamus niger 375
Hyoscyamus pusillus 375
Hypecoum 140
Hypecoum parviflorum 140
Hypecoum pendulum 140
Hypecoum trilobum 141
Hypericaceae 245
Hypericum 245
Hypericum elongatum 245
Hypericum perforatum 246
Hypericum scabrum 246
Hypogomphia 391
Hypogomphia turkestana 391
Hypogomphia turkestanica 391
Hyssopus 411
Hyssopus seravschanicus 411

I

Iberidella 292
Iberidella trinervia 292
Impatiens 346
Impatiens parviflora 346
Imperata 103
Imperata cylindrica 103
Inula 453
Inula britannica 453
Inula caspica 453
Inula glauca 453
Inula helenium 453
Inula macrophylla 453
Inula orientalis 453
Inula rhizocephala 453
Iridaceae 67
Iris 67
Iris linifoliiformis 67
Iris loczyi 67
Iris longiscapa 67
Iris maracandica 68
Iris narbutii 68
Iris parvula 69
Iris songarica 69
Iris tadshikorum 69
Isatis 277
Isatis brevipes 277
Isatis emarginata 277
Isatis minima 277
Isatis multicaulis 277
Isatis tinctoria 277
Isatis violascens 277
Isopyrum 146
Isopyrum anemonoides 146
Ixioliriaceae 66
Ixiolirion 66
Ixiolirion tataricum 66

J

Juglandaceae 242
Juglans 242

Juglans regia 242
Juncaceae 93
Juncaginaceae 50
Juncus 93
Juncus articulatus 93
Juncus bufonius 93
Juncus compressus 93
Juncus gerardii 93
Juncus heptopotamicus 93
Juncus hybridus 94
Juncus inflexus 94
Juncus jaxarticus 94
Juncus macrantherus 94
Juncus persicus subsp. *libanoticus* 94
Juncus rechingeri 94
Juncus schischkinii 94
Juncus sphaerocarpus 94
Juncus triglumis 94
Juncus turkestanicus 94
Juncus vvedenskyi 94
Juniperus 44
Juniperus polycarpos var. seravschanica 46
Juniperus pseudosabina 44
Juniperus semiglobosa 45
Jurinea 444
Jurinea abramowii 444
Jurinea androssovii 444
Jurinea ferganica 444
Jurinea helichrysifolia 444
Jurinea kokanica 445
Jurinea komarovii 445
Jurinea lasiopoda 445
Jurinea maxima 446
Jurinea olgae 446
Jurinea suffruticosa 447
Jurinea trautvetteriana 447
Jurinea zakirovii 447

K

Kalidium 336
Kalidium caspicum 336
Kalimeris 457
Kalimeris altaica 457
Karelinia 453
Karelinia caspia 453
Kickxia 377
Kickxia elatine 377
Kirilovia eriantha 334
Kobresia 99
Kobresia pamiroalaica 99
Kobresia persica 99
Kobresia stenocarpa 99
Kochia 334
Kochia iranica 334
Kochia prostrata 334
Kochia scoparia 334
Koeleria 115
Koeleria macrantha 115
Koelpinia 433
Koelpinia linearis 433
Koelpinia macrantha 434

Koelpinia tenuissima   434
Koelpinia turanica   434
Korshinskya   482
Korshinskya olgae   482
Kovalevskiella   424
Kovalevskiella zeravschanica   424
Kozlovia   480
Kozlovia paleacea   480
Krascheninnikovia   333
Krascheninnikovia ceratoides   333
Krascheninnikovia ewersmanniana   333
Kudrjaschevia   389
Kudrjaschevia jacubi   389
Kuhitangia   325
Kuhitangia knorringiana   325
Kuschakewiczia   370
Kuschakewiczia turkestanica   370

L

Labiatae   383
Lachnoloma   284
Lachnoloma lehmanii   284
Lachnophyllum   458
Lachnophyllum gossypinum   458
Lactuca   423
Lactuca altaica   424
Lactuca crambifolia   423
Lactuca dissecta   423
Lactuca glauciifolia   423
Lactuca orientalis   423
Lactuca serriola   424
Lactuca soongorica   424
Lactuca spinidens   424
Lactuca tatarica   424
Lactuca undulata   424
Ladyginia bucharica   491
Lagochilus   403
Lagochilus inebrians   403
Lagochilus olgae   404
Lagochilus proskorjakovii   404
Lagochilus seravschanicus   405
Lagotis   379
Lagotis korolkowii   379
Lallemantia   389
Lallemantia royleana   389
Lamiaceae   383
Lamium   401
Lamium album   401
Lamium amplexicaule   401
Laphangium   455
Laphangium luteoalbum   455
Lappula   365
Lappula brachycentra   365
Lappula consanguinea   366
Lappula marginata   366
Lappula microcarpa   365
Lappula myosotis   365
Lappula nuratavica   365
Lappula occultata   366
Lappula patula   366
Lappula sarawschanica   366
Lappula semialata   366

Lappula semiglabra   366
Lappula sessiliflora   366
Lappula sinaica   366
Lappula spinocarpos   366
Lappula squarrosa   366
Lathyrus   216
Lathyrus aphaca   216
Lathyrus asiaticus   216
Lathyrus cicera   216
Lathyrus inconspicuus   216
Lathyrus pratensis   216
Lathyrus sativus   216
Lathyrus tuberosus   217
Launaea   423
Launaea korovinii   423
Launaea procumbens   423
Leguminosae   172
Leiospora   281
Leiospora subscapigera   281
Lemna   48
Lemna gibba   48
Lemna minor   48
Lens   216
Lens culinaris subsp. orientalis   216
Lentibulariaceae   382
Leontice   144
Leontice eversmannii   144
Leontopodium   455
Leontopodium fedtschenkoanum   455
Leonurus   402
Leonurus glaucescens   402
Leonurus turkestanicus   402
Lepidium   289
Lepidium botschantsevianum   289
Lepidium draba subsp. chalepense   290
Lepidium ferganense   290
Lepidium lacerum   290
Lepidium latifolium   290
Lepidium lipskyi   290
Lepidium olgae   290
Lepidium orientale   291
Lepidium perfoliatum   291
Lepidolopha   472
Lepidolopha komarowii   472
Lepidolopha nuratavica   472
Lepidolopsis   464
Lepidolopsis turkestanica   464
Leptaleum   283
Leptaleum filifolium   283
Leptorhabdos   416
Leptorhabdos parviflora   416
Lepyrodiclis   317
Lepyrodiclis holosteoides   317
Lepyrodiclis stellarioides   317
Leymus   132
Leymus alaicus   132
Leymus angustus   132
Leymus multicaulis   133
Libanotis   489
Libanotis schrenkiana   489
Ligularia   452
Ligularia alpigena   452

Ligularia heterophylla   452
Ligularia thomsonii   452
Liliaceae   54
Liliopsida   47
Limonium   301
Limonium otolepis   301
Limonium reniforme   302
Limonium suffruticosum   302
Linaceae   256
Linaria   381
Linaria popovii   381
Linaria sessilis   381
Lindelofia   369
Lindelofia macrostyla   369
Lindelofia olgae   369
Lindelofia stylosa   369
Linum   256
Linum corymbulosum   256
Linum humile   257
Linum macrorhizum   256
Linum olgae   256
Linum usitatissimum   256
Lipskya   482
Lipskya insignis   482
Lipskyella   444
Lipskyella annua   444
Lithospermum   360
Lithospermum officinale   360
Litwinowia   284
Litwinowia tenuissima   284
Lloydia   64
Lloydia serotina   64
Loliolum   122
Loliolum subulatum   122
Lolium   126
Lolium perenne   126
Lolium persicum   126
Lolium temulentum   126
Lomatogonium   357
Lomatogonium carinthiacum   357
Londesia eriantha   334
Lonicera   473
Lonicera altmannii   474
Lonicera bracteolaris   473
Lonicera humilis   474
Lonicera korolkowii   474
Lonicera microphylla   475
Lonicera nummulariifolia   475
Lonicera olgae   475
Lonicera paradoxa   476
Lonicera semenovii   476
Lonicera simulatrix   476
Lonicera zeravschanica   476
Lophanthus   387
Lophanthus ouroumitanensis   387
Lophanthus schtschurowskianus   387
Lophanthus subnivalis   387
Lotus   178
Lotus krylovii   178
Lotus sergievskiae   178
Luzula   94
Luzula pallescens   94

Luzula spicata   94
Lycium   374
Lycium dasystemum   374
Lycium ruthenicum   374
Lycopus   412
Lycopus europaeus   412
Lysimachia   349
Lysimachia dubia   349
Lysimachia maritima   349
Lythraceae   261
Lythrum   261
Lythrum hyssopifolia   261
Lythrum nanum   261

M

Macrotomia euchroma   362
Magnoliopsida   135
Malus   222
Malus sieversii   222
Malva   270
Malva neglecta   270
Malvaceae   270
Marrubium   386
Marrubium anisodon   386
Marsilea   35
Marsilea quadrifolia   35
Marsileaceae   35
Matthiola   282
Matthiola chorassanica   282
Matthiola integrifolia   282
Matthiola obovata   282
Mattiastrum   368
Mattiastrum himalayense   368
Mausolea   464
Mausolea eriocarpa   464
Mazaceae   413
Mediasia   490
Mediasia macrophylla   490
Medicago   176
Medicago denticulata   176
Medicago lupulina   176
Medicago medicaginoides   176
Medicago minima   176
Medicago monantha   176
Medicago orbicularis   176
Medicago orthoceras   176
Medicago polymorpha   176
Medicago rigidula   177
Medicago sativa   177
Megacarpaea   291
Megacarpaea gigantea   291
Megacarpaea orbiculata   291
Melandrium album   322
Melandrium ruinarium   324
Melandrium turkestanicum   324
Melica   116
Melica altissima   116
Melica hohenackeri   116
Melica inaequiglumis   116
Melilotus   177
Melilotus albus   177
Melilotus officinalis   177

Melissitus   176
Melissitus adscendens   174
Melissitus aristatus   174
Melissitus gontscharovii   175
Melissitus pamiricus   176
Melissitus popovii   176
Meniocus linifolius   286
Mentha   412
Mentha longifolia subsp. asiatica   412
Mentha pamiroalaica   413
Merendera   53
Merendera robusta   53
Meristotropis triphylla   206
Mesostemma   315
Mesostemma karatavica   315
Microcephala   473
Microcephala lamellata   473
Microcephala turcomanica   473
Microthlaspi perfoliatum   292
Milium   108
Milium vernale   108
Minuartia   317
Minuartia hamata   317
Minuartia kryloviana   317
Minuartia meyeri   317
Monocots   47
Moraceae   242
Morus   242
Morus alba   242
Morus nigra   242
Myosotis   365
Myosotis alpestris   365
Myosotis caespitosa   365
Myosotis laxa subsp. caespitosa   365
Myosotis micrantha   365
Myosotis refracta   365
Myosotis stricta   365
Myricaria   297
Myricaria bracteata   297
Myricaria germanica   297
Myriophyllum   169
Myriophyllum spicatum   169

N

Najas   50
Najas marina   50
Najas minor   50
Nanophyton   344
Nanophyton erinaceum   344
Nanophyton saxatile   344
Nardurus krausei   123
Nasturtium   276
Nasturtium fontanum   276
Nasturtium officinale   276
Neotorularia   274
Neotorularia torulosa   274
Nepeta   387
Nepeta alatavica   387
Nepeta bracteata   387
Nepeta cataria   388
Nepeta kokanica   388
Nepeta mariae   388

Nepeta maussarifii   388
Nepeta micrantha   388
Nepeta nuda   388
Nepeta olgae   388
Nepeta pannonica   388
Nepeta podostachys   388
Nepeta pungens   388
Nepeta saturejoides   388
Nepeta ucranica   388
Neslia   294
Neslia apiculata   294
Neslia paniculata subsp. thracica   294
Neuroloma fruticulosum   279
Neuroloma nuratense   279
Neurotropis   293
Neurotropis platycarpa   293
Nicandra   376
Nicandra physaloides   376
Nigella   156
Nigella integrifolia   156
Nitrariaceae   264
Nonea   364
Nonea caspica   364
Nonea melanocarpa   365
Nonea pulla   365

O

Octoceras   284
Octoceras lehmannianum   284
Oedibasis   486
Oedibasis apiculata   486
Oedibasis tamerlanii   486
Oleaceae   377
Olimarabidopsis   275
Olimarabidopsis pumila   275
Olimarabidopsis umbrosa   275
Onagraceae   261
Onobrychis   208
Onobrychis chorassanica   208
Onobrychis echidna   209
Onobrychis gontscharovii   210
Onobrychis grandis   210
Onobrychis micrantha   210
Onobrychis pulchella   210
Onobrychis saravschanica   211
Ononis   174
Ononis spinosa subsp. antiquorum   174
Onopordum   435
Onopordum acantium   435
Onopordum leptolepis   436
Onosma   362
Onosma dichroanta   362
Onosma maracandica   362
Ophioglossaceae   34
Orchidaceae   65
Orchis   65
Orchis pseudolaxiflora   65
Origanum   411
Origanum tyttanthum   411
Origanum vulgare subsp. gracile   411
Orlaya   481
Orlaya grandiflora   481

Orobanchaceae 414
Orobanche 414
Orobanche aegyptiaca 414
Orobanche amoena 414
Orobanche cernua 414
Orobanche coelestis 414
Orobanche elatior 414
Orobanche gigantea 414
Orobanche hansii 414
Orobanche orientalis 414
Orobanche pallens 414
Orobanche sulphurea 415
Orthurus 228
Orthurus heterocarpus 228
Orthurus kokanicus 228
Oxyria 303
Oxyria digyna 303
Oxytropis 202
Oxytropis aspera 202
Oxytropis capusii 202
Oxytropis chesneyoides 202
Oxytropis humifusa 202
Oxytropis immersa 202
Oxytropis integripetala 202
Oxytropis kamelinii 202
Oxytropis lehmanni 202
Oxytropis leucocyanea 202
Oxytropis lipskyi 203
Oxytropis macrocarpa 203
Oxytropis michelsonii 204
Oxytropis microsphaera 204
Oxytropis platonychia 204
Oxytropis pseudorosea 204
Oxytropis riparia 205
Oxytropis rosea 205
Oxytropis savellanica 205
Oxytropis seravschanica 205
Oxytropis tachtensis 205
Oxytropis trichocalycina 205

P

Pachypterygium brevipes 277
Pachypterygium densiflorum 277
Paeonia 164
Paeonia hybrida 164
Paeonia tenuifolia 164
Paeoniaceae 164
Papaver 136
Papaver croceum 137
Papaver dubium 136
Papaver litwinowii 136
Papaver nudicaule 137
Papaver pavoninum 138
Papaveraceae 136
Paracaryum glochidiatum 368
Paracaryum himalayense 368
Paramicrorhynchus procumbens 423
Paraquilegia 147
Paraquilegia caespitosa 147
Parentucellia 416
Parentucellia flaviflora 416
Parietaria 242

Parietaria debilis 242
Parietaria lusitanica subsp. serbica 242
Parietaria micrantha 242
Parietaria serbica 242
Parnassia 245
Parnassia laxmannii 245
Parrya 279
Parrya fruticulosa 279
Parrya hispida 279
Parrya khorasanica 279
Parrya nuratensis 279
Parrya olgae 280
Parrya sarawschanica 281
Pedicularis 417
Pedicularis dolichorrhiza 417
Pedicularis krylovii 417
Pedicularis olgae 417
Pedicularis peduncularis 418
Pedicularis popovii 418
Pedicularis rhinanthoides 418
Pedicularis sarawschanica 418
Pedicularis waldheimii 418
Peganum 264
Peganum harmala 264
Pentanema 454
Pentanema albertoregelia 454
Pentanema divaricatum 454
Pentaphylloides 227
Pentaphylloides parvifolia 227
Perovskia 409
Perovskia angustifolia 409
Perovskia botschantzevii 409
Perovskia kudrjaschevii 409
Perovskia scrophulariifolia 409
Persicaria 313
Persicaria amphibia 313
Persicaria bistorta 313
Persicaria hydropiper 313
Persicaria lapathifolia 314
Persicaria maculata 314
Persicaria minor 314
Petrorhagia 326
Petrorhagia alpina 326
Phaecasium pulchrum 427
Phalaroides 105
Phalaroides arundinacea 105
Phelipanche aegyptiaca 414
Phelipanche coelestis 414
Phelipanche orientalis 414
Phelipanche pallens 414
Phleum 109
Phleum alpinum 109
Phleum exaratum 110
Phleum paniculatum 110
Phleum phleoides 111
Phleum pratense 111
Phlomis 398
Phlomis linearifolia 398
Phlomis nubilans 399
Phlomis olgae 400
Phlomis salicifolia 400
Phlomis thapsoides 400

Phlomoides 392
Phlomoides ambigua 392
Phlomoides anisochila 393
Phlomoides canescens 394
Phlomoides dszumrutensis 397
Phlomoides eriocalyx 395
Phlomoides kaufmanniana 395
Phlomoides labiosa 396
Phlomoides napuligera 396
Phlomoides oreophila 397
Phlomoides sogdiana 397
Phlomoides speciosa 397
Phlomoides uniflora 397
Phragmites 114
Phragmites australis 114
Picnomon 435
Picnomon acarna 435
Picris 422
Picris nuristanica 422
Pilosella 427
Pilosella echioides 427
Pilosella procera 428
Pimpinella 487
Pimpinella peregrina 487
Piptatherum 107
Piptatherum alpestre 107
Piptatherum ferganense 108
Piptatherum holciforme 108
Piptatherum laterale 108
Piptatherum latifolium 108
Piptatherum microcarpum 108
Piptatherum pamiralaicum 108
Piptatherum platyanthum 108
Piptatherum sogdianum 108
Piptatherum songaricum 108
Piptatherum vicarium 108
Pistacia 265
Pistacia vera 265
Plantaginaceae 377
Plantago 379
Plantago gentianoides 379
Plantago lagocephala 379
Plantago lanceolata 379
Plantago major 379
Platanaceae 164
Platanus 164
Platanus orientalis 164
Platycladus 44
Platycladus orientalis 44
Pleioneura 329
Pleioneura griffithiana 329
Plumbaginaceae 297
Poa 117
Poa alpina 117
Poa angustifolia 118
Poa annua 118
Poa bactriana 118
Poa bucharica 118
Poa bulbosa 98 118
Poa diaphora 120
Poa fragilis 120
Poa litvinoviana 120

Poa nemoralis   120
Poa palustris   120
Poa pratensis   120
Poa relaxa   120
Poa supina   120
Poa trivialis   120
Poaceae   103
Polycnemum   329
Polycnemum arvense   329
Polycnemum perenne   330
Polygala   217
Polygala hybrida   217
Polygalaceae   217
Polygonaceae   303
Polygonatum   90
Polygonatum sewerzowii   90
Polygonum   311
Polygonum acetosum   311
Polygonum amphibium   313
Polygonum argyrocoleon   311
Polygonum aviculare   311
Polygonum biaristatum   311
Polygonum coriarium   311
Polygonum fibrilliferum   311
Polygonum hissaricum   312
Polygonum inflexum   312
Polygonum molliiforme   312
Polygonum nitens   313
Polygonum paronychioides   312
Polygonum patulum   312
Polygonum polychnemoides   312
Polygonum pulvinatum   312
Polygonum rottboellioides   312
Polygonum subaphyllum   312
Polygonum vvedenskyi   312
Polypodiaceae   38
Polypodiophyta   33
Polypogon   111
Polypogon fugax   111
Polypogon monspleniensis   111
Polypogon semiverticillatus   111
Polypogon viridis   111
Populus   248
Populus afghanica   248
Populus alba   249
Populus euphratica   250
Populus nigra   250
Populus pruinosa   250
Populus talassica   250
Portulaca   345
Portulaca oleracea   345
Portulacaceae   345
Potamogeton   50
Potamogeton crispus   50
Potamogeton filiformis   50
Potamogeton lucens   50
Potamogeton natans   50
Potamogeton pectinatus   51
Potamogeton perfoliatus   51
Potamogeton pusillus   51
Potamogetonaceae   50
Potentilla   225

Potentilla asiae-mediae   225
Potentilla asiatica   225
Potentilla desertorum   225
Potentilla flabellata   225
Potentilla gelida   225
Potentilla hololeuca   226
Potentilla impolita   226
Potentilla inclinata   226
Potentilla orientalis   226
Potentilla pamiroalaica   226
Potentilla pedata   226
Potentilla reptans   227
Potentilla soongorica   227
Potentilla supina   227
Potentilla vvedenskyi   227
Poterium lasiocarpum   229
Poterium polygamum   229
Prangos   483
Prangos didyma   483
Prangos fedtschenkoi   483
Prangos ornata   483
Prangos pabularia   484
Primula   347
Primula algida   347
Primula fedtschenkoi   347
Primula iljinskii   347
Primula lactiflora   349
Primula matthioli   347
Primula olgae   349
Primula pamirica   349
Primula turkeviczii   349
Primulaceae   347
Prunella   391
Prunella vulgaris   391
Prunus   233
Prunus armeniaca   233
Prunus bucharica   233
Prunus cerasus   234
Prunus divaricata   234
Prunus domestica   234
Prunus dulcis   235
Prunus erythrocarpa   235
Prunus mahaleb   236
Prunus spinosissima   237
Prunus verrucosa   238
Pseudoclausia hispida   279
Pseudoclausia olgae   280
Pseudoclausia sarawschanica   281
Pseudoclausia turkestanica   279
Pseudohandelia   462
Pseudohandelia umbellifera   462
Pseudolinosyris   456
Pseudolinosyris grimmii   456
Pseudosedum   168
Pseudosedum bucharicum   168
Pseudosedum campanuliflorum   168
Pseudosedum lievenii   169
Pseudosedum longidentatum   169
Psilurus   126
Psilurus incurvus   126
Psoralea drupacea   178
Psychrogeton   458

Psychrogeton amorphoglossus   458
Psychrogeton aucherii   457
Psychrogeton leucophyllus   458
Psychrogeton pseuderigeron   458
Psylliostachys   303
Psylliostachys leptostachya   303
Psylliostachys suworowii   303
Psylliostachys × myosuroides   303
Pteridaceae   35
Puccinelia   121
Puccinelia distans   121
Puccinellia poecilantha   121
Pulicaria   454
Pulicaria dysenterica subsp.
   uliginosa   454
Pulicaria gnaphalodes   454
Pulicaria salviifolia   454
Pulicaria uliginosa   454
Pulsatilla   158
Pulsatilla campanella   158
Pycreus   99
Pycreus globosus   99
Pyrethrum   471
Pyrethrum galae   471
Pyrethrum parthenifolium   472
Pyrethrum pyrethroides   471
Pyrola   351
Pyrola rotundifolia   351
Pyrus   221
Pyrus communis   221
Pyrus korshinskyi   221
Pyrus regelii   221
Pyrus turcomanica   222

Q

Queria hispanica   317

R

Ranunculaceae   145
Ranunculus   159
Ranunculus arvensis   159
Ranunculus baldshuanicus   160
Ranunculus brevirostris   160
Ranunculus komarovii   160
Ranunculus linearilobus   160
Ranunculus mindshelkensis   160
Ranunculus natans   161
Ranunculus olgae   161
Ranunculus oxyspermus   161
Ranunculus paucidentatus   161
Ranunculus pinnatisectus   161
Ranunculus pulchellus   162
Ranunculus repens   162
Ranunculus rionii   162
Ranunculus rubrocalyx   162
Ranunculus rufosepalus   162
Ranunculus sceleratus   163
Ranunculus sewerzowii   163
Ranunculus songaricus   164
Ranunculus sphaerospermus   164
Ranunculus tenuilobus   164
Ranunculus trichophyllus   164

Reaumuria 295
Reaumuria alternifolia 295
Reaumuria fruticosa 295
Reseda 273
Reseda lutea 273
Reseda luteola 273
Resedaceae 273
Rhabdotheca korovinii 423
Rhamnaceae 240
Rhamnus 240
Rhamnus cathartica 240
Rhamnus coriacea 240
Rhaponticum 449
Rhaponticum repens 449
Rheum 305
Rheum macrocarpum 305
Rheum maximowiczii 305
Rheum turkestanicum 306
Rheum cordatum 305
Rhinactinidia 457
Rhinactinidia popovii 457
Rhodiola 166
Rhodiola heterodonta 166
Ribes 165
Ribes meyeri 165
Rindera 368
Rindera tetraspis 368
Rochelia 367
Rochelia bungei 367
Rochelia cardiosepala 367
Rochelia disperma subsp. retorta 367
Rochelia leiocarpa 367
Rochelia peduncularis 368
Roemeria 138
Roemeria hybrida 138
Roemeria refracta 139
Rorippa 276
Rorippa palustris 276
Rosa 229
Rosa beggeriana 229
Rosa canina 230
Rosa corymbifera 230
Rosa ecae 230
Rosa fedtschenkoana 230
Rosa hissarica 231
Rosa kokanica 231
Rosa lehmanniana 231
Rosa maracandica 232
Rosa nanothamnus 233
Rosa transturkestanica 233
Rosaceae 217
Rosularia 169
Rosularia paniculata 169
Rosularia radicosa 169
Rosularia subspicata 169
Rubia 354
Rubia regelii 354
Rubiaceae 351
Rubus 224
Rubus caesius 224
Rumex 304
Rumex acetosa 304

Rumex chalepensis 304
Rumex conglomeratus 304
Rumex crispus 304
Rumex dentatus *halacsyi* 304
Rumex drobovii 304
Rumex halacsyi 304
Rumex marschallianus 304
Rumex pamiricus 304
Rumex paulsenianus 304
Rumex rectinervis 304
Rumex syriacus 304
Rumex tianschanicus 304
Ruppia 51
Ruppia maritima 51
Ruppiaceae 51
Rutaceae 266

S

Saccharum 103
Saccharum spontaneum 103
Sageretia 240
Sageretia thea 240
Sagina 317
Sagina saginoides 317
Sagittaria 49
Sagittaria trifolia 49
Salicaceae 248
Salicornia 336
Salicornia europaea 336
Salix 251
Salix alba 251
Salix blakii 251
Salix olgae 251
Salix pycnostachya 251
Salix rosmarinifolia 251
Salix songarica 251
Salix wilhelmsiana 251
Salsola 337
Salsola arbuscula 337
Salsola arbusculiformis 337
Salsola collina 337
Salsola dendroides 337
Salsola foliosa 338
Salsola iberica 338
Salsola leptoclada 338
Salsola micranthera 338
Salsola orientalis 338
Salsola paletzkiana 338
Salsola paulsenii 338
Salsola richteri 338
Salsola sclerantha 339
Salsola titovii 339
Salsola turkestanica 339
Salsola vvedenskyi 339
Salvia 407
Salvia aequidens 407
Salvia deserta 407
Salvia glabricaulis 407
Salvia macrosiphon 407
Salvia sarawschanica 407
Salvia sclarea 407
Salvia spinosa 408

Salvia submutica 409
Salvia virgata 409
Salvinia 35
Salvinia natans 35
Salviniaceae 35
Samolus 349
Samolus valerandi 349
Sanguisorba 229
Sanguisorba minor 229
Sanguisorba minor subsp. balearica 229
Santalaceae 294
Sapindaceae 265
Saponaria griffithiana 329
Saussurea 444
Saussurea elegans 444
Saussurea salsa 444
Saxifraga 165
Saxifraga hirculus 165
Saxifraga sibirica 166
Saxifragaceae 165
Scabiosa 477
Scabiosa flavida 477
Scabiosa olivierii 477
Scabiosa rhodantha 477
Scabiosa songarica 477
Scandix 480
Scandix nodosa 480
Scandix pecten-veneris 480
Scandix stellata 480
Scapiarabis 276
Scapiarabis saxicola 276
Scariola orientalis 423
Schismus 117
Schismus arabicus 117
Schoenoplectus 95
Schoenoplectus litoralis 95
Schoenoplectus triqueter 95
Schoenus 99
Schoenus nigricans 99
Schrenkia 481
Schrenkia golickeana 481
Schrenkia pungens 482
Schrenkia vaginata 482
Schrenkiella 274
Schrenkiella parvula 274
Schtschurowskia 482
Schtschurowskia meifolia 482
Scirpus 95
Scirpus triquetriformis 95
Sclerochloa 117
Sclerochloa dura 117
Scorzonera 429
Scorzonera acanthoclada 429
Scorzonera bracteosa 429
Scorzonera circumflexa 429
Scorzonera hissarica 429
Scorzonera inconspicua 429
Scorzonera litwinowii 429
Scorzonera ovata 430
Scorzonera pusilla 431
Scorzonera songarica 431

Scorzonera tragopogonoides   431
Scrophularia   381
Scrophularia gontscharovii   381
Scrophularia griffithii   381
Scrophularia heucheriiflora   381
Scrophularia incisa   381
Scrophularia integrifolia   381
Scrophularia kiriloviana   382
Scrophularia leucoclada   382
Scrophularia scoparia   382
Scrophularia striata   382
Scrophularia umbrosa   382
Scrophularia vvedenskyi   382
Scrophularia xanthoglossa   382
Scrophulariaceae   380
Scutellaria   383
Scutellaria adenostegia   383
Scutellaria comosa   383
Scutellaria cordifrons   383
Scutellaria galericulata   384
Scutellaria glabrata   385
Scutellaria immaculata   385
Scutellaria intermedia   385
Scutellaria oxystegia   385
Scutellaria phyllostachya   386
Scutellaria ramosissima   386
Scutellaria schachristanica   386
Scutellaria squarrosa   386
Secale   128
Secale cereale   128
Secale segetale   128
Secale sylvestre   128
Sedum   167
Sedum hispanicum   167
Sedum pentapetalum   167
Sedum tetramerum   168
Semenovia   498
Semenovia dasycarpa   498
Semenovia heterodonta   498
Semenovia pimpinelloides   498
Senecio   451
Senecio franchetii   451
Senecio olgae   451
Senecio subdentatus   451
Sergia   421
Sergia regelii   421
Serratula   450
Serratula algida   450
Serratula lancifolia   450
Seseli   488
Seseli calycinum   488
Seseli korovinii   488
Seseli lehmannianum   488
Seseli mucronatum   488
Seseli schrenkianum   489
Seseli seravschanicum   488
Seseli tenuisectum   489
Seseli turbinatum   489
Setaria   104
Setaria lutescens   104
Setaria verticillata   104
Setaria viridis   105

Sibbaldia   227
Sibbaldia tetrandra   227
Sideritis   387
Sideritis montana   387
Silene   320
Silene brahuica   320
Silene claviformis   320
Silene conica   320
Silene coniflora   321
Silene conoidea   321
Silene graminifolia   322
Silene guntensis   322
Silene incurvifolia   322
Silene kuschakewiczii   322
Silene latifolia   322
Silene longicalycina   322
Silene longicarpophora   325
Silene nana   323
Silene nevskii   323
Silene obtusidentata   323
Silene paranadena   323
Silene plurifolia   324
Silene praelonga   324
Silene praemixta   324
Silene pugionifolia   324
Silene ruinarum   324
Silene schischkinii   322
Silene stenantha   324
Silene tachtensis   324
Silene turkestanica   324
Silene uralensis subsp. apetala   324
Silene vulgaris   324
Simaroubaceae   269
Sinapis   287
Sinapis alba   287
Sinapis arvensis   287
Sisymbrium   274
Sisymbrium altissimum   274
Sisymbrium brassiciforme   274
Sisymbrium loeselii   274
Sium   487
Sium sisaroideum   487
Smelowskia   275
Smelowskia sisymbrioides   275
Smirnowia   179
Smirnowia turkestana   179
Solanaceae   373
Solanum   373
Solanum americanum   373
Solanum asiae-mediae   373
Solanum kitagawae   373
Solanum olgae   373
Solenanthus   369
Solenanthus circinatus   369
Solenanthus turkestanicus   370
Solidago   456
Solidago kuhistanica   456
Sonchus   424
Sonchus arvensis   424
Sonchus asper   425
Sonchus oleraceus   425
Sonchus palustris   425

Sophiopsis sisymbrioides   275
Sophora   172
Sophora alopecuroides   172
Sophora lehmanni   173
Sophora pachycarpa   173
Sorbus   222
Sorbus persica   222
Sorbus tianschanica   223
Sorghum   104
Sorghum halepense   104
Sparganium   92
Sparganium stoloniferum   92
Spergularia   319
Spergularia diandra   319
Spergularia marina   319
Spergularia marina   319
Spergularia media   319
Spergularia rubra   319
Spinacia   332
Spinacia turkestanica   332
Spiraea   217
Spiraea hypericifolia   217
Spiraea pilosa   217
Spirorrhynchus sabulosus   277
Stachyopsis   400
Stachyopsis oblongata   400
Stachys   405
Stachys hissarica   405
Stachys setifera   405
Stellaria   314
Stellaria alsinoides   314
Stellaria brachypetala   314
Stellaria karatavica   315
Stellaria media   314
Stellaria neglecta   314
Stellaria pallida   314
Stellaria turkestanica   314
Steptorhamphus crambifolius   423
Stipa   106
Stipa arabica   106
Stipa capillata   106
Stipa caragana   106
Stipa caucasica   107
Stipa hohenackeriana   107
Stipa kirghisorum   107
Stipa kurdistanica   107
Stipa lessingiana   107
Stipa lingua   107
Stipa lipskyi   107
Stipa margelanica   107
Stipa orientalis   107
Stipa richteriana   107
Stipa sareptana   107
Stipa splendens   107
Stipa trichoides   107
Stipa turkestanica   107
Stipagrostis   105
Stipagrostis karelinii   105
Stipagrostis pennata   106
Stizolophus   449
Stizolophus balsamita   449
Streptoloma   283

Streptoloma desertorum 283
Strigosella 281
Strigosella africana 281
Strigosella brevipes 281
Strigosella grandiflora 281
Strigosella intermedia 281
Strigosella scorpioides 281
Strigosella stenopetala 281
Strigosella trichocarpa 281
Strigosella turkestanica 281
Stroganowia angustifolia 289
Stubendorffia lipskyi 290
Stubendorffia olgae 290
Stubendorffia orientalis 291
Suaeda 337
Suaeda acuminata 337
Suaeda altissima 337
Suaeda arcuata 337
Suaeda crassifolia 334
Suaeda dendroides 337
Suaeda heterophylla 337
Suaeda linifolia 337
Suaeda paradoxa 337
Suaeda physophora 337
Swertia 357
Swertia lactea 357
Symphyotrichum 458
Symphyotrichum ciliatum 458

T

Taeniatherum 130
Taeniatherum asperum 130
Taeniatherum caput-medusae 130
Takhtajaniantha 431
Takhtajaniantha pusilla 431
Tamaricaceae 295
Tamarix 295
Tamarix androssowii var. transcaucassica 295
Tamarix arceuthoides 295
Tamarix elongata 296
Tamarix hispida 296
Tamarix hohenackeri 297
Tamarix laxa 296
Tamarix ramosissima 297
Tamarix smyrnensis 297
Tanacetopsis 463
Tanacetopsis karataviensis 463
Tanacetopsis mucronata 464
Tanacetopsis urgutensis 464
Tanacetum 472
Tanacetum parthenifolium 472
Tanacetum pseudachillea 472
Taraxacum 425
Taraxacum bessarabicum 425
Taraxacum bicorne 425
Taraxacum brevirostre 425
Taraxacum contristans 425
Taraxacum ecornutum 425
Taraxacum elongatum 425
Taraxacum erostre 425
Taraxacum glaucivirens 425

Taraxacum maracandicum 426
Taraxacum marginatum 426
Taraxacum minutilobium 426
Taraxacum modestum 426
Taraxacum monochlamydeum 426
Taraxacum montanum 426
Taraxacum nevskii 426
Taraxacum nuratavicum 426
Taraxacum officinale 426
Taraxacum popovii 426
Taraxacum reflexum 426
Taraxacum repandum 426
Taraxacum sonchoides 426
Taraxacum strobilocephalum 426
Tauscheria 277
Tauscheria lasiocarpa 277
Tetracme 283
Tetracme quadricornis 283
Tetracme recurvata 283
Tetrataenium olgae 497
Teucrium 383
Teucrium scordium subsp. scordioides 383
Thalictrum 147
Thalictrum isopyroides 147
Thalictrum minus 147
Thalictrum sultanabadense 147
Thlaspi 292
Thlaspi arvense 292
Thlaspi kotschyanum 293
Thlaspi perfoliatum 292
Thymelaceae 272
Thymelaea 272
Thymelaea passerina 272
Thymus 411
Thymus seravschanicus 411
Thymus subnervosus 412
Torilis 480
Torilis arvensis 480
Torilis leptophylla 481
Trachomitum 358
Trachomitum scabrum 358
Tragopogon 432
Tragopogon capitatus 432
Tragopogon conduplicatus 432
Tragopogon kraschennikovii 433
Tragopogon kultiassovii 432
Tragopogon malikus 433
Tragopogon porrifolius subsp. longirostris 433
Tragopogon pseudomajor 433
Tragopogon turkestanicus 433
Tragopogon vvedenskyi 433
Tribulus 171
Tribulus terrestris 171
Trichantemis 471
Trichanthemis karataviensis 471
Trichochiton inconspicuum 282
Trichodesma 370
Trichodesma incanum 370
Trifolium 177
Trifolium campestre 177

Trifolium fragiferum 177
Trifolium neglectum 177
Trifolium pratense 177
Trifolium repens 178
Triglochin 50
Triglochin palustre 50
Trigon ella gontscharovii 175
Trigonella 174
Trigonella adscendens 174
Trigonella arcuata 176
Trigonella aristata 174
Trigonella cancellata 174
Trigonella geminiflora 175
Trigonella grandiflora 175
Trigonella noeana 176
Trigonella orthoceras 176
Trigonella popovii 176
Trigonella verae 176
Tripleurospermum 473
Tripleurospermum disciforme 473
Tripolium 457
Tripolium pannonicum subsp. tripolium 457
Trisetum 113
Trisetum spicatum 113
Trollius 149
Trollius komarovii 149
Tulipa 61
Tulipa affinis 61
Tulipa borszczowii 62
Tulipa buhseana 62
Tulipa lehmanniana 62
Tulipa micheliana 63
Turgenia 481
Turgenia latifolia 481
Turristis 276
Turristis glabra 276
Tussilago 452
Tussilago farfara 452
Tulipa 61
Tulipa affinis 61
Tulipa borszczowii 62
Tulipa buhseana 62
Tulipa dasystemon 62
Tulipa dasystemonoides 62
Tulipa korolkowii 62
Tulipa turkestanica 64
Typha 92
Typha angustata 92
Typha angustifolia 92
Typha latifolia 93
Typha laxmannii 93
Typha minima 93
Typhaceae 92
Tytthostemma alsinoides 314

U

Ulmaceae 241
Ulmus 241
Ulmus glabra 241
Ulmus laevis 241
Ulmus minor 241

INDEX OF LATIN NAMES

Ulmus pumila   241
Umbelliferae   479
Ungernia   76
Ungernia oligostroma   76
Urtica   242
Urtica dioica   242
Urticaceae   242
Utricularia   382
Utricularia minor   382

V

Vaccaria   326
Vaccaria hispanica   326
Valeriana   478
Valeriana chionophila   478
Valeriana fedtschenkoi   478
Valeriana ficarifolia   478
Valerianella   478
Valerianella coronata   478
Valerianella cymbocarpa   479
Valerianella dactylophylla   479
Valerianella muricata   479
Valerianella oxyrrhyncha   479
Valerianella plagiostephana   479
Valerianella szovitsiana   479
Valerianella turkestanica   479
Valerianella vvedenskyi   479
Vallisneria   49
Vallisneria spiralis   49
Velezia   329
Velezia rigida   329
Verbascum   380
Verbascum bactrianum   380
Verbascum blattaria   380
Verbascum erianthum   380
Verbascum songaricum   380
Verbascum turkestanicum   381
Verbena   382
Verbena officinalis   382

Verbenaceae   382
Veronica   377
Veronica anagallis-aquatica   377
Veronica anagalloides   377
Veronica arguteserrata   377
Veronica arvensis   377
Veronica beccabunga   378
Veronica biloba   378
Veronica campylopoda   378
Veronica capillipes   378
Veronica cardiocarpa   378
Veronica hederifolia   378
Veronica intercedens   378
Veronica oxycarpa   378
Veronica persica   378
Veronica polita   378
Veronica verna   378
Vexibia alopecuroides   172
Vexibia pachycarpa   173
Vicia   214
Vicia anatolica   214
Vicia angustifolia   215
Vicia ervilia   214
Vicia gracilior   215
Vicia hajastana   214
Vicia hyrcanica   214
Vicia kokanica   214
Vicia michauxii   214
Vicia peregrina   215
Vicia sativa subsp. *nigra*   215
Vicia subvillosa   215
Vicia tenuifolia   215
Vicia villosa   215
Vicoa albertoregelia   454
Viola   247
Viola alaica   247
Viola occulta   247
Viola suavis   248
Viola turkestanica   248

Violaceae   247
Vitaceae   170
Vitis   170
Vitis vinifera   170
Vulpia   122
Vulpia ciliata   122
Vulpia myuros   122
Vulpia persica   122
Vulpia unilateralis   123

W

Waldheimia   471
Waldheimia glabra   471
Waldheimia tomentosa   471

X

Xanthium   473
Xanthium spinosum   473
Xanthium strumarium   473
Xeranthemum   435
Xeranthemum longepapposum   435

Y

Youngia serawschanica   424

Z

Zannichellia   51
Zannichellia pedunculata   51
Ziziphora   410
Ziziphora clinopodioides   410
Ziziphora pamiroalaica   410
Ziziphora suffruticosa   410
Ziziphora tenuior   410
Zygophyllaceae   170
Zygophyllum   170
Zygophyllum jaxarticum   170
Zygophyllum macrophyllum   170
Zygophyllum miniatum   171
Zygophyllum oxianum   171